Lecture Notes in Computer Science 12982

More information about this subseries at https://link.springer.com/bookseries/7407

Jochen Koenemann · Britta Peis (Eds.)

Approximation and Online Algorithms

19th International Workshop, WAOA 2021
Lisbon, Portugal, September 6–10, 2021
Revised Selected Papers

 Springer

Editors
Jochen Koenemann
University of Waterloo
Waterloo, ON, Canada

Britta Peis
RWTH Aachen University
Aachen, Germany

ISSN 0302-9743   ISSN 1611-3349 (electronic)
Lecture Notes in Computer Science
ISBN 978-3-030-92701-1   ISBN 978-3-030-92702-8 (eBook)
https://doi.org/10.1007/978-3-030-92702-8

LNCS Sublibrary: SL1 – Theoretical Computer Science and General Issues

This Springer imprint is published by the registered company Springer Nature Switzerland AG
The registered company address is: Gewerbestrasse 11, 6330 Cham, Switzerland

Preface

The 19th Workshop on Approximation and Online Algorithms (WAOA 2021) focused on the design and analysis of algorithms for online and computationally hard problems. Both kinds of problems have a large number of applications in a variety of fields. Due to the COVID-19 pandemic, WAOA 2021 took place virtually in Lisbon, Portugal, during September 9–10, 2021, and was a success: it featured many interesting presentations and provided opportunity for stimulating interactions. WAOA 2021 was part of the ALGO 2021 event that also hosted ESA, ALGOCLOUD, ALGOSENSORS, and ATMOS.

Topics of interest for WAOA 2021 were as follows: graph algorithms, inapproximability results, network design, packing and covering, paradigms for the design and analysis of approximation and online algorithms, parameterized complexity, scheduling problems, algorithmic game theory, algorithmic trading, coloring and partitioning, competitive analysis, computational advertising, computational finance, cuts and connectivity, geometric problems, mechanism design, resource augmentation, and real-world applications.

In response to the call for papers we received 31 submissions. Each of the submissions was reviewed by at least three referees, and many of the submissions were reviewed by more than three referees. The submissions were mainly judged on originality, technical quality, and relevance to the topics of the conference. Based on the reviews, the Program Committee (PC) selected 16 papers. This volume contains final revised versions of these papers as well as an invited contribution by our invited speaker Daniel Lokshtanov. The EasyChair conference system was used to manage the electronic submissions, the review process, and the electronic Program Committee discussions. It made our task much easier.

We would like to thank all the authors who submitted papers to WAOA 2021 and all attendees of WAOA 2021, including the presenters of the accepted papers. A special thank you goes to the plenary invited speaker Daniel Lokshtanov for accepting our invitation despite the uncertainty involved due to the pandemic and giving a very nice talk. We would also like to thank the PC members and the external reviewers for their diligent work in evaluating the submissions and their contributions to the electronic discussions. Furthermore, we are grateful to all the local organizers of ALGO 2021, especially the general chair of the organizing committee, Arlindo Oliveira.

September 2021

Jochen Könemann
Britta Peis

Organization

General Chair

Arlindo Oliveira   University of Lisbon, Portugal

Program Committee Chairs

Jochen Könemann   University of Waterloo, Canada
Britta Peis   RWTH Aachen University, Germany

Steering Committee

Thomas Erlebach   University of Leicester, UK
Christos Kaklamanis   University of Patras, Greece
Nicole Megow   University of Bremen, Germany
Laura Sanità   Eindhoven University of Technology, The Netherlands
Martin Skutella   Technical University Berlin, Germany
Roberto Solis-Oba   University of Western Ontario, Canada
Klaus Jansen   University of Kiel, Germany

Program Committee

Jarosław Byrka   University of Wrocław, Poland
Parinya Chalermsook   Aalto University, Finland
Sándor Fekete   Technical University Braunschweig, Germany
Andreas Emil Feldmann   Charles University in Prague, Czech Republic
Zachary Friggstad   University of Alberta, Canada
Stefan Funke   University of Stuttgart, Germany
Naveen Garg   Indian Institute of Technology Delhi, India
Hung Le   University of Massachusetts Amherst, USA
Jochen Könemann   University of Waterloo, Canada
Asaf Levin   Technion Israel Institute of Technology, Israel
Daniel Lokshtanov   University of California, Santa Barbara, USA
Jannik Matuschke   KU Leuven, Belgium
Matthias Mnich   Hamburg University of Technology, Germany
Ben Moseley   Carnegie Mellon University, USA
Alantha Newman   Université Grenoble Alpes, France
Britta Peis   RWTH Aachen University, Germany
Heiko Röglin   University of Bonn, Germany
Chaitanya Swamy   University of Waterloo, Canada
Vera Traub   ETH Zürich, Switzerland
Marc Ütz   University of Twente, The Netherlands
Jose Verschae   Catholic University of Chile, Chile
Gerhard Woeginger   RWTH Aachen University, Germany

Additional Reviewers

Marek Adamczyk
Ioana Bercea
Martin Böhm
Cornelius Brand
Franziska Eberle
Leah Epstein
David Fischer
Waldo Gálvez
Julian Golak
Lukas Graf
Alexander Grigoriev
Ruben Hoeksma
Łukasz Jeż

Bart de Keijzer
Phillip Keldenich
Joe Mitchell
Anish Mukherjee
Christian Ortlieb
Kanstantsin Pashkovich
Malin Rau
Rebecca Reiffenhäuser
Vibha Sahlot
Jens M. Schmidt
Arne Schmidt
Zeev Nutov
Victor Verdugo

How to Navigate Through Obstacles (and How to Place Them) (Invited Talk)

Daniel Lokshtanov

Department of Computer Science, University of California at Santa Barbara, California, USA

Abstract. Consider an agent that has to move from a source point s to a destination point t through an environment. The environment can be the plane, multidimensional space, or a graph. Suppose now that the environment is littered with (possibly overlapping) obstacles that obstruct the path from s to t. What is the minimum number of obstacles that has to be removed so that the agent can move unobstructed from the source to the destination? This is a fundamental problem with applications ranging from sensor networks to robotics, and it has been intensively studied under different names. In this talk we will survey the state of the art for the problem, from the perspective of approximation algorithms, hardness of approximation, and parameterized algorithms. We will also consider the dual problem of placing the minimum number of obstacles between the source and the destination so that there is no un-obstructed path from the source to the destination.

How to Navigate Through Obstacles (and How to Place Them): Invited Talk

Contents

Approximation Algorithms
for Vertex-Connectivity Augmentation
on the Cycle

Waldo Gálvez[1] , Francisco Sanhueza-Matamala[2]([✉]), and José A. Soto[2,3]

[1] Department of Computer Science, Technical University of Munich,
Munich, Germany
galvez@in.tum.de
[2] Departamento de Ingeniería Matemática, Universidad de Chile, Santiago, Chile
{fsanhueza,jsoto}@dim.uchile.cl
[3] Centro de Modelamiento Matemático, IRL 2807 CNRS, Universidad de Chile,
Santiago, Chile

Abstract. Given a k-vertex-connected graph G and a set S of extra edges (links), the goal of the k-vertex-connectivity augmentation problem is to find a subset S' of S of minimum size such that adding S' to G makes it $(k + 1)$-vertex-connected. Unlike the edge-connectivity augmentation problem, research for the vertex-connectivity version has been sparse.

In this work we present the first polynomial time approximation algorithm that improves the known ratio of 2 for 2-vertex-connectivity augmentation, for the case in which G is a cycle. This is the first step for attacking the more general problem of augmenting a 2-connected graph.

Our algorithm is based on local search and attains an approximation ratio of 1.8703. To derive it, we prove novel results on the structure of minimal solutions.

Keywords: Approximation algorithms · Connectivity augmentation ·
Cycle augmentation · Network design

1 Introduction

In the field of **Survivable Network Design,** one of the main goals is to construct robust networks (e.g. transportation, telecommunication or electric power supply networks), meaning that they are resilient to failures of nodes or edges (for instance due to attacks or malfunctioning), at a low cost. A classical example is

Waldo Gálvez is supported by the European Research Council, Grant Agreement No. 691672, project APEG. Francisco Sanhueza-Matamala is partially supported by grants ANID-PFCHA/Magíster Nacional/2020- 22201780 and FONDECYT Regular 1190043. Francisco Sanhueza-Matamala and José A. Soto are partially supported by ANID via FONDECYT Regular 1181180 and PIA AFB170001.

© Springer Nature Switzerland AG 2021
J. Koenemann and B. Peis (Eds.): WAOA 2021, LNCS 12982, pp. 1–22, 2021.
https://doi.org/10.1007/978-3-030-92702-8_1

the **Connectivity Augmentation** problem, where we are given a k-connected[1] (k-edge-connected, resp.) undirected graph $G = (V, E)$ and a collection S of extra edges (*links*), and the goal is to select the smallest possible set of links $S' \subseteq S$ so that $G' = (V, E \cup S')$ becomes $(k+1)$-connected ($(k+1)$-edge-connected, resp.).

Most of the variants of Connectivity Augmentation are known to be NP-hard. In the case of augmenting the edge-connectivity of a graph, a classical result of Khuller and Vishkin [18] provides already a 2-approximation algorithm for this problem, and crucial breakthroughs have been developed in recent years (see [4,5]). However, results for the case of vertex-connectivity augmentation are more scarce and the developed techniques are arguably more involved.

In this work, we focus on the case of augmenting the vertex-connectivity of a given 2-connected graph by one. For this case, Auletta et al. [2] provide a 2-approximation for the problem even when links have different weights and the goal is to minimize the total weight of the solution, but no improvement on the problem has been developed ever since.

1.1   Our Results

In order to make progress on the latter problem, we study the case where the input 2-connected graph is a cycle, which we denote as the *cycle vertex-connectivity augmentation* problem (**cycle VCA**). Even for this simpler case, the best-known result is the aforementioned 2-approximation due to Auletta et al. [2], while the case of augmenting the edge-connectivity of a cycle by one admits much better approximation guarantees [5], even via simple iterative algorithms [14]. Similarly to the case of edge-connectivity, it is possible to prove that **cycle VCA** is APX-hard (see Sect. 5), so our focus will be on the design of approximation algorithms.

The following theorem summarizes the central result of this work.

Theorem 1. *There is a 1.8703-approximation for the cycle vertex-connectivity augmentation problem.*

Our algorithm consists basically of two phases: a first phase where we exhaust *locally efficient* choices of links to be added to the instance, and then a second phase to complete the solution constructed so far. Roughly speaking, the cost of the partial solution constructed in the first phase can be properly bounded but does not ensure feasibility, issue that the second phase addresses. The number of links required in this second phase may be large, but in that case, the fact that no locally efficient set of links can be added to the solution implies that any feasible solution must include a considerable amount of links.

Our algorithm is similar in spirit to the one presented by Gálvez et al. [14] for the case of augmenting the edge-connectivity of a cycle; however, much more sophisticated tools are required for its analysis. In particular, the results from

---

[1] For $k \in \mathbb{N}$, a k-connected graph is a graph $G = (V, E)$ satisfying that, for any $V' \subseteq V$ with $|V'| \leq k - 1$, G remains connected after the deletion of V'. If the definition holds when replacing nodes by edges, the graph is said to be k-edge-connected.

Gálvez et al. heavily rely on the notion of *contracting* links (i.e. merging the end-points of a link into a new super-node), as in the edge-connectivity case a solution is feasible if and only if iteratively contracting all its links turns the original cycle into a single super-node. This property, unfortunately, does not extend to the case of vertex-connectivity (see Fig. 1 for an example of a set of links whose addition makes the graph 3-edge-connected but not 3-connected), and consequently, we develop alternative methods to verify the feasibility of a solution. By properly adjusting the parameters involved and using linear programming formulations to obtain refined lower bounds for the optimal solution, we are able to prove the claimed approximation guarantees. To the best of our knowledge, this is the first improvement for a special case of the vertex-connectivity augmentation problem on 2-vertex-connected graphs since Auletta et al.'s results.

1.2 Related Results

As mentioned before, the edge-connectivity augmentation problem has received considerable attention. A main reason is that it can be reduced to augmenting the edge-connectivity of either a tree or a cactus[2] by one [10]. The first case is known in the literature as the **Tree Augmentation** problem and many results have been developed for it. The problem is known to be APX-hard [7,13,19], and several better-than-2 approximation algorithms have been developed using a wide range of techniques [1,6,8,11,12,16,20,21,24], being 1.393 the current best approximation ratio achieved by Cecchetto et al. [5]. Tree Augmentation has also been studied in the framework of Fixed-Parameter Tractability [3,23] and in presence of general edge weights. In the latter case, for a long time the best approximation ratio was 2 [13,15,17,18], but recently this approximation ratio has been improved in a major breakthrough due to Traub and Zenklusen [29]; prior to that, such an improvement had been achieved only for restricted cases [1, 5,9,12,16,27].

The second case is known as the **Cactus Augmentation** problem, which is also APX-hard even if the cactus is a cycle [14]. For a long time the best-known approximation guarantee was 2 due to Khuller and Vishkin [18]. Recently, Byrka et al. [4] broke this barrier providing a 1.91-approximation, which was further improved to 1.393 by Cecchetto et al. [5]; this is consequently the current best approximation ratio for edge-connectivity augmentation.

For the vertex-connectivity augmentation problem, it was first proved by Végh [30] that if all the possible chords are available as links, the problem can be solved in polynomial time for any k, but if the set of links is pre-specified then the problem becomes APX-hard [19]. It is worth noting that this result does not directly hold for the special case of cycles. For the special case of vertex-connectivity augmentation of a 2-connected graph, Auletta et al. [2] provide a 2-approximation, while for the case of k-connected graphs the current best approximation ratio is $4 + \varepsilon$ for any fixed k, due to Nutov [26]. Recently, for the

---

[2] A *cactus* is a 2-edge-connected graph where every edge belongs exactly to one cycle of the graph.

case of vertex-connectivity augmentation of 1-connected graphs, Nutov obtained a 1.91 approximation [25].

Organization of the Paper. Section 2 provides some preliminary definitions and useful tools. In Sect. 3 we describe our main algorithmic approach and formally prove that the approximation ratio of our algorithm is strictly better than 2. Then in Sect. 4 we describe how to enhance this algorithm and to refine the analysis so as to obtain the claimed approximation ratio and finally in Sect. 5 we prove that the problem is APX-hard. Due to space constraints, some proofs are deferred to the full version of the article.

2   Preliminaries

In this work, we only consider simple graphs of $n \geq 4$ vertices. We let C_n denote the cycle on the set $[n] = \{1, 2, \ldots, n\}$ of vertices. Due to the cyclic aspect of C_n, sometimes we may use $n+1$ to denote vertex 1. A **chord** is an edge between non (cyclically) consecutive vertices of C_n. Each chord ab divides the cycle into two nonempty *sides* V_1 and V_2 so that $V(C_n) = V_1 \dot\cup V_2 \dot\cup \{a, b\}$. If L is a set of chords, $V(L)$ will denote the set of vertices incident to at least one chord in L.

C_n is 2-connected and so is any graph obtained by adding chords to it. A *separating pair* of a 2-connected graph is a pair of vertices such that, if we remove them, the graph becomes disconnected. Thus, 3-connected graphs are exactly those 2-connected graphs without separating pairs. In what follows, a set L of chords is said to 3-connect the cycle if $C_n \cup L$ is 3-connected.

Definition 1. *Given a set of chords S of C_n such that $C_n \cup S$ is 3-connected, the goal of the **cycle vertex-connectivity augmentation** (**cycle VCA**) problem is to find a set $S' \subseteq S$ of minimum size that 3-connects the cycle. We say that (C_n, S) is an instance of **cycle VCA**.*

We reserve the word *link* to refer to chords in S. Sets of links that 3-connect the cycle are feasible solutions for **cycle VCA**. We start by introducing some required concepts to characterize these feasible solutions.

Definition 2. *Let a, b, c, d be distinct vertices in C_n satisfying $a < b$ and $c < d$ (in the standard ordering of $[n]$). We say that a chord ab **crosses** cd if $c < a < d < b$ or $a < c < b < d$.*

Equivalently, ab and cd cross if and only if c and d are on different sides of ab (chords sharing an extreme do not cross). This definition is actually natural: If we draw C_n in the plane as a circle, and ab and cd as straight open-ended line segments, then ab crosses cd if and only if the associated segments intersect.

Lemma 1. *$S' \subseteq S$ is feasible for **cycle VCA** if and only if every chord of C_n is crossed by some link of S'. As a consequence, if S' is a feasible solution, then every vertex of the cycle must be incident to some link and $|S'| \geq n/2$.*

Proof. The first claim follows since a chord of C_n is a separating pair of $C_n \cup S'$ if and only if it is not crossed by any link of S'. For the second claim, notice that for each vertex j there must be a link crossing the chord defined by vertices $(j-1)$ and $(j+1)$; since this link is incident to j, S' must be an edge-cover and the bound follows.

Since any set of links containing a feasible solution is feasible, it makes sense to study minimal feasible solutions. A theorem by Mader [22] states that every cycle Z of a k-connected graph G that is formed by critical-edges (i.e., edges e such that $G - e$ is not k-connected) contains a vertex whose degree in G is k. This can be translated into the following useful property.

Lemma 2. *Every minimal solution S' for* **cycle VCA** *is acyclic and thus has size at most $n - 1$. In particular, every minimal set that 3-connects the cycle defines a 2-approximate solution for this problem.*

Proof. Let C' be a cycle in a minimal solution S'. As all the vertices incident to C' have degree equal to four in $C_n \cup S'$, Mader's theorem above implies that at least one link of C' is not critical, contradicting the minimality of S'. Consequently S' is acyclic, and hence its size is at most $n - 1$. Using Lemma 1 the optimum solution has size at least $n/2 \geq |S'|/2$.

Lemma 2 suggests a simple greedy 2-approximation algorithm: Initialize S' as S, and as long as there exists a link e such that $S' - e$ is feasible, remove e from S'. The resulting set S' is a minimal solution and hence a 2-approximate solution.

2.1   Circle Components

The previous 2-approximation was built in a top-down way, i.e., deleting links from a feasible solution. In order to obtain a better than 2-approximation, we will instead build solutions in a bottom-up way. For that purpose, we use the concept of *circle components* (see Fig. 1).

Definition 3 [28]. *The* **circle graph** *defined by a set of chords L is the graph with vertex set L, where two chords are adjacent if and only if they cross.*

Definition 4. *A* **circle component** *is a set of links $L \subseteq S$ such that the circle graph defined by L is connected. A* **link path** *from vertex a to vertex b is a sequence $e_1 e_2 \ldots e_k$ of links such that a is an endpoint of e_1, b is an endpoint of e_k and each link crosses the next one in the sequence.*

Observe that if L is a circle component, then for every $a, b \in V(L)$ there is a link path in L from a to b (see Fig. 1). The next technical definition provides further useful concepts we will use along this work.

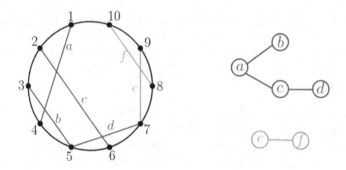

Fig. 1. Left: Two circle components in blue and red. With respect to the blue component, vertices 2, 3, 4, 5 and 6 are internal, vertices 1 and 7 are border and the remaining vertices are external; with respect to the red component, vertices 8 and 9 are internal while vertices 7 and 10 are border. The sequence *bac* is an example of a link path between 5 and 2. **Right:** The associated circle graph. (Color figure online)

Definition 5. *Let L be a circle component and $V(L) = \{v_1, v_2, \cdots, v_{|V(L)|}\}$ such that $v_1 < v_2 < \cdots < v_{|V(L)|}$ in the cycle. We say that v_i, v_{i+1} (indices considered modulo $|V(L)|$) are consecutive in $V(L)$.*

*Chords between non-consecutive vertices of $V(L)$ are called **internal chords of** L. Chords between consecutive vertices in $V(L)$ are called **border chords**. Vertices of $V(L)$ incident to some border chord are called **border vertices** and the rest of the vertices of $V(L)$ are called **internal vertices**. The vertices outside $V(L)$ are called **external vertices**.*

Notice that an internal chord of L may connect two internal vertices, one internal vertex and a border vertex or even two border vertices (if they are incident to different border chords). We also say that L crosses a chord ab of C_n if some link $e \in L$ crosses ab.

The following lemma contains the main characterization of feasibility we will use along this work.

Lemma 3. *Let $S' \subseteq S$ be an edge-cover of C_n. The set S' is a feasible solution for **cycle VCA** if and only if S' is a circle component.*

We need first the following technical lemmas.

Lemma 4. *Let ab and cd be chords of C_n and $W = e_1 \ldots e_k$ be a link path from a to b. If ab crosses cd then there is some chord in W that crosses cd.*

Proof. Let V_1 and V_2 be the sets of vertices of C_n on each side of the chord cd, so that $a \in V_1$ and $b \in V_2$. Observe also that $V_1 \cup \{c, d\}$ is an interval in the cycle with extremes c and d. Let e_i be the first chord in W with at least one endpoint in V_2 (this link exists since e_k has one endpoint in V_2). If $i = 1$ then e_1 crosses cd, so assume $i \geq 2$. Since the endpoints of e_{i-1} are in $V_1 \cup \{c, d\}$ which is an interval, one of the sides of e_{i-1} is fully contained in V_1. Therefore, as e_{i-1}

crosses e_i, one endpoint of e_i is also in V_1. We conclude that e_i has one endpoint in each side of cd and so e_i crosses cd.

Lemma 5. *Let L be a circle component and ab a chord. L crosses ab if and only if (1) ab is an internal chord of L or (2) ab crosses some border chord cd of L.*

Proof. We proceed by cases depending on ab.

Case 1: ab is an internal chord of L. Let V_1 and V_2 be the two sides defined by ab. Since ab is internal, $V(L)$ has vertices on both sides, so let $c \in V(L) \cap V_1$, $d \in V(L) \cap V_2$ and W be a link path in L from c to d. Since cd crosses ab, we conclude by using Lemma 4 that there exists a link of $W \subseteq L$ that crosses ab.

Case 2: ab crosses some border chord cd of L. Let W be a link path in L from c to d. By Lemma 4, some link in this path must cross ab.

Case 3: ab is a border chord of L Since a and b are consecutive in L there is no link in L with an extreme in one of the sides of ab and so L does not cross ab.

In Any Other Case. At least one extreme (say a) of the chord is external to L. Starting from a, let c (respectively d) be the first vertex of $V(L)$ that we encounter going clockwise (respectively counterclockwise) along the cycle. Then cd is a border chord of L. Let V_1 be the side of cd to which a belongs and note that $V_1 \cap V(L) = \emptyset$. Since ab does not cross cd (see case 2), we conclude that b is in the interval $V_1 \cup \{c, d\}$. Therefore, one side of ab is completely contained in V_1 and thus, this side does not contain any vertex of $V(L)$. It follows that no link in L crosses ab.

We now have all the ingredients to prove Lemma 3.

Proof (Proof of Lemma 3). Suppose that S' is a circle component. Due to Lemma 1, every chord ab of C_n is internal for S', and thus, by Lemma 5, ab is crossed by some link in S'. Lemma 1 implies that S' is feasible for **cycle VCA**.

For the converse, let S' be a feasible solution of **cycle VCA**. Suppose by contradiction that the circle graph G' of S' is not connected and let L be a connected component of G'. Note that L cannot be a singleton, since if $L = \{ab\}$ then by hypothesis there exists some chord of S' crossing ab, and so L would not be maximally connected in G'. It follows that L is a circle component.

Suppose first that L has a border chord ab. By Lemma 5, L does not cross ab. Since S' crosses ab, there exists some link $cd \in S' \setminus L$ such that cd crosses ab. Lemma 5 then implies that $L \cup \{cd\}$ is also a circle component, contradicting the maximality of L. We conclude that all chords with endpoints in $V(L)$ are internal chords of L. The only way for this to happen is that $V(L) = [n]$. But then, every chord of the cycle is internal to L. In particular, any $cd \in S'$ is crossed by L, and by maximality of L, $cd \in L$. We conclude that $S' = L$ and so S' is a circle component.

2.2 Minimal Completions

Another tool we will make use of along this work is a procedure to turn a set of circle components into a feasible solution while controlling the number of extra links.

Definition 6. *Let $F \subseteq S$ be a set of links. We call a set $Q \subseteq S$ a* **completion** *of F if $F \cup Q$ is feasible for* **cycle VCA**. *A* **minimal completion of F** *is a completion that is minimal for inclusion.*

We say that two circle components L_1 and L_2 cross if there exists $e_1 \in L_1$ and $e_2 \in L_2$ such that e_1 crosses e_2. The main idea of our algorithm in Sect. 3 is to build a collection $\mathcal{L}$ of circle components that, roughly speaking, use few links to cover many vertices and are pairwise non-crossing; then we add a minimal completion to obtain the desired feasible solution. A similar approach has been used by Gálvez et al. for the case of edge-connectivity [14], where the authors prove that any set of links can be completed by iteratively picking and contracting links so as to reduce the whole graph into a single super-node. As mentioned before, this procedure unfortunately does not work for the case of vertex-connectivity as now we require instead to merge the different circle components into a single circle component, which is in general a strictly stronger requirement. The following lemma provides a way to compute a minimal completion in the case of vertex-connectivity and also allows us to bound its size.

Lemma 6. *Let $L_1, \ldots, L_k$ be $k \geq 1$ non-crossing circle components, $|L_i| \geq 2$ for each $i = 1, \ldots, k$, and let $F = L_1 \cup \cdots \cup L_k$. Then every minimal completion Q of F uses at most $n - 3 - \sum_{i=1}^{k}(|V(L_i)| - 3)$ links. Furthermore, such a minimal completion can be computed in polynomial time.*

The proof consists of two steps: We first prove the claim for the simpler case of a single circle component, and then reduce the general case to this simpler one by introducing the notion of *zone graph* (see Fig. 2).

Step 1. Completing a single circle-component.

Lemma 7. *Let $L \subseteq S$ be a circle component. Then, every minimal completion Q of L has size at most $n - |V(L)|$.*

Proof. Let us first prove that there exists a completion Q of size $n - |V(L)|$. We do this by induction on $i = n - |V(L)|$. If $i = 0$ then L is an edge-cover and by Lemma 3, it is already feasible, so we can set $Q = \emptyset$. For $i \geq 1$ we notice that there must be a border chord ab of L. Since S itself is feasible, there must be a link $cd \in S$ that crosses ab. Then, by Lemma 5, $L' = L \cup \{cd\}$ is a component covering at least one more vertex than L. By induction, there is a completion Q' of L' with at most $n - |V(L')|$ links. Then $Q = Q' \cup \{cd\}$ is a completion of L, with $|Q| = |Q'| + 1 \leq 1 + n - |V(L')| \leq n - |V(L)|$, concluding the first proof.

Now, let Q be a minimal completion of L. Consider the new instance of **cycle VCA** $(C_n, L \cup Q)$. In this instance, Q is the only completion of L, and then following the previous argumentation we have that $|Q| \leq n - |V(L)|$.

It is worth noting that Lemma 7 provides an alternative proof that minimal solutions are 2-approximate not relying on Mader's theorem (see Lemma 2) as the following proposition states.

Proposition 1. *Every minimal completion of a single chord has at most $n - 3$ links, and every minimal solution for* **cycle VCA** *has size at most $n - 2$.*

Proof. Let Q be a minimal completion of a single link $\{e\}$. Since $Q \cup \{e\}$ is a solution for **cycle VCA**, there exists $f \in F$ that crosses e. Then $Q \setminus \{e, f\}$ is a minimal completion of the circle component $\{e, f\}$. Using Lemma 7, we get $|Q| = 1 + |Q \setminus \{f\}| \leq 1 + n - |V(\{e, f\})| = 1 + n - 4 = n - 3$.

For the second statement, we observe that if Q is a minimal solution for **cycle VCA** then $Q - e$ is a minimal completion of $\{e\}$ for any link $e \in Q$, and so by the previous paragraph $|Q| = 1 + |Q - e| \leq n - 2$.

Step 2. Completing a collection of arbitrary non-crossing circle components. As mentioned before, we will relate the case of arbitrary non-crossing circle components to the case of a single circle component via the following definitions.

Definition 7. *The* **open zone** *defined by a border chord ab of a circle component L is the interval of vertices $I(ab, L)$ corresponding to the side of ab that does not include any vertex of $V(L)$. The associated closed zone is $I(ab, L) \cup \{a, b\}$.*

The **zone graph** *defined by a border chord ab of a circle component L is the graph $Z(ab, L, S)$ obtained from $G = ([n], C_n \cup S)$ doing the following operations: (1) Erase all the links of L, (2) Add a link e between a and b and (3) Contract all the vertices in $C_n \setminus (I(ab, L) \cup \{a, b\})$ to a single vertex v_0.*

In $Z = Z(ab, L, S)$ we have removed all the loops and kept only one link from each parallel class. For every set $Q \subseteq S$ of chords in G, we let $\psi(Q)$ be the associated set of chords in Z, eliminating the links that become part of the external cycle. By this construction, we obtain a new *zone instance* $(C_{n'}, \psi(S) \cup \{ab\})$, associated to ab and L. Here we note that $n' = |I(ab, L)| + 1 < n$ which follows since L covers at least 4 vertices.

Let $\mathcal{L}$ be a collection of non-crossing circle components, and let $L \in \mathcal{L}$ be a fixed one. Let $P(L)$ be the border chords (P stands for perimeter) of L. Since they do not cross, every circle component $K \in \mathcal{L} \setminus \{L\}$ is completely contained in one closed zone of L (meaning that $V(K)$ is a subset of the closed zone), and every closed zone of L may contain further components of $\mathcal{L} \setminus \{L\}$. For each border chord $ab \in P(L)$, let $\mathcal{L}(ab, L)$ be the set of circle components in $\mathcal{L} \setminus \{L\}$ that are in the closed zone $I(ab, L) \cup \{a, b\}$, and let $F^{ab} = \bigcup_{K \in \mathcal{L}(ab, L)} K$ the sets of links that are contained in some circle component of the closed zone $I(ab, L) \cup \{a, b\}$. We can then prove the following.

Lemma 8. *Let $Q \subseteq S$ be a set of links. If Q is a completion of $F^{ab} \cup L$ in the original instance (C_n, S) then $\psi(Q)$ is a completion of $\psi(F^{ab}) \cup \{ab\}$ in the zone instance $(C_{n'}, \psi(S) \cup \{ab\})$. Conversely, if $\psi(Q)$ is a completion of $\psi(F^{ab}) \cup \{ab\}$ in the zone instance, then $Q \cup F^{ab}$ crosses all the chords of C_n with both extremes in $I(ab, L) \cup \{a, b\}$.*

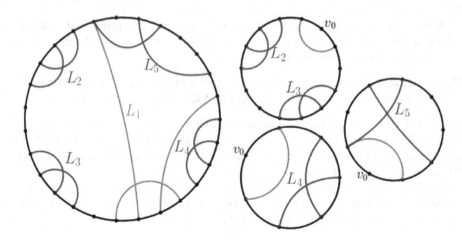

Fig. 2. Left: Graph with 5 components. L_1 defines 3 zones, which can have one component (L_4) Two or more components (On the zone that includes L_2 and L_3) and extra vertices (like the ones next to L_3). **Right:** The corresponding Zone Graphs

Proof. Let Q be a completion of L in the original instance and let cd be a chord in $C_{n'}$. If both c and d are uncontracted vertices (i.e., they are in the closed zone $I(ab, L) \cup \{a, b\}$), then they are not crossed by L in the original instance, and so they must be crossed by some link $e \in Q \cup F^{ab}$. Therefore, the respective $\psi(e) \in \psi(Q) \cup \psi(F^{ab})$ crosses cd on the zone instance. On the other hand, if one of the vertices of the chord cd is v_0 then cd is crossed by ab. In any case cd is crossed by $\psi(Q) \cup \psi(F^{ab}) \cup \{ab\}$.

For the converse, suppose that $\psi(Q)$ is a completion of $\psi(F^{ab}) \cup \{ab\}$ in the zone instance. We notice that all the vertices in $I(ab, L) \cup \{a, b\}$ are uncontracted vertices. Then for every chord cd in C_n between two vertices in that set, there is a link f' in $\psi(Q) \cup \psi(F^{ab})$ that crosses the chord in $C_{n'}$. Any link $f \in Q \cup F^{ab}$ with $\psi(f) = f'$ crosses the chord cd in the original instance.

We can now proceed with the proof of Lemma 6

Proof (Proof of Lemma 6). We will prove the claim by induction on n. Let Q be a minimal completion of $F = L_1 \cup \cdots \cup L_k$. The case $n = 4$ holds trivially since in that case $k = 1$ and L_1 covers all the vertices and so $\emptyset$ is the sought minimal completion.

Consider a border chord $ab \in P(L_1)$. Lemma 8 guarantees that $\psi(Q)$ is a completion of $\psi(F^{ab}) \cup \{ab\}$ in the associated zone instance.

Therefore, $\psi(Q) \cup \{ab\}$ is a completion of $\psi(F^{ab})$ in the same zone instance. So, there is a subset $Q'_{ab} \subseteq Q$ with $|Q'_{ab}| = |\psi(Q'_{ab})|$ such that either $\psi(Q'_{ab})$ or $\psi(Q'_{ab}) \cup \{ab\}$ are minimal completions of $\psi(F^{ab})$ in that zone instance (in any case, Q'_{ab} has size at most that of a minimal completion of $\psi(F_{ab})$ in the zone graph).

We have two cases here: let us consider first the case in which $\mathcal{L}(ab, L_1)$ is non-empty. Recall that the zone instance has $n_{ab} < n$ vertices, so we can apply the induction hypothesis to the zone instance to obtain that

$$|Q'_{ab}| \leq n_{ab} - 3 - \sum_{J \in \mathcal{L}(ab, L_1)} (|V(J)| - 3) = |I(ab, L_1)| - \sum_{J \in \mathcal{L}(ab, L_1)} (|V(J)| - 3|).$$

The other case is when there are no components in the zone defined by ab, i.e., $\mathcal{L}(ab, L_1)$ and F^{ab} are empty. In this case, $\psi(Q)$ is a completion of a single link $\{ab\}$ in the zone instance, and so there is a set $Q'_{ab} \subseteq Q$ with $|Q'_{ab}| = |\psi(Q'_{ab})|$ such that $\psi(Q'_{ab})$ is a minimal completion of $\{ab\}$. By Proposition 1, $|Q'_{ab}| = |\psi(Q'_{ab})| = n_{ab} - 3 = |I(ab, L_1)| = |I(ab, L_1)| - \sum_{J \in \mathcal{L}(ab, L_1)} (|V(J)| - 3)$ since the sum is empty.

Define $Q' = \bigcup_{ab \in P(L_1)} Q'_{ab}$. We claim that Q' is a completion of $F = L_1 \cup \cdots \cup L_k$ in the original instance. Indeed, let cd be an arbitrary chord of the cycle C_n and let us prove that $Q' \cup F$ crosses it. If cd is crossed by L_1 then we are done, so assume that it is not crossed by it. By Lemma 5, cd is neither an internal chord of L_1 nor it is a chord that crosses some border chord of L_1. The only possibility left is that cd connects two vertices in $I(ab, L_1) \cup \{a, b\}$ for some border chord $ab \in P(L_1)$. Lemma 8 implies that cd is crossed by a link in $Q'_{ab} \cup F^{ab} \subseteq Q' \cup F$, so we are done proving the claim.

Finally, as $Q' \subseteq Q$ and Q is a minimal completion of F we must have $Q' = Q$. Using the bounds above and the fact that the open zones $\{I(ab, L_1) : ab \in P(L_1)\}$ partitions $[n] \setminus V(L_1)$, we get

$$|Q| \leq \sum_{ab \in P(L_1)} |Q'_{ab}| \leq \sum_{ab \in P(L_1)} \left(|I(ab, L_1)| - \sum_{J \in \mathcal{L}(ab, L_1)} (|V(J)| - 3) \right)$$

$$= (n - |V(L_1)|) - \sum_{J \in \mathcal{L} \setminus \{L_1\}} (|V(J)| - 3) = n - 3 - \sum_{J \in \mathcal{L}} (|V(J)| - 3).$$

3   Local Search Algorithm

In this section we present our main algorithmic approach for **cycle VCA**. Let us first define special sets of links which will be fundamental for our algorithm.

Definition 8. *Let $F \subseteq S$ be a set of links. We say that F is **singleton-free** if the circle graph G' of F has no isolated vertices (singletons). The connected components of G' form a family $\mathcal{L}$ of non-crossing circle components. We define the **utility of F** as*

$$U(F) = -|F| + \sum_{J \in \mathcal{L}} (|V(J)| - 3).$$

Our approach is to initially find singleton-free sets in an increasing way (i.e., by only adding links) whose utility is high with respect to some increasing parameter (e.g. the number of covered vertices). As the following lemma shows, high utility sets indeed provide solutions with improved approximation guarantees.

Lemma 9. *If Q is a minimal completion of a singleton-free set F then $Q \cup F$ is a solution for* **cycle VCA** *with at most $n - 3 - U(F)$ links.*

Proof. By Lemma 6, Q has size at most $n - 3 - \sum_{J \in \mathcal{L}}(|V(J)| - 3) = n - 3 - |F| - U(F)$, and so $|Q| + |F| \leq n - 3 - U(F)$.

For any given $\alpha \in (1/2, 1]$ and any fixed step size $N_{\max} \in \mathbb{N}$ we define the following Local Search algorithm which works in two phases. In the first phase it constructs a singleton-free set of links F, by adding at most $N_{\max}$ links at a time and ensuring that in each iteration, the marginal utility gain is at least $(1 - \alpha)$ times the number of newly covered vertices. At the end of the phase, the utility of F is at least $(1 - \alpha)|V(F)|$. More in detail, the algorithm adds links in the first phase as long as the set F under construction is not $(\alpha, N_{\max})$-critical as defined below.

Definition 9. *Let $\alpha \in (1/2, 1]$ and let $N_{\max} \in \mathbb{N}$. A singleton-free set $F \subseteq S$ will be called $(\alpha, N_{\max})$-critical if there is no set $K \subseteq S \setminus F$ of size at most $N_{\max}$ such that $F \cup K$ is singleton-free and $U(F \cup K) - U(F) \geq (1 - \alpha)|V(F \cup K) \setminus V(F)|$.*

In the second phase, the algorithm finds a minimal completion Q of F and returns the pair (Q, F). The set $Q \cup F$ is the solution proposed.

Algorithm. Local Search algorithm

Input: Instance (C_n, S) of **cycle VCA**. $\alpha \in (1/2, 1)$, $N_{\max} \in \mathbb{N}$.

Phase 1 – Building a collection of non-crossing circle components.

1: $F \leftarrow \emptyset$.
2: **while** F is not $(\alpha, N_{\max})$-critical **do**
3:     Let $K \subseteq S \setminus F$ of size at most $N_{\max}$ such that $F \cup K$ is singleton-free and $U(F \cup K) - U(F) \geq (1 - \alpha)|V(F \cup K) \setminus V(F)|$;
4:     $F \leftarrow F \cup K$
5: **end while**

Phase 2 – Finding a completion

6: Find a minimal completion Q of F.
7: Return (Q, F)

To gain some intuition about the algorithm, suppose that at one iteration of Phase 1, the algorithm finds a set K of at most $N_{\max}$ links to add to the current solution F. Let $\mathcal{L}$ be the current circle components of F. The simplest case is when the set K crosses exactly one circle component $L \in \mathcal{L}$ and $V(K)$ does not intersect any other component of $\mathcal{L}$. In this case, the marginal utility gain $U(F \cup K) - U(F)$ is simply $-|K| + |V(L \cup K) \setminus V(L)| = -|K| + |V(F \cup K) \setminus V(F)|$. So, the algorithm adds K to F as long as $-|K| + |V(F \cup K) \setminus V(F)| \geq (1 - \alpha)|V(F \cup K) \setminus V(F)|$, or equivalently if the ratio from the number of new links K to the number of new covered vertices $|V(F \cup K) \setminus V(F)|$ is at most α. So, the algorithm only adds sets K of links that cover new vertices with few links.

The intuition behind adding sets K that cross (or touch) more than one component is more difficult to grasp, but as a general rule, connecting more components yields larger marginal utility gains. The following lemma provides a bound on the size of the returned solution with respect to the number of nodes covered by links from Phase 1.

Lemma 10. *For any fixed $N_{\max}$ and α, the Local Search algorithm runs in polynomial time and returns a pair (Q, F) such that $Q \cup F$ is feasible for* **cycle VCA***, and $|Q \cup F| \leq n - 3 - (1 - \alpha)|V(F)|$.*

Proof. Let $s = |S|$. Every iteration of the first phase may be achieved by trying all $s^{O(N_{\max})}$ possible subsets of $S \setminus L$ of size at most $N_{\max}$, hence taking polynomial time. Finding a minimal completion Q can also be done in polynomial time, by iteratively removing links from $S \setminus F$ that are not needed.

Let q be the number of iterations of the first phase and let F_i be the set F at the end of the i-th iteration (using $F_0 = \emptyset$). Then, by a telescopic sum and using the fact that $V(F_{i-1}) \subseteq V(F_i)$ for each i, we have

$$U(F) = \sum_{i=1}^{q} U(F_i) - U(F_{i-1}) \geq (1-\alpha) \sum_{i=1}^{q} |V(F_i) \setminus V(F_{i-1})| = (1-\alpha)|V(F)|.$$

By Lemma 9, $Q \cup F$ is feasible and has size at most $n - 3 - (1 - \alpha)|V(F)|$.

Roughly speaking, using that the optimal solution has size at least $n/2$ (by Lemma 1) the previous lemma guarantees good approximation factors as long as $|V(F)|$ is large enough (where F is the $(\alpha, N_{\max})$-critical set at the end of Phase 1). On the other hand, we will prove that whenever $|V(F)|$ is small there are stronger lower bounds on the size of the optimal solution (see Fig. 3).

3.1   Properties of Critical Sets

Let F be a fixed $(\alpha, N_{\max})$-critical set, with $N_{\max} \geq 1$ (for instance, the set F at end of Phase 1 of the algorithm). We also define $V' = V(C_n) \setminus V(F)$ and the set $\mathcal{L}$ of the non-crossing circle components associated to F. A first property we prove is that any link in $S \setminus F$ crosses at most one component $L \in \mathcal{L}$ and, if it does, it crosses at most one border chord of L. In particular, at least one endpoint will be in $V(L) \subseteq V(F)$. This intuitively implies that any relatively small set of remaining links in the instance cannot cover many vertices of the cycle and cross F simultaneously (as in particular at least one of the endpoints of each such link will be in $V(L) \subseteq V(F)$). In what follows $P(J)$ denotes the set of border chords of a circle component J, and $P(F) := \bigcup_{J \in \mathcal{L}} P(J)$ denotes the *perimeter* of F.

Lemma 11. *Any link $e \in S$ can cross at most one chord in $P(F)$. If both endpoints of e are in V', then e cannot cross any chord in $P(F)$.*

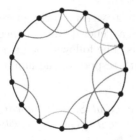

Fig. 3. Depiction of cases arising when applying the Local Search algorithm. Blue links correspond to F while orange links correspond to a minimal completion. **Left:** When $|V(F)|$ is large, many vertices are covered by sets of links of high utility which pay for the size of the minimal completion. **Right:** When $|V(F)|$ is small, since F is $(\alpha, N_{\max})$-critical, any set of links in $S \setminus F$ must have low utility, implying stronger lower bounds for $|OPT|$.

Proof. Suppose that e crosses at least one border chord in $P(F)$ then, by Lemma 5, it crosses at least one circle component of $\mathcal{L}$ and it can furthermore be proved that it has to cross exactly one (please refer to the full version for details). So let $L \in \mathcal{L}$ be the component crossed by e.

Note that e can cross at most 2 border chords of L, and the only way for this to happen is that the extremes of e are outside $V(L)$ and in different zones of L. In this case, $U(F \cup \{e\}) - U(F) = -1 + |V(L \cup \{e\})| - |V(L)| = 1 \geq |V(F \cup \{e\}) - V(F)|(1 - \alpha)$ which is not possible since F was $(\alpha, N_{\max})$-critical.

From here any link e in S can cross at most one chord in $P(F)$. Furthermore, if both endpoints of e are in V' and e crosses a chord in $P(F)$, then it would have to cross two by the previous paragraph, which is not possible. So in the latter case, e cannot cross any chord in P.

Another way to obtain lower bounds on the size of the optimal solution is to show that any maximal matching $M \subseteq \mathrm{OPT}[V']$ outside $V(F)$ must be large. We will argue about the number of links required to *connect* M as specified in the following definition.

Definition 10. *We say that a link $e \in S$ **connects** a set of links X if $X \cup \{e\}$ is circle component.*

Since F is $(\alpha, N_{\max})$-critical, M must be formed by a considerable amount of circle components which need to be connected so as to get a feasible solution; however, again since F is $(\alpha, N_{\max})$-critical, each one of the links from the optimal solution required to connect these circle components cannot cross too many links from M simultaneously. As a consequence, the number of such links cannot be too small, providing then a lower bound on the size of the optimal solution. The following lemma formalizes this previously described intuition and provides the required bounds to analyze the performance of the algorithm.

Lemma 12. *Suppose that $N_{\max} \geq \lceil (5 - 2\alpha)/(2\alpha - 1) \rceil + 1$. Let $M \subseteq S$ be any matching consisting of links with both endpoints in $V' := V(C_n) \setminus V(F)$. For any link $e \in S \setminus (M \cup F)$, the number of links from M that e connects is at most $\max\left\{ 0, \left\lceil \dfrac{5 - 2X(e) - V_F(e) - (4 - V_M(e) - V_F(e))\alpha}{2\alpha - 1} \right\rceil \right\}$, where $X(e)$ is the indicator that e crosses F, $V_M(e) = |V(\{e\}) \cap V(M)|$ and $V_F = |V(\{e\}) \cap V(F)|$. In particular, no link of $S \setminus (M \cup F)$ connects more than $\ell := \lceil (5 - 2\alpha)/(2\alpha - 1) \rceil$ links in M.*

Proof. Let $\gamma = \max\left\{ 0, \left\lceil \dfrac{5 - 2X(e) - V_F(e) - (4 - V_M(e) - V_F(e))\alpha}{2\alpha - 1} \right\rceil \right\}$ be the bound in the Lemma. Note that we always have $\gamma \leq \ell \leq N_{\max} - 1$.

Let M_e be the set of links of M that e connects (this set contains the set of links of M that e crosses). Suppose for the sake of contradiction that $|M_e| > \gamma$. If $|M_e| \geq N_{\max} - 1$ then redefine M_e as any subset of M_e with $N_{\max} - 1$ links that e connects (this can be done by pruning leaves in a covering tree of the circle graph of $M_e \cup \{e\}$). Let also $V_{M_e}(e) = |V(\{e\} \cap V(M_e)|$.

In any case we have $2 \leq \gamma + 2 \leq |M_e \cup \{e\}| \leq N_{\max}$. We also have that $F \cup M_e \cup \{e\}$ is singleton-free. Furthermore $|V(M_e \cup \{e\} \cup F) \setminus V(F)| = 2|M_e| + |V(\{e\}) \setminus V(M_e \cup F)| = 2|M_e| + 2 - V_{M_e}(e) - V_F(e)$. Then, since F is $(\alpha, N_{\max})$-critical, we have that

$$U(M_e \cup \{e\} \cup F) - U(F) < (1 - \alpha)(2|M_e| + 2 - V_{M_e}(e) - V_F(e)).$$

If e does not cross F then $M_e \cup \{e\}$ is a circle component not crossing F. Then $U(M_e \cup \{e\} \cup F) - U(F) = -|M_e \cup \{e\}| + |V(M_e \cup \{e\})| - 3 = -|M_e| - 1 + (2|M_e| + 2 - V_{M_e}(e)) - 3 = |M_e| - V_{M_e}(e) - 2$.

If e crosses F then it does on a single circle component L, and we know that exactly one endpoint of e is in $V(L)$. Let v be the other endpoint of e, and note that $V_{M_e}(e) = 1$ if $v \in V(M_e)$ and $V_{M_e}(e) = 0$ if $v \notin V(M_e)$. Therefore, $U(M_e \cup \{e\} \cup F) - U(F) = -|M_e \cup \{e\}| + |V(M_e \cup \{e\} \cup L)| - |V(L)| = -|M_e| - 1 + |\{v\} \cup V(M_e)| = -|M_e| - 1 + 2|M_e| + (1 - V_{M_e}(e)) = |M_e| - V_{M_e}(e)$.

Summarizing we always have

$$|M_e| - V_{M_e}(e) + 2 - 2X(e) < (1 - \alpha)(2|M_e| + 2 - V_{M_e}(e) - V_F(e))$$

Solving the inequality, we get $|M_e| < \frac{4 - 2X(e) - V_F(e) - \alpha(2 - V_{M_e}(e) - V_F(e))}{2\alpha - 1}$. Since $V_{M_e}(e) \leq V_M(e)$, the inequality above holds if we replace $V_{M_e}(e)$ by $V_M(e)$. Furthermore, since $|M_e|$ is an integer, this means that

$$|M_e| \leq \left\lceil \frac{4 - 2X(e) - V_F(e) - \alpha(2 - V_M(e) - V_F(e))}{2\alpha - 1} - 1 \right\rceil$$

$$= \left\lceil \frac{5 - 2X(e) - V_F(e) - \alpha(4 - V_M(e) - V_F(e))}{2\alpha - 1} \right\rceil = \gamma < |M_e|,$$

which is a contradiction.

3.2   Analysis of the Local Search Algorithm

Let F be an $(\alpha, N_{\max})$-critical set, OPT be an optimal solution for **cycle VCA** and M be a maximal matching from OPT$[V']$ (i.e. with both endpoints outside $V(F)$). Let ALG(F) be the set obtained by adding to F a minimal completion. To ease notation, let $m = |M|/n$, $f = |V(F)|/n$, opt $=$ OPT $/n$ and alg$(F) =$ $|$ALG$(F)|/n$. In particular, if (F, Q) is the pair obtained by our Local Search algorithm, then its approximation guarantee is $|$ALG$(F)|/|$OPT$| =$ alg$(F)/$opt. The next lemma provides a first better-than-2 bound on this guarantee.

Lemma 13. *If $N_{\max} \geq \ell + 1$, where $\ell = \lceil \frac{5-2\alpha}{2\alpha-1} \rceil$ then* opt $\geq \max\{1/2, 1 - f - m, m + \frac{m}{\ell}\}$. *Moreover, if $\alpha = 3/4$, then* alg$(F)/$opt $\leq 63/32 = 1.96875$.

Proof. By Lemma 1, opt $\geq 1/2$. To derive opt $\geq 1 - f - m$, note that there is no link $e \in$ OPT between two vertices of $V(C_n) \setminus (V(M) \cup V(F))$ because M is a maximal matching. Therefore, each vertex in $V(C_n) \setminus (V(M) \cup V(F))$ is covered by a different link in OPT $\setminus M$. We conclude that

$$\text{opt} \geq m + \frac{1}{n} \cdot |V(C_n) \setminus (V(M) \cup V(F))| = m + (1 - 2m - f) = 1 - f - m.$$

Since OPT is feasible, OPT $\setminus M$ needs to connect all the links of M. By Lemma 12, each link in $S \setminus (M \cup F)$ connects at most ℓ links of M. Then, OPT $\setminus M$ contains at least $|M|/\ell$ links. Therefore, opt $\geq m + m/\ell$, or equivalently $m \leq$ opt $\ell/(\ell+1)$. This proves the last bound for opt.

Now, by Lemma 10, alg$(F) \leq \alpha + (1-f-m)(1-\alpha) + m(1-\alpha)$. Using $1 \leq 2$opt, $(1-f-m) \leq$ opt and $m \leq$ opt $\ell/(\ell+1)$, we get alg$(F)/$opt $\leq (2\ell+1+\alpha)/(\ell+1)$. Among the values of $\alpha \in (1/2, 1]$ with $(5 - 2\alpha)/(2\alpha - 1)$ integer, the smallest bound occurs when $\alpha = 3/4$, or equivalently $\ell = 7$, giving the claimed guarantee (see Fig. 4).

4   Improving the Algorithm

In the previous section we proved that our Local Search algorithm is a $(63/32)$-approximation. We will now describe how to refine the analysis and enhance the algorithm so as to obtain the claimed 1.8703-approximation.

4.1   Refining the Analysis

We start by building a linear programming formulation that provides a stronger lower bound for $|$OPT$|$. To do so, we partition OPT into several subsets (one of them will be M), assigning a variable to each one of them representing its size, and provide relations and constraints these sizes should satisfy.

Let $A = A(F)$ and $B = B(F)$ be the internal and border vertices in $V(F)$ respectively. Let $C = V(M)$ and $D = V(C_n) \setminus (A \cup B \cup C)$ so that A, B, C and D partition $[n]$. Let OPT$_{ij} \subseteq$ OPT $\setminus M$, where $i, j \in \{A, B, C, D\}$, be the

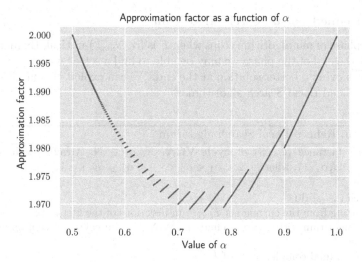

Fig. 4. Plot of the function $f(\alpha) = (2\ell + 1 + \alpha)/(\ell + 1)$ from the proof of Lemma 13. We can observe that the function is minimized when $\alpha = 3/4$.

links with an endpoint in i and the other in j. Our formulation creates variables for the sizes of each of these sets, allowing to encode the previously obtained constraints, such as requiring that every vertex in the cycle is incident to some link (Lemma 1), that any link from $\text{OPT}\setminus M$ can cross at most one link of the perimeter P (Lemma 11) and that M must be connected to the rest of the solution but links from $S \setminus F$ can cross a bounded number of links from M (Lemma 12). A careful analysis of the optimal solution for this formulation provides the following improved bound, please refer to the full version of the article for the details on the formulation and its analysis. In what follows, we define

$$f_\alpha := 3/(6 + 4\lceil (4 - 3\alpha)/(2\alpha - 1)\rceil + 2\lceil (2 - 3\alpha)/(2\alpha - 1)\rceil_+).$$

Lemma 14. *Let F be an $(\alpha, N_{\max})$-critical set with $N_{\max} \geq \lceil \frac{5-2\alpha}{2\alpha-1}\rceil + 1$. Then, it holds that $\text{alg}(F)/\text{opt} \leq 2 - 2(1 - \alpha)f_\alpha$.*

We can now prove a first improvement on the approximation ratio of the Local Search algorithm.

Theorem 2. *The approximation ratio of the Local Search Algorithm is at most $233/121 \approx 1.92562$.*

Proof. The expression f_α is an increasing step function of α with jumps and breakpoints on the countably infinite set $\mathcal{A}$ in which $(2 - 3\alpha)/(2\alpha - 1)$ or $(4 - 3\alpha)/(2\alpha - 1)$ are integers. Thus, $(1 - \alpha)f_\alpha$ is the area of the maximal rectangle with base $[\alpha, 1]$ lying below the plot of f_α, and our guarantee is 2 minus twice this area, which is maximized on the point $\alpha^* = 8/11 \in \mathcal{A}$ (see Fig. 5).

4.2   A Refined Local Search Algorithm

The first phase of our algorithm stops when F is $(\alpha, N_{\max})$-critical. By increasing α a bit at the end of the phase we may be able to find a set F' with larger utility than F, achieving a better solution at the end. We can exploit this intuition and obtain a refined Local Search algorithm.

Algorithm. Refined Local Search algorithm

Input: Instance (C_n, S) of **cycle VCA**. $N_{\max} \in \mathbb{N}$. Array of values $A = \{\alpha_1, \alpha_2, \cdots, \alpha_s, \alpha_{s+1}\}$, where $1/2 < \alpha_1 < \alpha_2 < \cdots < \alpha_s \le \alpha_{s+1} = 1$,

1: $F_0 \leftarrow \emptyset$.
2: **for each** $i \in [s]$ **do**
3:     Starting from the current set F_{i-1}, run iterations of the first phase of the Local Search algorithm with $\alpha = \alpha_i$ to find an $(\alpha_i, N_{\max})$-critical set F_i containing F_{i-1}.
4: **end for**
5: Find a minimal completion Q of F_s
6: Return (Q, F_s).

This allows to prove a generalized version of Lemma 14.

Lemma 15. *Let* ALG *be the output of the Refined Local Search algorithm for* $N_{\max} \ge \lceil \frac{5-2\alpha_1}{2\alpha_1-1} \rceil + 1$, *and let* $\varepsilon > 0$. *It holds that* $|\,\text{ALG}\,|/|\,\text{OPT}\,| \le 2 - 2\sum_{i=2}^{m+1}(\alpha_i - \alpha_{i-1})f_{\alpha_j}$.

Proof. Consider the sets $F_0, F_1, \ldots, F_s$ created by the Refined Local Search algorithm and denote $f^i = |F_i|/n$. For every $i \in [s]$, F_i was obtained from F_{i-1} by iteratively adding sets such that the marginal utility of that set is at least $(1 - \alpha_i)$ times the number of newly added vertices. We conclude that $U(F_i) - U(F_{i-1}) \ge (1 - \alpha_i)(|V(F_i)| - |V(F_{i-1})|)$. Using $U(F_0) = 0 = |V(F_0)|$, and that $\alpha_s = 1$ we conclude that $U(F_s) \ge \sum_{j=1}^s (1 - \alpha_j)(|V(F_s)| - |V(F_{i-1})|) = n\sum_{j=1}^s (\alpha_{j+1} - \alpha_j)f^j$.

Using that $3n = 3n\alpha_1 + 3n\sum_{j=1}^s (\alpha_{j+1} - \alpha_j)$ and Lemma 9 we get

$$\frac{|\,\text{ALG}\,|}{|\,\text{OPT}\,|} \le \frac{-2n + (3n - U(F_s))}{n\,\text{opt}} \le 2(-2 + 3\alpha_1) + \sum_{j=1}^s (\alpha_{j+1} - \alpha_j)\frac{(3 - f^j)}{\text{opt}}.$$

Note that if $f^j \ge f_{\alpha_j}$, then $(3 - f^j)/\text{opt} \le (3 - f_{\alpha_j})/\text{opt} \le 6 - 2f_{\alpha_j}$. On the other hand, if $f^j \le f_{\alpha_j}$, then as $f^j = |V(F_j)|/n$ and F_j is $(\alpha_j, N_{\max})$-critical, it is possible to prove that $(3 - f^j)/\text{opt} \le \Psi_\alpha(f^j) \le \Psi_\alpha(f_{\alpha_j}) = 6 - 2f_{\alpha_j}$ (please refer to the full version for the details).

Using these bounds on the expression for $|\,\text{ALG}\,|/|\,\text{OPT}\,|$ and that $6 = 6\alpha + 6\sum_{j=1}^s (\alpha_{j+1} - \alpha_j)$ we get,

$$\frac{|\,\text{ALG}\,|}{|\,\text{OPT}\,|} \le (6\alpha_1 - 4) + \sum_{j=1}^s (\alpha_{j+1} - \alpha_j)(6 - 2f_{\alpha_j}) = 2 - 2\sum_{j=1}^s (\alpha_{j+1} - \alpha_j)f_{\alpha_j}.$$

Now we can conclude our main theorem.

Proof (Proof of Theorem 1). Recall that $f_\alpha = 3/(6+4\lceil(4-3\alpha)/(2\alpha-1)\rceil+2\lceil(2-3\alpha)/(2\alpha-1)\rceil_+)$ is an increasing step function of α with jumps on the countably infinite set $\mathcal{A}$ in which $(2-3\alpha)/(2\alpha-1)$ or $(4-3\alpha)/(2\alpha-1)$ are integers. The function $\sum_{j=1}^{s}(\alpha_{j+1}-\alpha_j)f_{\alpha_j}$ is thus the area of the lower approximation by rectangles whose base are all the intervals $[\alpha_j, \alpha_{j+1}]$, and attains its maximum value $\int_{1/2}^{1} f_\alpha d\alpha$ by setting the values α_i as the sequence $\mathcal{A}$ of breakpoints of f_α (see Fig. 5). However, we require to only use finitely many points from $\mathcal{A}$; we cannot use points arbitrarily close to $1/2$ as this would require $N_{\max}$ to be arbitrarily large. By properly truncating the sequence, we can achieve in time $n^{O(1/\varepsilon)}$ a guarantee close to $2 - 2\int_{1/2}^{1} f_\alpha d\alpha \approx 1.87029$.

More in detail, we truncate the summation of Lemma 15 to the k-th term, obtaining the corresponding value of α by solving $\frac{4-3\alpha}{2\alpha-1} = k$, thus being $\alpha_k = \frac{4+k}{2k+3}$. We rewrite α_k as $1/2 + \hat{\varepsilon}$, obtaining $\hat{\varepsilon} = \frac{5}{4k+6}$. As f_α is an increasing function, we can bound the obtained approximation ratio by

$$2 - 2\int_{1/2+\hat{\varepsilon}}^{1} f_\alpha d\alpha \leq 2 - 2\int_{1/2}^{1} f_\alpha d\alpha + 2\int_{1/2}^{1/2+\hat{\varepsilon}} f_{\alpha_k} d\alpha = 1.87029 + 2\hat{\varepsilon}f_{\alpha_k}.$$

The value k can be chosen large enough so that the term $2\hat{\varepsilon}f_{\alpha_k}$ is small enough, leading to the claimed bound.

Regarding the running time, notice that every iteration of the first phase takes $|S|^{O(N_{\max})}$, but $|S| = O(n^2)$ and, for $\alpha = 1/2 + \varepsilon$, $N_{\max} \geq \lceil 2\frac{1}{\varepsilon} - 1\rceil = O(1/\varepsilon)$. Moreover, we have to repeat this iteration $\lceil(4-3(\frac{1}{2}+\varepsilon)/(2(\frac{1}{2}+\varepsilon)-1)\rceil = O(1/\varepsilon)$ times. Therefore, the total running time for the algorithm is $O(1/\varepsilon)n^{O(1/\varepsilon)} = n^{O(1/\varepsilon)}$.

5   Hardness of Approximation of Cycle VCA

In this section we prove that **cycle VCA** is APX-hard. The following theorem summarizes a hardness result from Gálvez et al. [14] in the context of edge-connectivity augmentation, stated in a way that will be useful for our purposes.

Theorem 3 [14]. *There exists a constant $\delta_0 > 0$ such that it is NP-hard to decide whether an instance of Cycle edge-connectivity augmentation admits a solution of size $\frac{n}{2}$, or no feasible solution for the instance has size at most $(1 + \delta_0)\frac{n}{2}$.*

Now we can prove that finding arbitrarily good solutions for our problem in polynomial time is hard as well.

Theorem 4. *There exists a constant $\delta > 0$ such that there is no $(1 + \delta)$-approximation for **cycle VCA** unless $P = NP$.*

Proof. We will show that if (C_n, S) is an instance of Cycle edge-connectivity augmentation, then

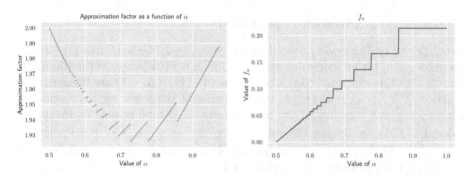

Fig. 5. Left: Plot of the function $f(\alpha) = 2 - 2(1 - \alpha)f_\alpha$ from Lemma 14. We can observe that this function is minimized when $\alpha = 8/11$. **Right:** Plot of f_α that illustrates the area under the curve that must be computed to derive the final approximation ratio according to the proof of Theorem 1.

1. If (C_n, S) admits a solution of size $n/2$, then the same input interpreted as an instance of **cycle VCA** admits a feasible solution of size $\frac{n}{2}$ too; and
2. If (C_n, S) does not admit a feasible solution of size at most $\left(1 - \delta_0\right)\frac{n}{2}$, then neither does the same input as an instance of **cycle VCA**.

The second statement follows directly since 3-vertex-connectivity implies 3-edge connectivity. For the first statement, we will prove that a feasible solution of size $n/2$ for Cycle edge-connectivity augmentation actually consists on a single circle component, being then feasible for **cycle VCA** thanks to Lemma 3. Indeed, if the solution has more than one circle component, no pair of components share border nodes as every node of the cycle is incident to exactly one link, and furthermore one of them has to be contiguous (i.e. a circle component L such that $V(L)$ is an interval on the cycle). However, in this case it is possible to disconnect the latter component by deleting the two edges of the cycle incident to its border nodes.

References

1. Adjiashvili, D.: Beating approximation factor two for weighted tree augmentation with bounded costs. ACM Trans. Algorithms **15**(2), 191–1926 (2019). https://doi.org/10.1145/3182395
2. Auletta, V., Dinitz, Y., Nutov, Z., Parente, D.: A 2-approximation algorithm for finding an optimum 3-vertex-connected spanning subgraph. J. Algorithms **32**(1), 21–30 (1999). https://doi.org/10.1006/jagm.1999.1006
3. Basavaraju, M., Fomin, F.V., Golovach, P., Misra, P., Ramanujan, M.S., Saurabh, S.: Parameterized algorithms to preserve connectivity. In: Esparza, J., Fraigniaud, P., Husfeldt, T., Koutsoupias, E. (eds.) ICALP 2014, Part I. LNCS, vol. 8572, pp. 800–811. Springer, Heidelberg (2014). https://doi.org/10.1007/978-3-662-43948-7_66
4. Byrka, J., Grandoni, F., Ameli, A.J.: Breaching the 2-approximation barrier for connectivity augmentation: a reduction to steiner tree. In: Proccedings of the 52nd

Annual ACM SIGACT Symposium on Theory of Computing (STOC), pp. 815–825. ACM (2020). https://doi.org/10.1145/3357713.3384301

5. Cecchetto, F., Traub, V., Zenklusen, R.: Bridging the gap between tree and connectivity augmentation: unified and stronger approaches. In: 53rd Annual ACM SIGACT Symposium on Theory of Computing (STOC), pp. 370–383. ACM (2021). https://doi.org/10.1145/3406325.3451086

6. Cheriyan, J., Gao, Z.: Approximating (unweighted) tree augmentation via lift-and-project, part I: stemless TAP. Algorithmica **80**(2), 530–559 (2017). https://doi.org/10.1007/s00453-016-0270-4

7. Cheriyan, J., Jordán, T., Ravi, R.: On 2-coverings and 2-packings of laminar families. In: Nešetřil, J. (ed.) ESA 1999. LNCS, vol. 1643, pp. 510–520. Springer, Heidelberg (1999). https://doi.org/10.1007/3-540-48481-7_44

8. Cheriyan, J., Karloff, H.J., Khandekar, R., Könemann, J.: On the integrality ratio for tree augmentation. Oper. Res. Lett. **36**(4), 399–401 (2008). https://doi.org/10.1016/j.orl.2008.01.009

9. Cohen, N., Nutov, Z.: A $(1 + \ln2)$-approximation algorithm for minimum-cost 2-edge-connectivity augmentation of trees with constant radius. Theor. Comput. Sci. **489–490**, 67–74 (2013). https://doi.org/10.1016/j.tcs.2013.04.004

10. Dinitz, E., Karnazov, A., Lomonosov, M.: On the structure of the system of minimum edge cuts of a graph. Stud. Discrete Optim. 290–306 (1976)

11. Even, G., Feldman, J., Kortsarz, G., Nutov, Z.: A 1.8 approximation algorithm for augmenting edge-connectivity of a graph from 1 to 2. ACM Trans. Algorithms **5**(2), 21:1-21:17 (2009). https://doi.org/10.1145/1497290.1497297

12. Fiorini, S., Groß, M., Könemann, J., Sanità, L.: Approximating weighted tree augmentation via Chvátal-Gomory cuts. In: Proceedings of the Twenty-Ninth Annual ACM-SIAM Symposium on Discrete Algorithms (SODA), pp. 817–831. SIAM (2018). https://doi.org/10.1137/1.9781611975031.53

13. Frederickson, G.N., JáJá, J.: Approximation algorithms for several graph augmentation problems. SIAM J. Comput. **10**(2), 270–283 (1981). https://doi.org/10.1137/0210019

14. Gálvez, W., Grandoni, F., Jabal Ameli, A., Sornat, K.: On the cycle augmentation problem: hardness and approximation algorithms. Theory Comput. Syst. **65**(6), 985–1008 (2021). https://doi.org/10.1007/s00224-020-10025-6

15. Goemans, M.X., Goldberg, A.V., Plotkin, S.A., Shmoys, D.B., Tardos, É., Williamson, D.P.: Improved approximation algorithms for network design problems. In: Proceedings of the Fifth Annual ACM-SIAM Symposium on Discrete Algorithms (SODA), pp. 223–232. ACM/SIAM (1994). http://dl.acm.org/citation.cfm?id=314464.314497

16. Grandoni, F., Kalaitzis, C., Zenklusen, R.: Improved approximation for tree augmentation: saving by rewiring. In: Proceedings of the 50th Annual ACM SIGACT Symposium on Theory of Computing (STOC), pp. 632–645. ACM (2018). https://doi.org/10.1145/3188745.3188898

17. Khuller, S., Thurimella, R.: Approximation algorithms for graph augmentation. J. Algorithms **14**(2), 214–225 (1993). https://doi.org/10.1006/jagm.1993.1010

18. Khuller, S., Vishkin, U.: Biconnectivity approximations and graph carvings. J. ACM **41**(2), 214–235 (1994). https://doi.org/10.1145/174652.174654

19. Kortsarz, G., Krauthgamer, R., Lee, J.R.: Hardness of approximation for vertex-connectivity network design problems. SIAM J. Comput. **33**(3), 704–720 (2004). https://doi.org/10.1137/S0097539702416736

20. Kortsarz, G., Nutov, Z.: A simplified 1.5-approximation algorithm for augmenting edge-connectivity of a graph from 1 to 2. ACM Trans. Algorithms **12**(2), 23:1-23:20 (2016). https://doi.org/10.1145/2786981

21. Kortsarz, G., Nutov, Z.: LP-relaxations for tree augmentation. Discret. Appl. Math. **239**, 94–105 (2018). https://doi.org/10.1016/j.dam.2017.12.033

22. Mader, W.: Ecken vom gradn in minimalenn-fach zusammenhängenden graphen. Archiv der Mathematik **23**(1), 219–224 (1972)

23. Marx, D., Végh, L.A.: Fixed-parameter algorithms for minimum-cost edge-connectivity augmentation. ACM Trans. Algorithms **11**(4), 271–2724 (2015). https://doi.org/10.1145/2700210

24. Nagamochi, H.: An approximation for finding a smallest 2-edge-connected subgraph containing a specified spanning tree. Discret. Appl. Math. **126**(1), 83–113 (2003). https://doi.org/10.1016/S0166-218X(02)00218-4

25. Nutov, Z.: 2-node-connectivity network design. CoRR abs/2002.04048 (2020). https://arxiv.org/abs/2002.04048

26. Nutov, Z.: A $4 + \varepsilon$ approximation for k-connected subgraphs. In: Proceedings of the 2020 ACM-SIAM Symposium on Discrete Algorithms (SODA), pp. 1000–1009. SIAM (2020). https://doi.org/10.1137/1.9781611975994.60

27. Nutov, Z.: On the tree augmentation problem. Algorithmica **83**(2), 553–575 (2020). https://doi.org/10.1007/s00453-020-00765-9

28. Spinrad, J.P.: Recognition of circle graphs. J. Algorithms **16**(2), 264–282 (1994). https://doi.org/10.1006/jagm.1994.1012

29. Traub, V., Zenklusen, R.: A better-than-2 approximation for weighted tree augmentation. CoRR abs/2104.07114 (2021). https://arxiv.org/abs/2104.07114

30. Végh, L.A.: Augmenting undirected node-connectivity by one. SIAM J. Discret. Math. **25**(2), 695–718 (2011). https://doi.org/10.1137/100787507

Constant Factor Approximation for Tracking Paths and Fault Tolerant Feedback Vertex Set

Václav Blažej, Pratibha Choudhary[(✉)], Dušan Knop, Jan Matyáš Křišťan, Ondřej Suchý, and Tomáš Valla

Department of Theoretical Computer Science, Faculty of Information Technology, Czech Technical University in Prague, Prague, Czech Republic
{blazeva1,pratibha.choudhary,dusan.knop,kristja6, ondrej.suchy,tomas.valla}@fit.cvut.cz

Abstract. Consider a vertex-weighted graph G with a source s and a target t. TRACKING PATHS requires finding a minimum weight set of vertices (*trackers*) such that the sequence of trackers in each path from s to t is unique. In this work, we derive a factor 66-approximation algorithm for TRACKING PATHS in weighted graphs and a factor 4-approximation algorithm if the input is unweighted. This is the first constant factor approximation for this problem. While doing so, we also study approximation of the closely related r-FAULT TOLERANT FEEDBACK VERTEX SET problem. There, for a fixed integer r and a given vertex-weighted graph G, the task is to find a minimum weight set of vertices intersecting every cycle of G in at least $r+1$ vertices. We give a factor $\mathcal{O}(r^2)$ approximation algorithm for r-FAULT TOLERANT FEEDBACK VERTEX SET if r is a constant.

1 Introduction

In this paper, we study the TRACKING PATHS problem, which involves finding a minimum weight set of vertices in a vertex-weighted simple graph G that can track moving objects in a network along the way from a source s to a target t. A set of vertices T is a *tracking set* if for any (simple) s-t path P in G the sequence $\overrightarrow{T_P}$ of the subset of vertices from T that appear in P, in their order along P, uniquely identifies the path P. That is, if $\overrightarrow{T_{P_1}} \neq \overrightarrow{T_{P_2}}$ holds for any two distinct s-t paths P_1 and P_2. Formally, the problem is the following.

> TRACKING PATHS
> **Input:** Undirected graph G with positive integer weights $w \colon V(G) \to \mathbb{N}$, a source s, and a target t.
> **Output:** A minimum weight tracking set $T \subseteq V(G)$.

The authors acknowledge the support of the OP VVV MEYS funded project CZ.02.1.01/0.0/0.0/16_019/0000765 "Research Center for Informatics". This work was supported by the Grant Agency of the Czech Technical University in Prague, grant No. SGS20/208/OHK3/3T/18.

J. Koenemann and B. Peis (Eds.): WAOA 2021, LNCS 12982, pp. 23–38, 2021.
https://doi.org/10.1007/978-3-030-92702-8_2

In the current age of information, social media networks have an important role in information exchange and dissemination. However, due to the unregulated nature of this exchange, spreading of rumours and fake news pose serious challenges in terms of authenticity of information [10,32]. Identifying and studying patterns of rumor spreading in social media poses a lot of challenges due to huge amounts of data in constant movement in large networks [36]. Tracing the sequence of channels (people, agents, ...) through which rumors spread can make it easier to contain the spread of such unwanted messages [27,37]. A basic approach would require tracing the complete route traversed by each message in the network. Here an optimum tracking set can serve as a resource-efficient solution for tracing the spread of rumors and dissolving them. Furthermore, TRACKING PATHS finds applications in tracking traffic movement in transport networks and tracing object movement in wireless sensor networks [1,30].

The graph theoretic version of the problem was introduced by Banik et al. [6], wherein the authors studied the unweighted (i.e., $w(v) = 1$ for all $v \in V(G)$) shortest path variant of the problem, namely TRACKING SHORTEST PATHS (i.e., the set T is required to uniquely identify each of the shortest s-t paths). They showed that this problem is NP-hard, even to approximate it within a factor of 1.1557. They also show that TRACKING SHORTEST PATHS admits a 2-approximation algorithm for planar graphs. Later parameterized complexity of TRACKING PATHS was studied in [4], where the problem was also proven to be NP-hard.

To the best of our knowledge, Eppstein et al. [17] were the first to study approximation algorithms for the unweighted TRACKING PATHS. They gave a 4-approximation algorithm when the input graph is planar. Recently this result was extended by Goodrich et al. [20] to a $(1 + \varepsilon)$-approximation algorithm for H-minor free graphs and an $\mathcal{O}(\log \mathrm{OPT})$-approximation algorithm for general (unweighted) graphs [20]. They also gave an $\mathcal{O}(\log n)$-approximation algorithm for TRACKING PATHS. The existence of constant factor application algorithm was posed as an open problem by Eppstein et al. [17]. In this paper we answer this affirmatively.

Theorem 1. *There is a 66-approximation algorithm for* TRACKING PATHS *in weighted graphs and a 4-approximation algorithm if the input is unweighted.*

There exists an interesting connection between FEEDBACK VERTEX SET (FVS) and TRACKING PATHS. Before we discuss this in more detail, we introduce FVS and its fault tolerant variant. Formally, for a given vertex-weighted graph $G = (V, E)$, FVS requires finding a minimum weight set of vertices $S \subseteq V$, referred to as a *feedback vertex set* (fvs), such that the graph induced by the vertex set $V \setminus S$ does not have any cycles. FVS is a classical NP-hard problem [25] that has been thoroughly studied in graph theory. An *r-fault tolerant feedback vertex set* (*r-ftfvs*) is a set of vertices that intersect with each cycle in the graph in at least $r + 1$ vertices; finding a minimum weight r-ftfvs is the r-FAULT TOLERANT FEEDBACK VERTEX SET problem [31]. Note that if a graph has a cycle of length less than or equal to r, then it cannot have an r-ftfvs.

Relation Between FVS *and* TRACKING PATHS. For a graph G with source s and destination t, if each vertex and edge participates in an s-t path, then we refer to G as a preprocessed graph. It is known that in a preprocessed graph, a tracking set is also a feedback vertex set [4]. Thus, the weight of a minimum feedback vertex set serves as a lower bound for the weight of a tracking set in preprocessed graphs. This lower bound has proven to be helpful in the analysis of TRACKING PATHS. However, approximating TRACKING PATHS has been challenging since the size of a tracking set can be arbitrarily larger than that of a minimum fvs.

Further, it is known [11,17] that in a graph G, if a set of vertices T contains at least three vertices from each cycle in G, then T is a tracking set for G. Thus, a 2-fault tolerant feedback vertex set is also a tracking set. In this paper, we borrow inspiration from this concept to derive a polynomial time algorithm to compute an approximate tracking set. In particular, we start with finding an fvs for the input graph G, and then identify cycles that need more vertices as trackers additional to the ones selected as feedback vertices.

Observe that a feedback vertex set is indeed a 0-fault tolerant fvs. Misra [31] gave a 3-approximation algorithm for the problem of finding a 1-fault tolerant fvs in unweighted graphs and 34-approximation algorithm for weighted graphs. In this paper, we give an approximation algorithm for finding an r-fault tolerant feedback vertex set, where r is a constant. We do this by using the MULTICUT IN FORESTS problem (see Sect. 2) as an auxiliary problem. Misra [31] pointed out that the complexity of r-FAULT TOLERANT FEEDBACK VERTEX SET is not known for $r \geq 2$ and asked for an approximation algorithm.

Theorem 2. *There is an $\mathcal{O}(r^2)$-approximation algorithm for r-FAULT TOLERANT FEEDBACK VERTEX SET in weighted graphs, where r is a constant.*

It is worth mentioning that our approach relies on explicit enumeration of certain cycles in the input graph G. This can be done in polynomial time if r is a constant (see Observation 9). Thus, it remains open how to approximate the r-FAULT TOLERANT FEEDBACK VERTEX SET if r depends on the size of the input.

Motivation for (Fault Tolerant) FVS. The FVS problem is motivated by applications in deadlock recovery [18,34], Bayesian inference [7], VLSI design [24], and other areas. Fault tolerant solutions are crucial to real world applications that are prone to failure of nodes in a network or entities in a system [33]. In the case of FVS, the failure corresponds to not being able to eliminate the node from the network.

Related Work. There has been a lot of heuristic based work on the problem of tracking moving objects in a network [29,35,38]. Parameterized complexity of TRACKING SHORTEST PATHS and TRACKING PATHS was studied in [4,5,8,12, 13,17]. FEEDBACK VERTEX SET is known to admit a 2-approximation algorithm which is tight under UGC [3,14]. The best known parameterized algorithm for FVS runs in $2.7^k \cdot n^{\mathcal{O}(1)}$ time, where k is the size of the solution [28]. It is worth noting that Misra [31] uses MULTICUT IN FORESTS as a subroutine as well. The edge version is known to admit an LP formulation whose matrix is totally unimodular [19] if the family of paths is non-crossing; thus, it is solvable in polynomial time. A related problem is the d-HURDLE MULTIWAY CUT for which a factor 2-approximation algorithm is known (and it is again tight under UGC) [15].

Preliminaries and Notations. We refer to [16] for the standard graph theory terminology. All paths that we consider in this work are simple paths. For a graph G, we use $V(G)$ to denote its vertex set. For a set $S \subseteq V(G)$ we use $G \setminus S$ to denote a graph that results from the deletion of vertex set S and the edges incident to $V(S)$. For a weight function $w: V(G) \to \mathbb{N}$, let $w(S)$ for $S \subseteq V(G)$ denote the sum of respective elements, i.e., $\sum_{v \in S} w(v)$. An unweighted version of the problem is obtained by assigning all vertices the same weight. In fact, we can also omit the weights in this case. We write vectors in boldface and their entries in normal font, i.e., x_3 is the third entry of a vector $\boldsymbol{x}$. If we apply minimum to two vectors, then it is applied entry-wise. We write $\mathbf{1}^n$ for the vector of n ones; we omit the superscript if the dimension is clear from the context.

2   Vertex Multicut in Forests

In this section, we gather polynomial time (approximation) algorithms for solving VERTEX MULTICUT IN FORESTS. The algorithms are used later in the subsequent sections to derive the approximation algorithms for r-FAULT TOLERANT FVS and TRACKING PATHS.

MULTICUT IN FORESTS (MULTICUT IN FORESTS)
Input: A forest $F = (V, E)$, weight function $w: V \to \mathbb{N}$, terminal pairs $(s_1, t_1), \ldots, (s_\ell, t_\ell)$.
Output: A minimum weight set of vertices $S \subseteq V$ such that in the graph $F \setminus S$ there is no path between s_i and t_i for $i \in [\ell]$.

In this work, we consider the UNRESTRICTED version of MULTICUT IN FORESTS i.e., a solution set is allowed to contain vertices from the terminal pairs. We will consider a set of paths $\mathcal{P}$ instead of a set of terminal pairs. It is not hard to see that these two versions are equivalent on forests; if the terminals in a pair belong to different trees we can discard the pair since there does not exist any path between them, otherwise there exists a unique path between each pair of terminals, which the solution S should intersect.

While MCF is NP-hard [9], the unweighted version is polynomial time solvable [9,23]. However, recently the following strong version of constant factor approximability of the problem was shown (in fact, the result holds for all chordal graphs, not just for forests).

Consider the natural ILP formulation of the problem, where we introduce a binary variable $x_v \in \{0, 1\}$ for each vertex $v \in V$, describing whether the corresponding vertex should or should not be taken into a solution. At least one vertex must be taken from each path from $\mathcal{P}$ to construct a feasible solution. Consider the following LP relaxation of the natural ILP.

$$\text{Minimize} \quad \sum_{v \in V} w(v) \cdot x_v$$

$$\text{subject to} \quad \sum_{v \in V(P)} x_v \geq 1 \quad \forall P \in \mathcal{P} \qquad \text{(LP}_{\text{MCF}})$$

$$0 \leq x_v \leq 1 \quad \forall v \in V.$$

Proposition 3 (Agrawal et al. [2, Lemma 5.1]). *For a given instance of* MULTICUT IN FORESTS *one can find a solution S such that $w(S) \leq 32\text{OPT}$ in polynomial time, where* OPT *is the objective value of an optimal solution to the corresponding* (LP$_{\text{MCF}}$).

We also need the following result of similar nature for unweighted forests, strengthening the polynomial time solvability of the problem.

Lemma 4. *If all the weights are equal, then there is an integral optimal solution for* (LP$_{\text{MCF}}$). *Furthermore, such a solution can be found in polynomial time.*

Proof. It is enough to show the lemma for each tree T of F separately.

Root the tree T in an arbitrary vertex r. Among all optimal solutions to the LP, consider the solution $\boldsymbol{x}$ that minimizes $\sum_{v \in V(T)} x_v \cdot \text{dist}(v, r)$.

Assume for the sake of contradiction, that $\boldsymbol{x}$ is not integral. Let u be a vertex with $0 < x_u < 1$ and the maximum distance from the root r. Note that, in particular, for all descendants v of u it holds that $x_v \in \{0, 1\}$. Let us first assume that $u \neq r$ and let p be the parent of u. We define $\widehat{\boldsymbol{x}}$ as

$$\widehat{x}_v = \begin{cases} 0 & \text{if } v = u, \\ \min\{x_p + x_u, 1\} & \text{if } v = p, \\ x_v & \text{otherwise.} \end{cases}$$

We claim that $\widehat{\boldsymbol{x}}$ is a solution to (LP$_{\text{MCF}}$). Obviously, $0 \leq \widehat{x}_v \leq 1$ for every $v \in V$. Let $P \in \mathcal{P}$. If $u \notin V(P)$, then $\sum_{v \in V(P)} \widehat{x}_v \geq \sum_{v \in V(P)} x_v \geq 1$. If $u \in V(P)$, then either the path P is fully contained in the subtree of F rooted in u, or $p \in V(P)$. In the first case, we have $\sum_{v \in V(P)} \widehat{x}_v \geq \left(\sum_{v \in V(P)} x_v \right) - x_u \geq 1 - x_u > 0$. Since all the summands on the left hand side are integral by the choice of u, it follows that $\sum_{v \in V(P)} \widehat{x}_v \geq 1$. In the later case, if $\widehat{x}_p = 1$, then $\sum_{v \in V(P)} \widehat{x}_v \geq 1$ and otherwise $\sum_{v \in V(P)} \widehat{x}_v = \left(\sum_{v \in V(P)} x_v \right) + x_u - x_u = \sum_{v \in V(P)} x_v \geq 1$.

Hence, $\widehat{\boldsymbol{x}}$ is a solution to (LP$_{\text{MCF}}$). Furthermore, we have that $\sum_{v \in V} w(v) \cdot \widehat{x}_v \leq \sum_{v \in V} w(v) \cdot x_v$ (since all the weights are equal) and $\sum_{v \in V} x_v \cdot \text{dist}(v, r) > \sum_{v \in V} \widehat{x}_v \cdot \text{dist}(v, r)$ which is a contradiction. The case $u = r$ can be proved using a similar argument for $\widehat{\boldsymbol{x}}$ such that $\widehat{x}_u = 0$ and $\widehat{x}_v = x_v$ if $v \neq u$.

The second part of the lemma follows from polynomial time algorithm for MCF [9], since any optimal solution to the instance of MCF represents an optimal solution for the corresponding (LP$_{\text{MCF}}$) by the first part of the lemma. □

For the rest of the paper we use μ to denote the best LP relative approximation ratio achievable for MCF with respect to LP relative approximation. That is $\mu \leq 32$ for weighted instances and $\mu = 1$ for unweighted instances.

3 Approximate r-Fault Tolerant Feedback Vertex Set

In this section, we give an algorithm for computing an approximate r-fault tolerant feedback vertex set for undirected weighted graphs, for any fixed integer $r \geq 2$. Recall that an r-fault tolerant feedback vertex set is a set of vertices that contains at least $r + 1$ vertices from each cycle in the graph. A polynomial time algorithm for computing a constant factor approximate 1-fault tolerant feedback vertex set was given by Misra [31]. The factor can be easily observed to be $2 + \mu$, where μ is the best possible approximation ratio for MCF. We start in Sect. 3.1 by giving an algorithm for finding 2-fault tolerant feedback vertex set. Later, in Sect. 3.2, we show how this technique can be generalized to give an r-fault tolerant feedback vertex set for any $r \geq 3$.

3.1 Two-Fault Tolerant FVS

> 2-FAULT TOLERANT FEEDBACK VERTEX SET
> **Input:** Undirected graph G and a weight function $w \colon V(G) \to \mathbb{N}$.
> **Output:** A minimum weight set of vertices $S \subseteq V(G)$ such that for each cycle C in G, it holds that $|V(C) \cap S| \geq 3$.

Let $G = (V, E)$ be the input graph. First, we compute a $(2 + \mu)$-approximate 1-fault tolerant feedback vertex set S for G using the algorithm of Misra [31]. Note that S contains at least two vertices from each cycle in G. Our goal is to compute a vertex set that contains at least three vertices from each cycle in G. Hence, for each cycle C in G for which $|V(C) \cap S| = 2$, we need to pick at least one more vertex from C into our solution. We first identify, which pairs of vertices from S are involved in such cycles. This can be done in polynomial time, by considering each pair of vertices $\{a, b\} \in S$ and checking the graph $G_{ab} = G \setminus (S \setminus \{a, b\})$ for cycles. If no such G_{ab} contains a cycle, then we return S as a 2-fault tolerant feedback vertex set. Otherwise, there exists at least one cycle in G such that S contains exactly two vertices from it. Observe that even though S does not contain three vertices from each cycle in G, S might contain at least three vertices from some cycles in G. We shall ignore such cycles.

If a cycle C in G intersects with S at vertices a and b, then there exist two vertex-disjoint paths P_1 and P_2 between a and b in G_{ab}, such that $V(P_1) \cup V(P_2) = V(C)$. In order to find a 2-fault tolerant fvs that extends S, we need to ensure that at least one vertex from $(V(P_1) \cup V(P_2)) \setminus \{a, b\}$ is included in the solution. Observe that paths P_1 and P_2 are a pair of paths between the vertices a, b in the graph $G \setminus (S \setminus \{a, b\})$, which implies that the subpaths between the neighbors of a and b are paths in the graph $G \setminus S$, which is a forest. As each pair of paths is uniquely determined by the neighbors of a and b on P_1 and P_2, there are at most $\mathcal{O}(n^4)$ such pairs in total and all of them can be found in $\mathcal{O}(n^5)$ time. We create a family $\mathcal{P}$ of all such pairs of vertex disjoint paths between each pair of vertices in S. More precisely, for each pair P_1', P_2' of such paths of length at least 2 we obtain P_1 and P_2 by removing a and b from P_1' and P_2', respectively. Then we add the pair $\{P_1, P_2\}$ to $\mathcal{P}$. If a and b are adjacent, then for each a-b-path P_1' of length at least 2 in $G \setminus (S \setminus \{a, b\})$, we add to $\mathcal{P}$ the pair $\{P_1, P_1\}$, where P_1 is obtained from P_1' by removing a and b.

Next, we use the following linear program to identify which path among each pair should be selected, from which a vertex would be picked in order to be included in the solution.

$$\text{Minimize} \quad \sum_{v \in V \setminus S} w(v) \cdot x_v$$

$$\text{subject to} \quad \sum_{v \in V(P_1) \cup V(P_2)} x_v \geq 1 \quad \forall \{P_1, P_2\} \in \mathcal{P} \qquad (\text{LP}_{\text{pairs}})$$

$$0 \leq x_v \leq 1 \qquad \forall v \in V \setminus S.$$

We solve the above linear program in polynomial time using the Ellipsoid method [21] (see also [22, Chapter 3]). Let $\boldsymbol{x}^*$ be an optimal solution for the above LP and $\text{OPT}_{\boldsymbol{x}^*}$ be its value.

Observation 5. *Let S^* be a 2-fault tolerant fvs in G (not necessarily extending S) and let $\text{OPT}_{\boldsymbol{x}^*}$ be the optimum value of $(\text{LP}_{\text{pairs}})$. Then, $\text{OPT}_{\boldsymbol{x}^*} \leq w(S^*)$.*

Proof. We claim that the vector $\widehat{\boldsymbol{x}}$ defined for $v \in V \setminus S$ as

$$\widehat{x}_v = \begin{cases} 1 & \text{if } v \in S^* \\ 0 & \text{otherwise} \end{cases}$$

constitutes a solution to $(\text{LP}_{\text{pairs}})$. In order to see this let $\{P_1, P_2\} \in \mathcal{P}$ be a pair of paths and let C be the cycle formed by these paths together with some $a, b \in S$. Since S^* is a 2-fault tolerant fvs, we have $|S^* \cap V(C)| \geq 3$ and thus $|S^* \cap (V(C) \setminus S)| = \sum_{v \in V(P_1) \cup V(P_2)} x_v \geq 1$ as needed. Hence, $\text{OPT}_{\boldsymbol{x}^*} \leq w(S^*)$. $\square$

Next, we create a set of paths $\mathcal{P}_{\boldsymbol{x}^*}$. For each path $P \in \bigcup_{\{P_1, P_2\} \in \mathcal{P}} \{P_1, P_2\}$, we include P in $\mathcal{P}_{\boldsymbol{x}^*}$ if $\sum_{v \in V(P)} x_v^* \geq \frac{1}{2}$ holds. Through this process, we are selecting the paths from which we will include at least one vertex in our solution.

Finally, we create an instance $(G \setminus S, w|_{(V \setminus S)}, \mathcal{P}_{\boldsymbol{x}^*})$ of MULTICUT IN FORESTS. Consider the corresponding LP relaxation.

$$\text{Minimize} \quad \sum_{v \in V \setminus S} w(v) \cdot y_v$$

$$\text{subject to} \quad \sum_{v \in V(P)} y_v \geq 1 \quad \forall P \in \mathcal{P}_{\boldsymbol{x}^*} \qquad (\text{LP}_{\text{paths}})$$

$$0 \leq y_v \leq 1 \quad \forall v \in V \setminus S.$$

Lemma 6. *Let* x^* *be an optimal solution of* (LP$_{\text{pairs}}$) *and let* OPT$_{x^*}$ *be its objective value. Then,* $y = \min\{1, 2x^*\}$ *is a solution to* (LP$_{\text{paths}}$). *In particular,* OPT$_{y^*} \leq 2 \cdot$ OPT$_{x^*}$ *holds, where* OPT$_{y^*}$ *is the value of an optimal solution to* (LP$_{\text{paths}}$).

Proof. Recall that we have $\sum_{v \in V(P)} x_v^* \geq \frac{1}{2}$ for every path $P \in \mathcal{P}_{x^*}$, by the definition of $\mathcal{P}_{x^*}$. Thus, we have $\sum_{v \in V(P)} y_v = \sum_{v \in V(P)} \min\{1, 2 \cdot x_v^*\} \geq \min\{1, \sum_{v \in V(P)} 2 \cdot x_v^*\} \geq 1$ for all $P \in \mathcal{P}_{x^*}$ and clearly $0 \leq y_v \leq 1$ for all $v \in V \setminus S$. We conclude that y is a solution to (LP$_{\text{paths}}$). □

We now show how to combine the $(2 + \mu)$-approximate 1-fault tolerant fvs with an approximate solution for MULTICUT IN FORESTS to obtain a $(2 + 3\mu)$-approximate 2-fault tolerant fvs.

Lemma 7. *Let* S *be an* α-*approximate 1-fault tolerant fvs and let* y *be an integral solution to* (LP$_{\text{paths}}$) *of weight at most* μ *times the weight of an optimal solution. Then,* $S' = S \cup \{v \in V \setminus S \mid y_v = 1\}$ *is an* $(\alpha + 2\mu)$-*approximate 2-fault tolerant fvs.*

Proof. Let S^* be an optimal 2-fault tolerant fvs. We know that $w(S) \leq \alpha \cdot w(S^*)$. By Observation 5 there is a solution x to (LP$_{\text{pairs}}$) with $\sum_{v \in V \setminus S} w(v) \cdot x_v \leq w(S^*)$. Thus, by Lemma 6 we have $\sum_{v \in V \setminus S} w(v) \cdot y_v \leq 2\mu \cdot w(S^*)$. In total we get

$$w(S') = w\left(S \cup \{v \in V \setminus S \mid y_v = 1\}\right)$$
$$= w(S) + \sum_{v \in V \setminus S} w(v) \cdot y_v$$
$$\leq \alpha \cdot w(S^*) + 2\mu \cdot w(S^*)$$
$$= (\alpha + 2\mu)w(S^*).$$

It is not hard to see that S' is a 2-fault tolerant fvs. Indeed, if S contains at least three vertices in a cycle of the input graph, so does S'. Thus, we can focus on a cycle C with $V(C) \cap S = \{a, b\}$. If this is the case, then the conditions of (LP$_{\text{pairs}}$) imply that in y it holds $y_v = 1$ for some $v \in V(C) \setminus \{a, b\}$ (follows from the construction of (LP$_{\text{paths}}$)). Therefore, we have $|V(C) \cap S'| \geq 3$. □

Corollary 8. *There is a 5-approximation algorithm for unweighted* 2-FAULT TOLERANT FVS *and 98-approximation algorithm for weighted* 2-FAULT TOLERANT FVS.

Proof. We begin with the $(2 + \mu)$-approximation algorithm for 1-FAULT TOLERANT FVS by Misra [31]. In polynomial time we construct (LP$_{\text{pairs}}$) and obtain an optimal solution x^* for it. Based on that we construct $\mathcal{P}_{x^*}$ and (LP$_{\text{paths}}$) in polynomial time. By Proposition 3 or Lemma 4 one can in polynomial time find an integral solution to (LP$_{\text{paths}}$) of weight at most μ times the weight of an optimal solution. By Lemma 7 this solution combined with the initial 1-fault tolerant fvs gives $(2 + 3\mu)$-approximate 2-fault tolerant fvs. The algorithm works in polynomial time as it uses polynomial-time routines. □

3.2   Higher Fault Tolerant FVS

Now we explain the procedure to scale up the algorithm from Sect. 3.1 to compute an r-fault tolerant fvs for $r \geq 3$.

---

r-FAULT TOLERANT FEEDBACK VERTEX SET
Input: Undirected graph G, weight function $w \colon V \to \mathbb{N}$.
Output: A minimum weight set of vertices $S \subseteq V(G)$ such that for each cycle C in G, it holds that $|V(C) \cap S| \geq r + 1$.

For the rest of the section we assume that $r \geq 3$ is a fixed constant. We follow a recursive process to compute an r-fault tolerant fvs. We start with an approximate solution S for $(r-1)$-FAULT TOLERANT FVS. Note that S contains at least r vertices from each cycle in G. Similar to the process in algorithm in Sect. 3.1, here we identify every group of r vertices that are involved in a cycle C in G, such that $|V(C) \cap S| = r$. Such cycles can be found by checking whether the graph $G_X = G \setminus (S \setminus X)$, for $X \subseteq S$ such that $|X| = r$, contains a cycle. If no such G_X contains a cycle, then S is an r-fault tolerant fvs, and we return it as a solution. Else, we focus on cycles which contain exactly r vertices from S.

We create a family $\mathcal{P}$ of path sets in the following way. For each cycle C that contains exactly r vertices of S, labeled in cyclic order along C as vertices $\{v_1, v_2, \ldots, v_r\}$, we consider the r paths, say $P_1', P_2', \ldots, P_r'$, where P_i' starts in v_i and ends in v_{i+1} (modulo r) and P_i' is a subpath of C. We remove paths with only two vertices, and we shorten the rest by removing their end vertices, leaving us with paths $P_1, \ldots, P_s$, where $s \leq r$ (as some paths may have been removed). If the set of paths $\{P_1, \ldots, P_s\}$ is non-empty, then we add it to the family $\mathcal{P}$. If the set $\{P_1, \ldots, P_s\}$ is empty, then we have found a cycle of length r and we report that there is no r-fault tolerant feedback vertex set for G as we cannot choose at least $r + 1$ vertices on a cycle of length r.

Observation 9. *Construction of $\mathcal{P}$ can be done in $n^{\mathcal{O}(r)}$ time.*

Proof. There are no more than $\binom{n}{r}$ subsets of S of order r. Each such subset X can be a part of many different cycles. To find all such cycles we take each of its $r!$ orderings $v_1, \ldots, v_r$ and we fix a predecessor and a successor (taken out of the respective vertex neighborhood) of each v_i. This constitutes at most n^{2r} different possibilities for each ordering of X. There is at most one path from the successor of v_i to the predecessor of v_{i+1} (modulo r) in $(G \setminus S)$ as it is a forest. As the paths are now fixed, we just need to check whether they form a cycle by checking that the paths are vertex disjoint, which can be done in polynomial time. Altogether, we have $\binom{n}{r} r! n^{2r + \mathcal{O}(r)} = n^{\mathcal{O}(r)}$ time. $\square$

Once the family $\mathcal{P}$ is computed, we solve the following linear program using the Ellipsoid method in polynomial time. Let $\boldsymbol{x}^*$ be its optimal solution and OPT_{x^*} its value.

$$\text{Minimize} \quad \sum_{v \in V \setminus S} w(v) \cdot x_v$$

$$\text{subject to} \quad \sum_{v \in \bigcup_{i=1}^{s} V(P_i)} x_v \geq 1 \quad \forall \{P_1, \ldots, P_s\} \in \mathcal{P} \qquad (\text{LP}_{s\text{-tuples}})$$

$$0 \leq x_v \leq 1 \qquad \forall v \in V \setminus S.$$

Similarly to Observation 5 we get the following.

Observation 10. *Let S^* be an r-fault tolerant fvs in G and let OPT_{x^*} be the optimum value of* (LP$_{s\text{-tuples}}$). *Then,* $\text{OPT}_{x^*} \leq w(S^*)$.

Next, we create a set of paths $\mathcal{P}_{x^*}$. For each path $P \in \bigcup_{\{P_1,\ldots,P_s\} \in \mathcal{P}} \{P_1, \ldots, P_s\}$, we include P in $\mathcal{P}_{x^*}$ if $\sum_{v \in V(P)} x_v^* \geq \frac{1}{r}$. As in Sect. 3.1, we create an instance $(G \setminus S, w|_{(V \setminus S)}, \mathcal{P}_{x^*})$ of MULTICUT IN FORESTS and consider its LP relaxation (LP$_{\text{paths}}$).

Lemma 11. *Let S be an $(r-1)$-fault tolerant fvs. Let x^* be an optimal solution of* (LP$_{s\text{-tuples}}$) *and let OPT_{x^*} be its objective value. Then, $y = \min\{1, r \cdot x^*\}$ is a solution to* (LP$_{\text{paths}}$) *for $\mathcal{P}_{x^*}$. In particular, $\text{OPT} \leq r \cdot \text{OPT}_{x^*}$ holds, where OPT is the value of an optimal solution to* (LP$_{\text{paths}}$).

Proof. Recall that we have $\sum_{v \in V(P)} x_v^* \geq \frac{1}{r}$ for each path $P \in \mathcal{P}_{x^*}$. Thus, we have $\sum_{v \in V(P)} y_v = \sum_{v \in V(P)} \min\{1, r \cdot x_v^*\} \geq \min\{1, \sum_{v \in V(P)} r \cdot x_v^*\} \geq 1$ for all $P \in \mathcal{P}$ and clearly $0 \leq y_v \leq 1$ for all $v \in V \setminus S$. We conclude that y is a solution to (LP$_{\text{paths}}$). $\square$

Lemma 12. *Let r be a constant. Let S be an α-approximate $(r-1)$-fault tolerant fvs and let y be a solution to* (LP$_{\text{MCF}}$) *induced by $\mathcal{P}_{x^*}$ with weight at most μ times the optimal. Then, $S' = S \cup \{v \in V \setminus S \mid y_v = 1\}$ is an $(\alpha + \mu \cdot r)$-approximate r-fault tolerant fvs.*

Proof. Let S^* be an optimal r-fault tolerant fvs. We know that $w(S) \leq \alpha \cdot w(S^*)$. By Observation 10 there is a solution x to (LP$_{s\text{-tuples}}$) with $\sum_{v \in V \setminus S} w(v) \cdot x_v \leq w(S^*)$. Thus, by Lemma 11 we have $\sum_{v \in V \setminus S} w(v) \cdot y_v \leq \mu \cdot r \cdot w(S^*)$. In total we get

$$w(S') = w\left(S \cup \{v \in V \setminus S \mid y_v = 1\}\right)$$
$$= w(S) + \sum_{v \in V \setminus S} w(v) \cdot y_v$$
$$\leq \alpha \cdot w(S^*) + \mu \cdot r \cdot w(S^*)$$
$$= (\alpha + \mu r) w(S^*).$$

Let us prove that S' is r-fault tolerant. If S contains at least $r + 1$ vertices in a cycle of the input graph, so does S'. Thus, we can focus on a cycle C with $V(C) \cap S = \{v_1, v_2, \ldots, v_r\}$. If this is the case, then the conditions of (LP$_{s\text{-tuples}}$) imply that in y it holds $y_v = 1$ for some $v \in V(C) \setminus \{v_1, v_2, \ldots, v_r\}$ (follows from the construction of (LP$_{\text{paths}}$)). Therefore, we have $|V(C) \cap S'| \geq r + 1$ for every such C. Hence, S' is an r-fault tolerant fvs. $\square$

Theorem 13 (precise version of Theorem 2). *Let r be a constant. There is a $(2 + \sum_{i=1}^{r} i \cdot \mu)$-approximation algorithm for r-*FAULT TOLERANT FEEDBACK VERTEX SET.*

Proof. The 1-FAULT TOLERANT FVS problem has a $(2 + \mu)$-approximation algorithm due to Misra [31]. By Lemma 12 we add $\mu \cdot r$ to the approximation factor when devising r-fault tolerant fvs from $(r - 1)$-fault tolerant fvs it follows that for an arbitrary r we have $(2 + \mu \cdot \sum_{i=1}^{r} i)$-approximation algorithm for finding an r-fault tolerant fvs.

The algorithm by Misra [31] gives the $(2 + \mu)$-approximation for 1-FAULT TOLERANT FVS in polynomial time. We showed how to devise an s-fault tolerant fvs from an $(s-1)$-fault tolerant fvs. We use $r-1$ such steps to incrementally increase the fault tolerance of the fvs. In step of devising s-fault tolerant fvs we use $n^{\mathcal{O}(s)}$ time to find $\mathcal{P}$ as seen in Observation 9, then we construct and solve (LP$_{s\text{-tuples}}$) in polynomial time using the Ellipsoid method and then construct (LP$_{\text{MCF}}$) and solve it using either Proposition 3 or Lemma 4 in polynomial time. As r is a constant we conclude that our $(2+\mu\cdot\sum_{i=2}^{r} i)$-approximation algorithm for r-FAULT TOLERANT FVS has polynomial time complexity. $\qquad\square$

4   Approximate Tracking Set

In this section, we give a constant factor approximation algorithm for TRACKING PATHS.

Let $G = (V, E)$ be the input graph and s and t the source and the target. We start by applying the following reduction rule on G. This can clearly be done in polynomial time.

Reduction Rule 1 (Banik et al. [4]). *If there exists a vertex or an edge that does not participate in any s-t path, then delete it.*

We use the term *reduced graph* to denote a graph that has been preprocessed using Reduction Rule 1. For the sake of simplicity, after the application of reduction rule, we continue to refer to the reduced graph as G.

Next, we describe *local source-destination pair (local s-t pair)*, a concept that has served as crucial for developing efficient algorithms for TRACKING PATHS [4, 11,13,17]. For a subgraph $G' \subseteq G$, and vertices $a, b \in V(G')$, we say that a, b is a *local s-t pair* for G' if

1. there exists a path in G from s to a, say P_{sa},
2. there exists a path in G from b to t, say P_{bt},
3. $V(P_{sa}) \cap V(P_{bt}) = \emptyset$, and
4. $V(P_{sa}) \cap V(G') = \{a\}$ and $V(P_{bt}) \cap V(G') = \{b\}$.

Note that a subgraph can have more than one local source-destination pair. It can be verified in $\mathcal{O}(n^2)$ time whether a pair of vertices $a, b \in V(G')$ form a local source-destination pair for G' by checking if there exist disjoint paths from s to a and b to t in the graph $G \setminus (V(G') \setminus \{a,b\})$, using the disjoint path algorithm from [26].

Observation 14. *Let G be a graph and let G' be a subgraph of G. We can verify in polynomial time whether $a, b \in V(G')$ is a local s-t pair for G'.*

We recall the following lemma from previous work.

Lemma 15 ([11, Lemma 2])**.** *In a graph G, if $T \subseteq V(G)$ is not a tracking set for G, then there exist two s-t paths with the same sequence of trackers, and they form a cycle C in G, such that C has a local source a and a local destination b, and $T \cap (V(C) \setminus \{a, b\}) = \emptyset$.*

Eppstein et al. [17] mentioned that a 2-fault tolerant feedback vertex set is always a tracking set. Here we use a variation of this idea to compute an approximate tracking set. Specifically, we start with a 2-approximate feedback vertex set and then identify the cycles that contain only one or two feedback vertices. We check if these cycles need more vertices as trackers and we use (LP$_\text{pairs}$) and (LP$_\text{MCF}$) explained in the previous section to add them.

Now we present the algorithm for computing a $2(1+\mu)$-approximate tracking set in polynomial time. We start by computing a 2-approximate feedback vertex set S on the reduced graph G using the algorithm by Bafna et al. [3]. We first check whether S is a tracking set for G by using the tracking set verification algorithm given in [4]. If it is a tracking set, we return S as the solution, otherwise we proceed further.

If S is not a tracking set, we will find vertices on which to place additional trackers in the following way. First we identify cycles C such that $|V(C) \cap S| = 1$. Each such cycle C can be obtained by taking a vertex $a \in S$ together with a path between a pair of its neighbors in $G \setminus S$. For each vertex $b \in V(C) \setminus \{a\}$ we check whether a, b (or b, a) is a local s-t pair for C. If this is the case, then we distinguish two cases. If a and b are adjacent on C, then let P_1 be the path $C \setminus \{a, b\}$. We add to $\mathcal{P}$ the pair $\{P_1, P_1\}$. If a and b are non-adjacent on C, then let P_1' and P_2' be the two paths between a and b forming the cycle C. We obtain P_1 and P_2 by removing a and b from P_1' and P_2', respectively. Then we add the pair $\{P_1, P_2\}$ to $\mathcal{P}$.

If a cycle C in G intersects with S in vertices a and b, then there exist two vertex-disjoint paths P_1' and P_2' between a and b, such that $V(P_1') \cup V(P_2') = V(C)$. Hence, each such cycle C is uniquely determined by the neighbors of a and b on P_1' and P_2'. If, furthermore, a, b (or b, a) is a local s-t pair for C, then we add some pair to $\mathcal{P}$. In particular, if both P_1', P_2' are of length at least 2 we obtain P_1 and P_2 by removing a and b from P_1' and P_2', respectively. Then we add the pair $\{P_1, P_2\}$ to $\mathcal{P}$. If one of the paths, say P_2', is of length 1 (i.e., a and b are adjacent on C), we add to $\mathcal{P}$ the pair $\{P_1, P_1\}$, where P_1 is obtained from P_1' by removing a and b.

Similarly to Observation 9, we have at most n^2 candidate cycles with a single vertex of S and for each of them we have at most n candidates on b. We have n^4 cycles with two vertices of S. For each of them, we check, whether a, b (or b, a) is a local s-t pair in $\mathcal{O}(n^2)$ time. Hence, $\mathcal{P}$ can be obtained in $\mathcal{O}(n^6)$ time.

Now we use (LP$_\text{pairs}$) with $\mathcal{P}$ to identify the paths on which we want to place at least one additional tracker. Let $\boldsymbol{x}^*$ be an optimal solution of (LP$_\text{pairs}$), which can be obtained in polynomial time using the Ellipsoid method. We construct $\mathcal{P}_{x^*}$ as the set of all paths $P \in \mathcal{P}$ such that $\sum_{v \in V(P)} x_v^* \geq \frac{1}{2}$.

We first show the following observation.

Observation 16. *Let G be a reduced graph and T^* be a tracking set for G. Let* OPT_{x^*} *be the optimum value of* (LP$_{\mathrm{pairs}}$). *Then* $\mathrm{OPT}_{x^*} \leq w(T^*)$.

Proof. Let $\widehat{x}$ be a vector such that for all $v \in V(G) \setminus S$

$$\widehat{x}_v = \begin{cases} 1 & \text{if } v \in T^* \\ 0 & \text{otherwise.} \end{cases}$$

We show that $\widehat{x}$ is a solution to (LP$_{\mathrm{pairs}}$). Suppose that $\sum_{v \in V(P_i) \cup V(P_j)} x_v < 1$ for some $\{P_i, P_j\} \in \mathcal{P}$, therefore $T^* \cap (V(P_i) \cup V(P_j)) = \emptyset$. Let a, b be the vertices such that $G[V(P_i) \cup V(P_j) \cup \{a, b\}]$ contains a cycle C, such that a, b are a local s-t pair for C. Such a, b must exist because of the way $\mathcal{P}$ was constructed.

Since a, b is a local s-t pair, there exist two distinct paths $\widehat{P}_1$ and $\widehat{P}_2$ such that $\overrightarrow{T^*_{\widehat{P}_1}} = \overrightarrow{T^*_{\widehat{P}_2}}$: Both reach a from s and reach t from b but one reaches b from a by P_i and the other one by P_j. This contradicts the assumption that T^* is a tracking set. □

To decide which additional vertices to include in the solution, we compute the minimum multicut in forests on $(G \setminus S, w|_{(V \setminus S)}, \mathcal{P}_{x^*})$ using (LP$_{\mathrm{MCF}}$). Let y^* be an integral solution of (LP$_{\mathrm{MCF}}$) of weight at most μ times the optimal one. By Proposition 3 or Lemma 4 such a solution can be obtained in polynomial time. Let X be the set of vertices such that $y_v = 1$. Then we claim that $S \cup X$ is a $2(1 + \mu)$-approximate solution to the TRACKING PATHS.

Lemma 17. *Let S be a 2-approximate fvs in G and let y be a solution to* (LP$_{\mathrm{MCF}}$) *induced by $\mathcal{P}_{x^*}$ with weight at most μ times the optimal. The set* $T = S \cup \{v \in V \setminus S \mid y_v = 1\}$ *is a $2(1 + \mu)$-approximate solution to* TRACKING PATHS *on G.*

Proof. Let $X = \{v \in V \setminus S \mid y_v = 1\}$. First we show that $T = S \cup X$ is a tracking set. Suppose there exists a pair of distinct s-t paths P_1, P_2 in G that are not distinguished by T, i.e., $\overrightarrow{T_{P_1}} = \overrightarrow{T_{P_2}}$. Then, by Lemma 15, there exists a cycle C such that $V(C) \subseteq V(P_1) \cup V(P_2)$ and a local s-t pair $a, b \in V(C)$ such that $T \cap V(C) \setminus \{a, b\} = \emptyset$. But then C must have been one of the cycles enumerated by the algorithm when constructing $\mathcal{P}$ and therefore there must be a $v \in (V(C) \setminus \{a, b\}) \cap X$, contradicting the choice of C.

Now we show that $w(T)$ is at most $2(1 + \mu)$ times the weight of a minimum tracking set in G. Let t^* be the weight of the minimum tracking set on G and f^* be the weight of a minimum feedback vertex set in G. Furthermore, let OPT_{x^*} be the objective value of an optimal solution to (LP$_{\mathrm{pairs}}$) and OPT_{y^*} be the objective value of a solution to (LP$_{\mathrm{MCF}}$).

We claim that $|S| \leq 2f^* \leq 2t^*$. The first inequality follows from the fact that S is a 2-approximate feedback vertex set for G. The second inequality follows from the fact that every tracking set is also a feedback vertex set [4]. It also holds $w(X) \leq \mu \mathrm{OPT}_{y^*} \leq 2\mu \cdot \mathrm{OPT}_{x^*} \leq 2\mu \cdot t^*$. Here the second inequality follows from Lemma 6, while the third one follows from Observation 16.

Together this gives us $w(T) = w(S) + w(X) \leq 2t^* + 2\mu \cdot t^* = 2(1 + \mu) \cdot t^*$. $\square$

The result is summed up as follows.

Theorem 18 (precise version of Theorem 1). *There exists a* $2(1 + \mu)$-*approximation algorithm for* TRACKING PATHS, *in particular there is a 4-approximation algorithm for unweighted graphs and 66-approximation algorithm for weighted graphs.*

References

1. Abid, A., Khan, F., Hayat, M., Khan, W.: Real-time object tracking in wireless sensor network. In 2017 10th International Conference on Electrical and Electronics Engineering (ELECO), pp. 1103–1107 (2017)
2. Agrawal, A., Lokshtanov, D., Misra, P., Saurabh, S., Zehavi, M.: Polylogarithmic approximation algorithms for weighted-$\mathcal{F}$-deletion problems. ACM Trans. Algorithms **16**(4), 51:1-51:38 (2020)
3. Bafna, V., Berman, P., Fujito, T.: A 2-approximation algorithm for the undirected feedback vertex set problem. SIAM J. Discret. Math. **12**(3), 289–297 (1999)
4. Banik, A., Choudhary, P., Lokshtanov, D., Raman, V., Saurabh, S.: A polynomial sized kernel for tracking paths problem. Algorithmica **82**(1), 41–63 (2020)
5. Banik, A., Choudhary, P., Raman, V., Saurabh, S.: Fixed-parameter tractable algorithms for tracking shortest paths. Theor. Comput. Sci. **846**, 1–13 (2020)
6. Banik, A., Katz, M.J., Packer, E., Simakov, M.: Tracking paths. Discret. Appl. Math. **282**, 22–34 (2020)
7. Bar-Yehuda, R., Geiger, D., Naor, J., Roth, R.M.: Approximation algorithms for the feedback vertex set problem with applications to constraint satisfaction and Bayesian inference. SIAM J. Comput. **27**(4), 942–959 (1998)
8. Bilò, D., Gualà, L., Leucci, S., Proietti, G.: Tracking routes in communication networks. Theor. Comput. Sci. **844**, 1–15 (2020)
9. Călinescu, G., Fernandes, C.G., Reed, B.A.: Multicuts in unweighted graphs and digraphs with bounded degree and bounded tree-width. J. Algorithms **48**(2), 333–359 (2003)
10. Chierichetti, F., Lattanzi, S., Panconesi, A.: Rumor spreading in social networks. Theor. Comput. Sci. **412**(24), 2602–2610 (2011)
11. Choudhary, P.: Polynomial time algorithms for tracking path problems. In: Combinatorial Algorithms - 31st International Workshop, IWOCA 2020, Bordeaux, France, 8–10 June 2020, Proceedings, pp. 166–179 (2020)
12. Choudhary, P., Raman, V.: Improved kernels for tracking path problems. CoRR, abs/2001.03161 (2020)
13. Choudhary, P., Raman, V.: Structural parameterizations of tracking paths problem. In: Proceedings of the 21st Italian Conference on Theoretical Computer Science, Ischia, Italy, 14–16 September 2020, volume 2756 of CEUR Workshop Proceedings, pp. 15–27. CEUR-WS.org (2020)
14. Chudak, F.A., Goemans, M.X., Hochbaum, D.S., Williamson, D.P.: A primal-dual interpretation of two 2-approximation algorithms for the feedback vertex set problem in undirected graphs. Oper. Res. Lett. **22**(4–5), 111–118 (1998)

15. Dean, B.C., Griffis, A., Parekh, O., Whitley, A.A.: Approximation algorithms for k-hurdle problems. Algorithmica **59**(1), 81–93 (2011)
16. Diestel, R.: Graph Theory. GTM, vol. 173, 5th edn. Springer, Heidelberg (2017). https://doi.org/10.1007/978-3-662-53622-3
17. Eppstein, D., Goodrich, M.T., Liu, J.A., Matias, P.: Tracking paths in planar graphs. In: 30th International Symposium on Algorithms and Computation, ISAAC 2019, 8–11 December 2019, Shanghai University of Finance and Economics, Shanghai, China, pp. 54:1–54:17 (2019)
18. Gardarin, G., Spaccapietra, S.: Integrity of data bases: a general lockout algorithm with deadlock avoidance. In: IFIP Working Conference on Modelling in Data Base Management Systems, pp. 395–412 (1976)
19. Golovin, D., Nagarajan, V., Singh, M.: Approximating the k-multicut problem. In: Proceedings of the Seventeenth Annual ACM-SIAM Symposium on Discrete Algorithms, SODA 2006, Miami, Florida, USA, 22–26 January 2006, pp. 621–630. ACM Press (2006)
20. Goodrich, M.T., Gupta, S., Khodabandeh, H., Matias, P.: How to catch marathon cheaters: New approximation algorithms for tracking paths. CoRR, abs/2104.12337 (2021)
21. Grötschel, M., Lovász, L., Schrijver, A.: The ellipsoid method and its consequences in combinatorial optimization. Combinatorica **1**(2), 169–197 (1981)
22. Grötschel, M., Lovász, L., Schrijver, A.: Geometric Algorithms and Combinatorial Optimization. Algorithms and Combinatorics, vol. 2. Springer, Heidelberg (1988). https://doi.org/10.1007/978-3-642-97881-4
23. Guo, J., Hüffner, F., Kenar, E., Niedermeier, R., Uhlmann, J.: Complexity and exact algorithms for vertex multicut in interval and bounded treewidth graphs. Eur. J. Oper. Res. **186**(2), 542–553 (2008)
24. Hudli, A.V., Hudli, R.V.: Finding small feedback vertex sets for VLSI circuits. Microprocess. Microsyst. **18**(7), 393–400 (1994)
25. Karp, R.M.: Reducibility among combinatorial problems. In: Miller, R.E., Thatcher, J.W. (eds.) Proceedings of a symposium on the Complexity of Computer Computations, held 20–22 March 1972, at the IBM Thomas J. Watson Research Center, Yorktown Heights, New York, USA, The IBM Research Symposia Series, pp. 85–103. Plenum Press, New York (1972)
26. Kawarabayashi, K., Kobayashi, Y., Reed, B.: The disjoint paths problem in quadratic time. J. Comb. Theory Ser. B **102**(2), 424–435 (2012)
27. Koohikamali, M., Kim, D.J.: Rumor and truth spreading patterns on social network sites during social crisis: big data analytics approach. In: Sugumaran, V., Yoon, V., Shaw, M.J. (eds.) WEB 2015. LNBIP, vol. 258, pp. 166–170. Springer, Cham (2016). https://doi.org/10.1007/978-3-319-45408-5_15
28. Li, J., Nederlof, J.: Detecting feedback vertex sets of size k in $O^*(2.7^k)$ time. In: Chawla, S. (ed.) Proceedings of the 2020 ACM-SIAM Symposium on Discrete Algorithms, SODA 2020, Salt Lake City, UT, USA, 5–8 January 2020, pp. 971–989. SIAM (2020)
29. Lin, C.-Y., Peng, W.-C., Tseng, Y.-C.: Efficient in-network moving object tracking in wireless sensor networks. IEEE Trans. Mob. Comput. **5**(8), 1044–1056 (2006)
30. Manley, E.D., Al Nahas, H., Deogun, J.S.: Localization and tracking in sensor systems. In: IEEE International Conference on Sensor Networks, Ubiquitous, and Trustworthy Computing (SUTC'06), vol. 2, pp. 237–242 (2006)
31. Misra, P.: On fault tolerant feedback vertex set. CoRR, abs/2009.06063 (2020)
32. Nekovee, M., Moreno, Y., Bianconi, G., Marsili, M.: Theory of rumour spreading in complex social networks. Physica Stat. Mech. Appl. **374**(1), 457–470 (2007)

33. Parter, M.: Fault-tolerant logical network structures. Bull. EATCS, **118**, 1–3 (2016)
34. Siberschatz, A., Galvin, P.B.: Operating System Concepts, 4th edn. Addison-Wesley Longman Publishing Co., Inc., Boston (1993)
35. Tanaka, T., Eum, S., Ata, S., Murata, M.: Design and implementation of tracking system for moving objects in information-centric networking. In: 2019 22nd Conference on Innovation in Clouds, Internet and Networks and Workshops (ICIN), pp. 302–306 (2019)
36. Varshney, C., Jain, S.C., Tripathi, V.: An overview of rumour detection based on social media. In: 2020 11th International Conference on Computing, Communication and Networking Technologies (ICCCNT), pp. 1–6 (2020)
37. Wang, S., Terano, T.: Detecting rumor patterns in streaming social media. In: 2015 IEEE International Conference on Big Data (Big Data), pp. 2709–2715 (2015)
38. Zhou, Y., Maskell, S.: Detecting and tracking small moving objects in wide area motion imagery (WAMI) using convolutional neural networks (CNNs). In: 22th International Conference on Information Fusion, FUSION 2019, Ottawa, ON, Canada, 2–5 July 2019, pp. 1–8. IEEE (2019)

An Improved Approximation Bound for Minimum Weight Dominating Set on Graphs of Bounded Arboricity

Hao Sun[(✉)]

University of Waterloo, Waterloo, ON N2L 3G1, Canada
hao.sun@uwaterloo.ca

Abstract. We consider the minimum weighted dominating set problem on graphs having bounded arboricity. We show that the natural LP has an integrality gap of at most one more than the arboricity of our graph via a primal-dual algorithm. This is nearly the best approximation possible, as Bansal and Umboh have shown it is NP-hard to approximate dominating sets to within one less than the arboricity of the graph.

Keywords: Dominating set · Bounded arboricity · Approximation algorithms

1 Introduction

In the minimum weighted dominating set (MWDS) problem, we are given a graph $G = (V, E)$ with weights $w_v, \forall v \in V$, and wish to find a minimum weight set D of vertices for which each vertex $v \in V$ is either in D, or has a neighbour in D. When all weights are 1 we call this the minimum dominating set (MDS) problem. One can see that MWDS is a special case of weighted set cover. Hence, by applying the greedy algorithm for weighted set cover, one can obtain a H_n approximation for MWDS, where $n := |V|$ and H_n is the n-th harmonic number. Bansal and Umboh [2] made the observation that it is NP-hard to approximate MDS to within $(1 - \epsilon) \ln n$ by reducing it to set cover and using a hardness result for set cover proven in [3].

Baker [1] showed that MDS in planar graphs admits a PTAS. Fomin et al. [5] extended this to an EPTAS on H minor free graphs for any fixed graph H.

Given that planar graphs are sparse, it may seem natural to generalize the previous results on MDS to sparse graphs. Lenzen and Wattenhofer [8] observed that MDS remains hard on graphs of low average degree and unit weights by the following reduction. Given any graph on n nodes, add a star on $n^2 - n$ nodes. The resulting graph G' has average degree at most 2 and MDS one more than that of G. Hence, approximating the minimum dominating set in H is as hard as approximating the minimum dominating set in G.

Supported by University of Waterloo.

J. Koenemann and B. Peis (Eds.): WAOA 2021, LNCS 12982, pp. 39–47, 2021.
https://doi.org/10.1007/978-3-030-92702-8_3

Lenzen and Wattenhofer [8] proposed studying MDS on a class of graphs that informally speaking have a local sparsity property. A graph has *arboricity* a if a is the minimum number of edge-disjoint forests into which its edges can be partitioned. It is well known that a graph has arboricity a, if and only if each subgraph induced by a subset of vertices $S \subset V$ has at most $a(|S| - 1)$ edges. In this sense, bounded arboricity is equivalent to local sparsity. This fact will also be important in our analysis. We use a-MWDS and a-MDS to refer to MWDS and MDS in graphs of arboricity a.

Lenzen and Wattenhofer presented a distributed $\mathcal{O}(a^2)$ approximation algorithm for a-MDS. Bansal and Umboh [2] improved this by giving a $3a$-approximation for a-MDS by rounding the natural LP relaxation. They also show that it is NP-hard to approximate a-MDS to within $a - 1 - \epsilon$ for any $\epsilon > 0$. Dvorak [4] showed that the algorithm of Bansal and Umboh [2] actually gives a $(2a + 1)$-approximation for a-MDS. We present an $(a + 1)$-approximation algorithm for a-MWDS using the primal-dual method.

Our main result is the following:

Theorem 1. *There is an $(a + 1)$-approximation algorithm for a-MWDS.*

Our analysis actually requires a slightly weaker condition than arboricity a, namely that for our graph G, $|E(G[S])| \le a|S|$ $\forall S \subset V(G)$. We use an integer programming formulation and apply the primal-dual method [6], which we outline as follows.

We start with the empty solution set $S = \emptyset$. While S is not feasible, we increment the dual variables y_v for v not in or adjacent to a vertex of S and add nodes u for which $w_u = \sum_{t \in \{u\} \cup N(u)} y_t$ to S, where $N(u)$ are the neighbours of u. At the end of the algorithm we apply the standard *reverse delete* procedure. Here, we consider all nodes in S in reverse order of addition to S, and we delete such a node if the feasibility of S is maintained.

Under the standard primal-dual method terminology [6], each dual variable y_v "pays" for solution vertices that are either v or in the neighbourhood of v. The intuition behind our algorithm is to note that graphs of arboricity a have average degree at most $2a$. Informally, this leads to each dual variable paying for at most $2a + 1$ solution nodes on average. To clarify, we actually need that at any time during the algorithm, the graph induced by the undominated nodes has average degree at most $2a$. Another observation we make is that for each node v we keep in S there is a *witness* node u for which v is the only node of $\{u\} \cup N(u)$ in S. This allows us to tighten our analysis to an $(a + 1)$-approximation.

In contrast, the algorithm of Bansal and Umboh [2] works by first taking the set H of vertices whose LP value is at least $\frac{1}{3a}$ and then taking the remaining set $U := V \setminus (H \cup N(H))$ of undominated vertices. H has cost at most $3a$ times $\sum_{v \in H} c_v x_v$ where x_v is the LP value for node v in the a-MWDS LP (P_{MWDS}). To bound the cost of U, they then use the fact the graph has arboricity a to obtain an orientation of the graph so that each node has at most a out neighbours. They analyze the size of U relative to $\sum_{v \in V \setminus H} x_v$ (their analysis requires unit weights) by effectively having each node of $V \setminus H$ "pay" its cost to each out neighbour. The final step of their proof is to observe that each node

$u \in U$ satisfies $\sum_{v \in \delta_{in}(u)} x_v \geq \frac{1}{3}$, where $\delta_{in}(u)$ are the in neighbours of u. Combining this observation with the fact that each node of $V \backslash H$ pays towards at most a other nodes yields a $3a$-approximation.

As previously mentioned [4] showed this algorithm actually yields a $(2a+1)$-approximation. This analysis is almost tight as there are instances, where the algorithm yields a solution of size (at least) 2a-1 times the optimum. We defer the example to the appendix.

2 Approximation Algorithm

2.1 Preliminaries

Denote by $N_H(v) := \{u \in V(H) : vu \in E(H)\}$ the neighbourhood of v in graph H, $N_H(S) := \cup_{v \in S} N_H(v)$, $N(S) := N_G(S)$ and $N[v] := \{v\} \cup N(v)$ the closed neighbourhood of v. Consider the following natural LP relaxation for MWDS and its dual. Given a graph G, for each $v \in V(G)$, the LP has a variable x_v indicating whether v is part of the minimum dominating set.

$$
\begin{array}{ll}
\min & \sum_{v \in V} w_v x_v \qquad (P_{\mathrm{MWDS}}) \\
\text{s.t.} & \sum_{u \in N[v]} x_u \geq 1 \quad \forall\, v \in V \\
& x \geq 0
\end{array}
$$

$$
\begin{array}{ll}
\max & \sum_{v \in V} y_v \qquad (D_{\mathrm{MWDS}}) \\
\text{s.t.} & \sum_{u:v \in N[u]} y_u \leq w_v \quad \forall\, v \in V \\
& x \geq 0
\end{array}
$$

We will show that the LP (P_{MWDS}) has integrality gap at most $a+1$ in graphs of arboricity a. Our algorithm is based on the primal-dual method. We describe the ideas of our primal-dual approach. Given a feasible dual solution y to (D_{MWDS}), define the *residual cost* of node $v \in V$ to be $w_v - \sum_{u:v \in N[u]} y_u$. Our primal-dual method starts with a trivial feasible dual solution $y = 0$, and the empty, infeasible "solution set" $S = \emptyset$.

Next, in each iteration, we increase y_v for all v in $V(G) \backslash (S \cup N(S))$, while maintaining dual feasibility, and until some dual constraint $\sum_{u:v \in N(u) \cup \{u\}} y_u \leq w_v$ becomes *tight* in the increase process, that is, the residual cost of the corresponding node v becomes 0.

When this happens, we add the node to S. As soon as S is feasible our algorithm terminates the first phase and implements a *reverse-delete* procedure. At this time, the algorithm will consider all nodes in S in reverse order of being added to S, and deletes such a node if the feasibility of S is preserved.

2.2 Analysis of Our Algorithm

We now show our algorithm is an $(a+1)$-approximation. To do so we need the following lemma of [6].

Algorithm 1: MinWeightDominatingSet (G, c)

Input : A graph $G = (V, E)$ with non-negative node-costs w_v, for each $v \in V$.
Output: A dominating set S of G.
$S = \emptyset$
while S *is not a dominating set of G* **do**
    Increment all dual variables y_v for $v \in V\backslash(S \cup N(S))$ uniformly. Add all
    nodes that became tight to S.
end
Reverse-Deletion:
    Let $s_1, s_2, .., s_l$ be nodes of S in the order they were added.
    for $t = l$ **downto** 1 **do**
        if $S\backslash\{s_t\}$ *is feasible* **then**
            | $S \leftarrow S\backslash\{s_t\}$
        end
    end

---

return S

Lemma 1. *([6] adapted for (P_{MWDS})): Suppose S and y are solutions to MWDS on G and the LP (D_{MWDS}) output by our primal-dual algorithm such that the following holds.*

1. *y is obtained starting with the initial feasible solution $y := 0$ and incrementing some set of dual variables $\{y_v : v \in W_t\}$ uniformly and maintaining feasibility of y for iterations $t = 1, 2, .., l$ for some $l \in \mathbb{N}$.*
2. *For each iteration $t \in \{1, 2, 3, .., l\}$, the set $\{y_v : v \in W_t\}$ of incremented dual variables satisfies $\sum_{v \in W_t} |S \cap N[v]| \leq \beta|W_t|$. Intuitively, the number of nodes of S each dual variable y_v "pays" for, $|S \cap N[v]|$ is at most β on average.*
3. *$\forall v \in S, \sum_{u \in N[v]} y_u = c_v$.*

Then S is a β-approximation. In fact, the characteristic vector $\hat{x} := \mathbb{1}^S$ (that is, $\hat{x}_v = 1$ for $v \in S$ and $\hat{x}_v = 0$ for $v \notin S$) satisfies $\sum_{v \in V} w_v \hat{x}_v \leq \beta \sum_{v \in V} y_v$.

Theorem 2. *Algorithm 1 is an $(a + 1)$-approximation on graphs of arboricity a.*

Proof. Let S' be the set of nodes returned by our primal-dual algorithm. We follow the standard method of analyzing the primal and dual increase rates.

Using Lemma 1 it suffices to prove that during any iteration t, the set W_t of nodes whose dual variables are incremented satisfies

$$\sum_{u \in W_t} |S' \cap N[u]| \leq (a + 1)|W_t|.$$

We illustrate the intuition of our proof as follows. Graphs G of arboricity a have at most $a|V(G)|$ edges, so the average degree of G is at most $2a$. Suppose each node of G had degree $2a$. Then for all $u \in V(G)$, $|S' \cap N[u]| \leq 1 + 2a$ which shows that S' is a $2a+1$-approximation. This analysis does not work in general because it may be the case that incremented dual variables correspond exclusively to high degree nodes. We will show that because S' is "minimal" in the sense that Algorithm 1 performed a reverse deletion step, it follows that although the nodes corresponding to incremented dual variables may have high average degree, they are not adjacent to too many nodes of S' on average. Minimality of S' also means that for each node u of S' there is another node $v \in V(G)$ called a witness, for which $N[v] \cap S' = \{u\}$. Informally speaking, we show y_v for witnesses v will pay for only one solution node. Intuitively, these y_v which pay for a single solution node help to bring the average number of solution nodes a dual variable pays for down to $a + 1$.

Consider some iteration t during the algorithm. Let S be the solution set in the current iteration, so $\{y_v : v \in V \backslash (S \cup N(S))\}$ is the set of incremented dual variables. Thus W_t, the nodes corresponding to the dual variables incremented during this iteration, is equal to $V \backslash (S \cup N(S))$. To simplify notation, we will refer to W_t simply by W. Define $S'_W := W \cap S'$, the set of solution nodes in W and $S'_{N(W)} := (S' \cap N(W)) \backslash W$, the set of solution nodes outside W adjacent to nodes of W. Using the convention introduced in Theorem 1, $S'_W \cup S'_{N(W)}$ are the nodes "paid for" during this iteration.

The minimality of S' will help us show Algorithm 1 is an $(a + 1)$-approximation.

Definition 1. *[7] For $u \in S'$ we call a node $v \in V$ such that $N[v] \cap S' = \{u\}$ a witness node for u.*

Proposition 3 *Each node $u \in S'_W \cup S'_{N(W)}$ has a witness node in W.*

Proof. Let S'' be the solution our algorithm has before it begins its reverse-deletion step. Consider the iteration of our reverse deletion procedure during which node u was considered for deletion. Let S''' be the solution set the algorithm keeps at this time. Since u was added to our solution after all nodes of S, no nodes of S have been removed from the current solution set. So $S \subset S'''$. Since we did not remove u from our solution, there is a node $w_u \in V$ such that $N[w_u] \cap S''' = \{u\}$. Since $N[w] \cap S \neq \emptyset$ for any $w \in S \cup N(S)$, $w_u \in W$.

For $s \in S'_W \cup S'_{N(W)}$ let $p(s)$ be a witness node for s in W. Denote $\hat{S}_W := \{s \in S'_W : p(s) \neq s\}$, the set of nodes in S'_W that are not their own witness and suppose $\hat{S}_W = \{s_1, s_2, .., s_l\}$. For brevity, we let $p_t := p(s_t)$.

Let $A := \{p_1, p_2, .., p_l\}$ be the set of witnesses for nodes of $\hat{S}_W$, $\tilde{A} := \{p(s) : s \in S'_{N(W)}\}$ the set of witnesses for $S'_{N(W)}$ and $B = W \backslash (A \cup S'_W \cup \tilde{A})$ be the nodes of W which are neither solution nodes nor witnesses (see Fig. 1). Let us make the following observations.

Lemma 2. *The following holds:*

1. *For $v \in S'_W \backslash \hat{S}_W$, $S' \cap N[v] = \{v\}$.*
2. *For $v \in A \cup \tilde{A}$, $|S' \cap N[v]| = 1$.*

3. $|A| = |\hat{S}_w|$, $|\tilde{A}| = |S'_{N(W)}|$, *in other words A is in one to one correspondence with $\hat{S}_w$ and $\tilde{A}$ with $S'_{N(W)}$.*

Proof. For items 1, 2, each node $v \in A \cup \tilde{A} \cup (S'_W \backslash \hat{S}_W)$ is a witness and hence $N[v]$ contains a single solution vertex so $|S' \cap N[v]| = 1$. For $v \in S'_W \backslash \hat{S}_W$, v is a solution node so $S' \cap N[v] = \{v\}$. For item 3, A (resp. $\tilde{A}$) is the set of witnesses for $\hat{S}_w$ (resp. $S'_{N(W)}$) and hence A and $\hat{S}_w$ (resp. $\tilde{A}$ and $S'_{N(W)}$) are in one to one correspondence with each other.

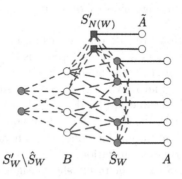

Fig. 1. Partition of W into $S'_W \backslash \hat{S}_W, B, \hat{S}_W, \tilde{A}$ and A. Nodes of S'_W are colored in red. The square blue nodes forming $S'_{N(W)}$, are solution nodes outside W but adjacent to some node inside W. (Color figure online)

Intuitively, $|N(v) \cap X|$ "counts" the number of edges with one endpoint v and the other in X. In this sense, $\sum_{v \in B} |N(v) \cap S'_W|$ counts edges between B and S'_W once, while $\sum_{q \in S'_W} |(N(q)) \cap S'_W|$ counts edges within S'_W twice. In the same line, $2\sum_{v \in B} |N(v) \cap S'_W| + \sum_{q \in S'_W} |(N(q)) \cap S'_W|$ counts each edge of $E(G[B \cup S'_W]) \backslash E \cap (B \times B)$ twice. Thus

$$2\sum_{v \in B} |N(v) \cap S'_W| + \sum_{q \in S'_W} (|(N(q)) \cap S'_W| + 1)$$
$$= 2|E(G(B \cup S'_W))| + |S'_W| - 2|E \cap B \times B| \qquad (1)$$
$$\leq 2|E(G(B \cup S'_W))| + |S'_W|.$$

Note from Lemma 2, $S' \cap N[v] = \{v\}$. Thus, $\sum_{q \in (S'_W \backslash \hat{S}_W)} (|N(q) \cap S'_W| + 1) = |S'_W \backslash \hat{S}_W|$. This also implies $\sum_{q \in \hat{S}_W} (|N(q) \cap S'_W| + 1) = \sum_{q \in \hat{S}_W} (|N(q) \cap \hat{S}_W| + 1)$.

Note $\sum_{q \in \hat{S}_W} |N(q) \cap \hat{S}_W|$ counts each edge of $G[\hat{S}_W]$ twice, so $\sum_{q \in \hat{S}_W} (|N(q) \cap \hat{S}_W| + 1) = 2|E(G[\hat{S}_W])| + |\hat{S}_W|$.
Thus,

$$\sum_{q \in S'_W} (|N(q) \cap S'_W| + 1)$$
$$= \sum_{q \in \hat{S}_W} (|N(q) \cap \hat{S}_W| + 1) + \sum_{q \in (S'_W \backslash \hat{S}_W)} (|N(q) \cap S'_W| + 1) \qquad (2)$$
$$= 2|E(G[\hat{S}_W])| + |\hat{S}_W| + |S'_W \backslash \hat{S}_W|.$$

Adding (1) and (2) and dividing by 2 we get

$$\sum_{v \in B} |N(v) \cap S'_W| + \sum_{q \in S'_W} (|N(q) \cap S'_W| + 1)$$
$$\leq \tfrac{1}{2}(2|E(G[B \cup S'_W])| + 2|E(G[\hat{S}_W])| + 2|\hat{S}_W|) + |S'_W \backslash \hat{S}_W|. \qquad (3)$$
$$= |E(G[B \cup S'_W])| + |E(G[\hat{S}_W])| + |S'_W|.$$

To recap, we wish to bound $\sum_{v \in W} |N[v] \cap S'|$. Note that for $v \in W$, if $v \in B$, then $|S'_W \cap N[v]| = |N(v) \cap S'_W|$. For $v \in \hat{S}_W$, $|S'_W \cap N[v]| = |N(v) \cap \hat{S}_W| + 1$. We will next bound $\sum_{v \in B \cup S'_W} (|N[v] \cap (S'_{N(W)} \cup S'_W)| - |N[v] \cap S'_W|)$, which will get us a bound on $\sum_{v \in B \cup S'_W} |N[v] \cap S'|$, the number of nodes dual variables indexed by $B \cup S'_W$ pay for. Using bounded arboricity we will bound the number of edges in subgraphs of G. Afterwards we will bound $\sum_{v \in A \cup \tilde{A}} |N[v] \cap S'|$, which will bound $\sum_{v \in W} |N[v] \cap S'|$ completing the analysis.

Recall $S'_{N(W)} := (S' \cap N(W)) \backslash W$. Let us account for the amount nodes of W pay towards $S'_{N(W)}$, their neighbours outside W which are solution nodes. Each node of $\tilde{A}$ only pays towards the node of $S'_{N(W)}$ it is a witness of.

Note $\sum_{v \in B \cup S'_W} (|N[v] \cap (S'_{N(W)} \cup S'_W)| - |N[v] \cap S'_W|)$ counts every edge of $G[B \cup S'_W \cup S'_{N(W)}]$ not in $G[B \cup S'_W]$, thus

$$\sum_{v \in B \cup S'_W} (|N[v] \cap (S'_{N(W)} \cup S'_W)| - |N[v] \cap S'_W|)$$
$$= |E(G[B \cup S'_W \cup S'_{N(W)}])| - |E(G[B \cup S'_W])|.$$

Adding this to the (3) we obtain:

$$\sum_{v \in B \cup S'_W} |N[v] \cap S'_W|$$
$$+ \sum_{v \in B \cup S'_W} (|(N[v] \cap (S'_{N(W)} \cup S'_W)| - |N[v] \cap S'_W|)$$
$$\leq |E(G[B \cup S'_W])| + |E(G[\hat{S}_W])| + |S'_W| + |E(G[B \cup S'_W \cup S'_{N(W)}])| - |E(G[B \cup S'_W])|.$$

Which simplifies to $|E(G[B \cup S'_W \cup S'_{N(W)}])| + |E(G[\hat{S}_W])| + |S'_W|$. So

$$\sum_{v \in B \cup S'_W} |(N[v]) \cap (S'_{N(W)} \cup S'_W)| \leq |E(G[B \cup S'_W \cup S'_{N(W)}])| + |E(G[\hat{S}_W])| + |S'_W|.$$
$$(4)$$

We now have a bound on the number of nodes that dual variables for $B \cup S'_W$ pay for. Let us bound the right hand side of the above equation. From the fact that G has arboricity a, $|E(G[X])| \leq a|X|$ for any $X \subset V$ (in fact this weaker condition is sufficient for our analysis). Thus, the right hand side of the above inequality is at most

$$a(|B| + |S'_W| + |S'_{N(W)}|) + a|\hat{S}_W| + |S'_W|$$
$$\leq a|B| + (2a+1)|\hat{S}_W| + (a+1)|S'_W \backslash \hat{S}_W| + a|S'_{N(W)}|. \qquad (5)$$

So combining (4) and (5), we obtain

$$\sum_{v \in B \cup S'_W} |N[v] \cap (S'_{N(W)} \cup S'_W)| \leq a|B| + (2a+1)|\hat{S}_W| + (a+1)|S'_W \backslash \hat{S}_W| + a|S'_{N(W)}|.$$
$$(6)$$

For $v \in A \cup \tilde{A}$, $|S' \cap N[v]| = 1$.

Also note $|A| = |\hat{S}_W|$, $|\tilde{A}| = |S'_{N(W)}|$, for each $v \in W$ we have $S' \cap N[v] = (S'_W \cup S'_{N(W)}) \cap N[v]$ and W is the disjoint union of $A, \tilde{A}, B$ and S'_W.

Putting these observations together, we get the total amount the dual variables "pay" for is

$\sum_{v \in W} |S' \cap N[v]|$
$= \sum_{v \in B} |N(v) \cap S'| + \sum_{q \in S'_W} (|N(q) \cap S'| + 1) + \sum_{q \in A} (|N(q) \cap S'| + 1) \sum_{q \in \tilde{A}} (|N(q) \cap S'| + 1)$
$= \sum_{v \in B} |N(v) \cap S'| + \sum_{q \in S'_W} (|N(q) \cap S'| + 1) + |A| + |\tilde{A}|.$

From (6), this is at most

$$a|B| + (2a+1)|\hat{S}_W| + (a+1)|S'_W \setminus \hat{S}_W| + a|S'_{N(W)}| + |A| + |\tilde{A}|.$$

Noting that $|A| = |\hat{S}_W|$ and $|\tilde{A}| = |S'_{N(W)}|$, the previous line is at most

$$a|B| + (2a+2)(|\hat{S}_W|) + (a+1)|S'_W \setminus \hat{S}_W| + (a+1)|\tilde{A}|$$
$$\leq a|B| + (a+1)(|\hat{S}_W| + |A|) + (a+1)|S'_W \setminus \hat{S}_W| + (a+1)|\tilde{A}|$$
$$\leq (a+1)|W|.$$

Since the set of incremented dual variables is $\{y_v : v \in W\}$, by Theorem 1, this shows our algorithm is an $(a+1)$-approximation.

Appendix 1.A   Example of Bansal and Umboh Algorithm Yielding $2a - 1$ Approximation

The following example shows the Bansal and Umboh Algorithm returning a solution $2a - 1 - o(1)$ times the optimum. Let $a \in \mathbb{N}$ and $n = k(a-1) + 1$. Construct graph G with vertex set $\{v_1, v_2, .., v_{k(a-1)+1}\}$ and edge set $\{v_i v_j : 0 \leq i - j \leq a - 1\} \cup \{v_i v_j : 0 \leq i + k(a-1) + 1 - j \leq a - 1\}$. One can see that $|E(G(U))| \leq a(|U| - 1)$ for any $U \subset V$ and hence G has arboricity at most a. For $i > k(a-1) + 1$ and $i < 0$, define $v_i = v_{i \pmod{k(a-1)+1}}$.

Consider the Bansal and Umboh algorithm on G with unit weights. We claim $x = \frac{1}{2a-1}\mathbb{1}$ is the optimal LP solution for the a-MWDS LP on G with unit weights. Note that the LP cost $\sum_{v \in V(G)} x_v$ of x is $\frac{k(a-1)+1}{2a-1}$. Adding up $\sum_{i=-a+1}^{a-1} v_{i+j} \geq 1$ for $j = 1, 2, .., k(a-1) + 1$ we get

$$(2a - 1) \sum_{v \in V(G)} x_v \geq k(a-1) + 1$$

so $\sum_{v \in V(G)} x_v \geq \frac{k(a-1)+1}{2a-1}$ and that equality can only hold if $\sum_{i=-a+1}^{a-1} v_{i+j} = 1$ for all i. It thus follows the unique optimal solution satisfies $x_{v_i} = x_{v_j}$ $\forall i, j \in [k(a-1) + 1]$ and is thus $\frac{1}{2a-1}\mathbb{1}$.

The algorithm of Bansal and Umboh would thus take every vertex of G, which has a cost of $k(a-1) + 1$. However, the vertices $\{v_{(2a-1)i} : 1 \leq i \leq \lceil \frac{k(a-1)+1}{2a-1} \rceil\}$ are a solution of cost $\lceil \frac{k(a-1)+1}{2a-11} \rceil \approx \frac{k(a-1)+1}{2a-1}$ for large k. Hence the solution returned by the algorithm is $2a - 1 - o(1)$ times the optimum.

References

1. Baker, B.S.: Approximation algorithms for np-complete problems on planar graphs. J. ACM **41**(1), 153–180 (1994)
2. Bansal, N., Umboh, S.W.: Tight approximation bounds for dominating set on graphs of bounded arboricity. Inf. Process. Lett. **122**, 21–24 (2017)
3. Dinur, I., Steurer, D.: Analytical approach to parallel repetition. In: Proceedings of the Forty-Sixth Annual ACM Symposium on Theory of Computing, STOC 2014, pp. 624–633, New York, NY, USA. Association for Computing Machinery (2014)
4. Dvorak, Z.: On distance r-dominating and 2r-independent sets in sparse graphs. J. Graph Theory **91**, 10 (2017)
5. Fomin, F., Lokshtanov, D., Raman, V., Saurabh, S.: Bidimensionality and EPTAS. In: Proceedings of the Annual ACM-SIAM Symposium on Discrete Algorithms, May 2010
6. Goemans, M.X., Williamson, D.P.: The Primal-Dual Method for Approximation Algorithms and Its Application to Network Design Problems, pp. 144–191. PWS Publishing Co., USA (1996)
7. Goemans, M.X., Williamson, D.P.: Primal-dual approximation algorithms for feedback problems in planar graphs. Combinatorica **18**, 37–59 (1998)
8. Lenzen, C., Wattenhofer, R.: Minimum dominating set approximation in graphs of bounded arboricity. In: Lynch, N.A., Shvartsman, A.A. (eds.) DISC 2010. LNCS, vol. 6343, pp. 510–524. Springer, Heidelberg (2010). https://doi.org/10.1007/978-3-642-15763-9_48

Tight Inapproximability of Minimum Maximal Matching on Bipartite Graphs and Related Problems

Szymon Dudycz[1], Pasin Manurangsi[2], and Jan Marcinkowski[1](✉)

[1] University of Wrocław, Wrocław, Poland
{szymon.dudycz,jan.marcinkowski}@cs.uni.wroc.pl
[2] Google Research, Mountain View, USA
pasin@google.com

Abstract. We study the *Minimum Maximal Matching* problem, where we are asked to find in a graph the smallest matching that cannot be extended. We show that this problem is hard to approximate with any constant smaller than 2 even in bipartite graphs, assuming either of two stronger variants of Unique Games Conjecture. The bound also holds for computationally equivalent *Minimum Edge Dominating Set*. Our lower bound matches the approximation provided by a trivial algorithm.

Our results imply conditional hardness of approximating Maximum Stable Matching with Ties and Incomplete Lists with a constant better than $\frac{3}{2}$, which also matches the best known approximation algorithm.

Keywords: Matching · Hardness of approximation

1 Introduction

Minimum Maximal Matching (MMM) and Minimum Edge Dominating Set (EDS) are two well-studied algorithmic graph problems. In MMM, we are asked to find the smallest (or lightest in the weighted version) matching that cannot be extended—which means that for every edge in the graph at least one of its endpoints is matched. In EDS, we are asked to find the smallest *edge dominating set*, which is a subset D of edges of a graph such that every edge of the graph either is a member of D or is adjacent to at least one edge in D.

Minimum Maximal Matching and Minimum Edge Dominating Set in unweighted (cardinality) variants have been known to have equivalent optimum solutions and were both proven to be NP-hard to solve optimally even when restricted to bipartite graphs of maximum degree 3 [25]. This equivalence does not carry into the variants with weights on edges, where we are asked to find the lightest Maximal Matching (resp. Edge Dominating Set).[1]

---

[1] To illustrate that imagine a cycle of four edges with weights $(1, 1, 100, 100)$. The lightest EDS weighs 2, but MMM must contain two opposite edges and weighs 101.

Supported by Polish NSC grants 2018/29/B/ST6/02633 and 2015/18/E/ST6/00456.

J. Koenemann and B. Peis (Eds.): WAOA 2021, LNCS 12982, pp. 48–64, 2021.
https://doi.org/10.1007/978-3-030-92702-8_4

Since any two maximal matchings M_1, M_2 satisfy $|M_1| \leqslant 2 \cdot |M_2|$, we can deduce a simple 2-approximation algorithm for the cardinality (i.e. unweighted) versions of MMM and EDS: output any maximal matching. The weighted version of EDS was shown to be approximable within a factor of $\frac{21}{10}$ [4], and then 2 [11,21]. Efforts have been put into improving upon the naive 2-approximation algorithm for the unweighted problem. Better-than-two constant has been achieved for dense graphs [24]. Furthermore, a local search approximation algorithm was constructed with approximation ratio of $2 - \Omega\left(\frac{\log n}{n}\right)$ for the general unweighted graphs [12].

On the hardness side, Chlebík and Chlebíková proved that it is NP-hard to approximate EDS within factor better than $\frac{7}{6}$ [6]. The result was later improved to[2] 1.207 [10]; under the *Unique Games Conjecture* [15], the same work [10] provided a stronger inapproximability ratio of $\frac{3}{2}$. Recently, Dudycz, Lewandowski and Marcinkowski proved that assuming *Unique Games Conjecture* it is hard to approximate Minimum Maximal Matching (and hence also EDS) within constant factor better than 2 in general, unweighted graphs [9].

Our Result. In this paper we focus on MMM (and, equivalently, EDS) in bipartite graphs. We provide two reductions, starting from different extensions of Unique Games Conjecture, from which it follows that MMM is hard to approximate within factor $2 - \epsilon$ even when restricted to bipartite graphs. Precisely, we prove the following theorems:

Theorem 1. *Assuming* Strong Unique Games Conjecture *(Conjecture 2), there is no polynomial-time algorithm that approximates MMM (or equivalently EDS) on unweighted, bipartite graphs within factor $2 - \varepsilon$ for any constant $\varepsilon > 0$, unless $P = NP$.*

Theorem 2. *Assuming* Small Set Expansion Hypothesis *(Conjecture 3), there is no polynomial-time algorithm that approximates MMM (or equivalently EDS) on unweighted, bipartite graphs within factor $2 - \varepsilon$ for any constant $\varepsilon > 0$, unless $P = NP$.*

For a related problem of finding Maximum Stable Matching with Ties and Incomplete Lists (Max SMTI), Huang et al. showed that, if MMM is hard to approximate with a constant better than 2 in bipartite graphs that contain almost perfect matching, then Max SMTI is hard to approximate with a constant better than $\frac{3}{2}$, which matches the best known algorithm for this problem [14]. Both our reductions satisfy that requirement and thus this (conditional) tight gap is a corollary of our results.

---

[2] The inapproximation ratio claimed in the original work [10] is 1.18, which follows from a reduction from Vertex Cover. However, due to the recent improvement in NP-hardness of approximating the latter problem [7,8,16], the inapproximation ratio for EDS improves to 1.207.

The Assumptions. Our results rely on two respective computational complexity conjectures. When proving hardness of approximation, we introduce and reduce between *gap problems*, in which the algorithm is promised that the incoming instance will come from one of two sets separated by a large gap in a specific measure. A good approximation algorithm with regard to this measure would need to distinguish between instances coming from these sets. The conjectures in this field are hence often concerned with hardness of this kind of gap problems.

Since the two conjectures we use are variants of the *Unique Games Conjecture (UGC)*, let us start by stating the UGC itself. To do so, we need the following promise problem:

Problem 1 (Unique Label Cover, $ULC(\epsilon, R)$).

Given a bipartite bi-regular graph $G = (U \uplus V, E)$, a set of labels $[R] = \{0, \ldots, R-1\}$, and for each $e \in E$ a permutation $\pi_e : [R] \to [R]$ distinguish between two cases:

- (YES) There exists a subset $V' \subset V$ of size $|V'| \geqslant (1-\epsilon)|V|$ and a labelling $l : U \cup V \to [R]$ such that for every edge $(u, v) \in E$ where $v \in V'$, $\pi_{(u,v)}(l(u)) = l(v)$.
- (NO) For every labelling $l : U \cup V \to [R]$ the number of satisfied edges (u, v), i.e. $\pi_{(u,v)}(l(u)) = l(v)$ is smaller than $\epsilon|E|$.

The Unique Games Conjecture [15] can then be stated as a hardness of ULC[3]:

Conjecture 1 (UGC [15]). For every $\epsilon > 0$, there exists R such that $ULC(\epsilon, R)$ is NP-hard.

One of the conjectures we rely on is the so-called *Strong Unique Games Conjecture*[4], which is a strengthening of UGC, where in the (NO) case the graph G is also promised to be a "small set expander"; a precise definition is given below.

Problem 2 (ULC with weak expansion, $ULCE(\epsilon, R, \delta)$). Given a bipartite bi-regular graph $G = (U \uplus V, E)$, a set of labels $[R]$, and for each $e \in E$ a permutation $\pi_e : [R] \to [R]$ distinguish between two cases:

- (YES) There exists a subset $V' \subset V$ of size $|V'| \geqslant (1-\epsilon)|V|$ and a labelling $l : U \cup V \to [R]$ such that for every edge $(u, v) \in E$ where $v \in V'$, $\pi_{(u,v)}(l(u)) = l(v)$.
- (NO) For every labelling $l : U \cup V \to [R]$ the number of satisfied edges (u, v), i.e. $\pi_{(u,v)}(l(u)) = l(v)$ is smaller than $\epsilon|E|$. **Moreover**, for every $S \subset V$ of size $|S| = \delta|V|$, $|\Gamma(S)| \geqslant (1 - \delta)|V|$, where $\Gamma(S)$ is a neighbourhood of the set S in the graph G.

---

[3] This formulation is slightly different that the original one from [15], but the two formulations are known to be equivalent [17].

[4] Originally, Strong UGC was used to refer to an equivalent formulation of UGC [17]. Here we follow the nomenclature of [3], which refers to Conjecture 2 as Strong UGC.

Conjecture 2 (Strong UGC [2]). For every $\epsilon, \delta > 0$, there exists R such that $\mathrm{ULCE}(\epsilon, R, \delta)$ is NP-hard.

Another conjecture we base our results on is the *Small Set Expansion Hypothesis (SSEH)*, which concerns the hardness of finding well-separated parts of a graph. Formally, in a d-regular weighted graph $G = (V, E)$, an edge expansion of a subset S of vertices is defined as $\Phi(S) = \frac{E(S, V \setminus S)}{d \cdot \min\{|S|, |V \setminus S|\}}$, where $E(S, V \setminus S)$ is a total weight of edges crossing the cut $(S, V \setminus S)$. The small set expansion problem parametrised by $\delta, \eta \in (0, 1)$ is defined as follows.

Problem 3 ($SSE(\delta, \eta)$). Given a regular, edge-weighted graph $G = (V, E, w)$ distinguish between two cases:

- (YES) There exists $S \subset V$ of size $|S| = \delta|V|$ such that $\Phi(S) \leqslant \eta$.
- (NO) For every $S \subset V$ of size $|S| = \delta|V|$, $\Phi(S) \geqslant 1 - \eta$.

The SSEH postulates the hardness of this problem:

Conjecture 3 (SSEH [22]). For every $\eta > 0$ exists $\delta = \delta(\eta) \in (0, \frac{1}{2}]$ such that $SSE(\delta, \eta)$ is NP-hard.

In the same work that introduced SSEH [22], Raghavendra and Steurer showed that SSEH can be equivalently stated as a hardness of ULC where the NO case additionally requires a certain condition on the edge expansion of small sets of the graph[5].

We remark that both Strong UGC and SSEH have been used as starting points for many hardness of approximation results [1–3, 5, 13, 19, 20, 22, 23]. Furthermore, some evidences supporting the UGC–such as the SDP integrality gap of [18]– also apply to these hypotheses.

Our Method. Similarly to the hardness of MMM in general graphs, our proofs in this paper take advantage of a close relationship between the INDEPENDENT SET problem and MMM. In short, the set of vertices left unmatched by a maximal matching must form an independent set. In [17] it was proven that it is NP-hard (under UGC) to distinguish graphs with (YES) independent set of size $(\frac{1}{2} - \epsilon)|V|$ from (NO) those with no independent set of size $\epsilon|V|$. In [9] this reduction was modified by only adding edges, so that in the (YES) case there was a perfect matching between vertices not included in the independent set. This strategy aims for simplicity, as typically, when proving correctness of gap reductions, analysing the (NO) case is significantly more difficult than the (YES) case. When we only add edges, no independent set can be created, so there is no need to look at the (NO) part at all.

In bipartite graphs, we are aiming to employ the same strategy. However, while maximum independent sets in bipartite graphs can be found in polynomial time, we notice that vertices left out by any maximal matching must form

---

[5] This is similar to Strong UGC except that Strong UGC deals with *vertex* expansion whereas SSEH concerns with *edge* expansion.

a *balanced bi-independent set*—an independent set with the same number of vertices on both sides of bipartition. We must hence look at existing hardness of approximation proofs for balanced bi-independent set [3,20][6] and carefully add edges to them.

2   Part I: From Strong UGC To MMM

In this section we will review the reduction from Unique Label Cover with Weak Expansion to Bi-Independent Set problem, that was presented in [3]. Then we will present our modification and prove its correctness. The main building block of their reduction is the graph T that we are going to use as a gadget. It will be composed of identical layers and parametrised by a prime number $q > 2$ (see Fig. 1).

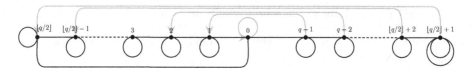

Fig. 1. The gadget T_q for a prime number $q > 2$ (black) is an undirected 3-regular graph. The modified gadget T_q' is obtained by adding directed arcs from i to $q - i$ for $i \notin Lo_q$ (grey).

Definition 1 (T_q). *The vertex set is* $[q] = \{0, 1, \ldots, q - 1\}$. *There are three kinds of (undirected) edges.*

- $\{i, i + 1 \pmod{q}\}$ *for* $i \neq \lfloor q/2 \rfloor$.
- *A self-loop* $\{i, i\}$ *for* $i \neq 0$ *and an additional one for* $\lfloor q/2 \rfloor + 1$.
- *An edge* $\{0, \lfloor q/2 \rfloor\}$.

The resulting graph is 3-regular and connected.

Let additionally define the sets $Hi_q = \{\lfloor q/2 \rfloor + 1, \ldots, q - 1\}$ and $Lo_q = \{1, \ldots, \lfloor q/2 \rfloor\}$. 0 belongs to neither. As one can see, there are no edges in T_q between Hi_q and Lo_q, so they form a balanced bi-independent set (even though T_q is not bipartite). The graph $T = T_q^{\otimes R}$ has the vertex set $[q]^R$ (so every vertex is an R-element vector of numbers from $[q]$, every coordinate corresponds to one label colour in Unique Label Cover) and two vertices x and y are connected, if x_i and y_i are connected in T_q for every coordinate $i \in R$. The reduction from [3] is now defined as follows.[7]

---

[6] These results talk about *Balanced Biclique* which is more natural, but since bi-independent set fits our needs more, we will silently take graph complement when recalling their reductions.

[7] The character 'o' is a function composition operator. Since a vector can be viewed as a function mapping indices to values, for a permutation π and a vector x, $\pi \circ x$ is the vector with permuted elements.

Reduction 3. *Input: An* ULCE(ϵ, R, δ) *instance* $G = \langle (U_G \uplus V_G, E_G),$ $\{\pi_e\}_{e \in E}\rangle$.
Output: A bipartite graph $H = \langle U_H \uplus V_H, E_H \rangle$.

Let $U_H = V_H = V_G \times [q]^R$, *so every vertex in* H *corresponds to one vertex in* V_G *and* R *numbers in* $[q]$. *For any* $u \in U_G$ *and two of its neighbours* v_1, v_2 *in* G, *we add an edge between* $(v_1, x) \in U_H$ *and* $(v_2, y) \in V_H$ *in* E_H *iff* $\pi^{-1}_{(u,v_1)} \circ x$ *is connected to* $\pi^{-1}_{(u,v_2)} \circ y$ *in* T *(x and y's coordinates are permuted to the point of view of vertex* u*).*[7]

It is easy to see that if the input graph came from the (YES) case, then there is a bi-independent set of size $(\frac{1}{2} - \epsilon - \frac{1}{q})|U_H \uplus V_H|$ in H. To find it pick the labelling l and the set $V' \subset V_G$ of vertices fully satisfied by the labelling l. For any $v \in V'$ and x such that $x(l(v)) \in Lo_q$ pick (v, x) into the left side of the bi-independent set. Similarly, pick $(v, y) \in V_H$ if $y(l(v)) \in Hi_q$. The reader can see, that in fact two symmetrical bi-independent sets exist in H. The (NO) case is analysed in the original paper.

Lemma 1 (soundness, [3, Lemma 14 reinterpreted]). *For every constant* γ, *there exist constants* ϵ, δ *such that if* G *came from the NO case of* ULCE(ϵ, R, δ) *then* H *has no balanced bi-independent larger than* $\epsilon|U_H \uplus V_H|$.

Now it is time to introduce our modification. Let us define a new gadget T'_q by adding directed arcs $\{(i, q - i)\}_{i \notin Lo_q}$ (see Fig. 1). The undirected edge between two vertices is replaced by a pair of two directed opposite arcs, the self-loop is replaced by a single arc. Consequently, the graph $T' = T'^{\otimes R}_q$ is also directed. Our reduction is similar to the one above.

Reduction 4. *Input: An* ULCE(ϵ, R, δ) *instance* $G = \langle (U_G \uplus V_G, E_G),$ $\{\pi_e\}_{e \in E}\rangle$.
Output: A bipartite graph $H' = \langle U_H \uplus V_H, E'_H \rangle$.

Let $U_H = V_H = V_G \times [q]^R$, *like before. For any* $u \in U_G$ *and two of its neighbours* v_1, v_2 *in* G, *we add an edge between* $(v_1, x) \in U_H$ *and* $(v_2, y) \in V_H$ *in* E_H *iff there is an arc from* $\pi^{-1}_{(u,v_1)} \circ x$ *to* $\pi^{-1}_{(u,v_2)} \circ y$ *in* T'.

Lemma 2 (completeness). *If the graph* G *comes from the (YES) case of* ULCE, *there is a bi-independent set* $BIS \subset U_H \uplus V_H$ *of size* $|BIS| \geqslant (\frac{1}{2} - \epsilon - \frac{1}{q})|U_H \uplus V_H|$ *in* H'. *Moreover, there is a perfect matching in* H' *between* $U_H \setminus BIS$ *and* $V_H \setminus BIS$.

Proof. Take l and V' as in Problem 2. For every $v \in V'$ and $x \in [q]^R$ such that $x(l(v)) \in Lo_q$, we pick the vertex $(v, x) \in U_H$ into the set BIS. Similarly, for every $v \in V'$ and y with $y(l(v)) \in Hi_q$, we pick $(v, y) \in V_H$ into BIS. To see that BIS is indeed a bi-independent set pick any $(v_1, x) \in U_H \cap BIS$ and $(v_2, y) \in V_H \cap BIS$. Since $v_1, v_2 \in V'$, for every u in their common neighbourhood in G, $\pi_{(u,v_1)}(l(u)) = l(v_1)$ and $\pi_{(u,v_2)}(l(u)) = l(v_2)$. There is no arc from $x(l(v_1)) \in Lo_q$

to $y(l(v_2)) \in \mathcal{H}i_q$ in T'_q, so there is no edge between $(v_1, x) \in U_H$ and $(v_2, y) \in V_H$. The size of $\mathcal{BIS}$ is $|\mathcal{BIS}| \geqslant (1 - \epsilon)|V_G|(\frac{\lfloor q/2 \rfloor}{q})q^R \cdot 2 \geqslant (\frac{1}{2} - \epsilon - \frac{1}{q})|U_H \uplus V_H|$.

Now we define a perfect matching $\mathcal{M}$ between the vertices not in $\mathcal{BIS}$. First, for every $v \in V'$ we will match the vertex $(v, x) \in U_H$ where $x(l(v)) \notin \mathcal{L}o_q$ with $(v, x') \in V_H$ where

$$x'(i) = \begin{cases} x(i), & \text{when } i \neq l(v) \\ q - x(i), & \text{when } i = l(v) \end{cases}$$

For any vertex $v \in V_G \setminus V'$ let $\mathcal{M}((v, x)) = (v, x)$. This edge exists since there was a loop (i, i) in T'_q for every $i \in [q]$. □

With Lemmas 2 and 1 we can conclude.

Corollary 1. *Assuming Strong Unique Games Conjecture, for any $\epsilon > 0$, it is NP-hard given a bipartite graph $G = (U \uplus V, E)$ (with $|U| = |V|$) to distinguish between two cases.*

- *(YES) There is a maximal matching in G of size $(\frac{1}{2} - \epsilon)|U|$.*
- *(NO) There is no bi-independent set of size $\epsilon|U \uplus V|$, so every maximal matching must have at least $(1 - \epsilon)|U|$ edges.*

which gives us Theorem 1, as the $(2 - \epsilon)$-approximation algorithm faced with a (YES) instance would produce a solution of size at most $(1 - 2\epsilon)|U|$ unequivocally qualifying it as a (YES) case.

3   Part II: Hardness based on SSEH

Our second reduction starts from the Small Set Expansion problem. It is a modification of the proof of hardness of approximation of MBB from [20]. It is more complicated than the one presented in Sect. 2. The reduction is two-step, with an intermediate problem called Max UnCut Hypergraph Bisection (MUCHB) defined below.

Problem 4 (Max UnCut Hypergraph Bisection). Given a hypergraph $H = (V_H, E_H)$ decide between cases:

- (YES) There is a bisection[8] (V_H^0, V_H^1) of V_H s.t. $|E_H(V_H^0)|, |E_H(V_H^1)| \geqslant (\frac{1}{2} - \epsilon)|E_H|$.
- (NO) For every set $T \subset V_H$ of size $|T| \leqslant \frac{|V_H|}{2}$, $|E_H(T)| \leqslant \epsilon|E_H|$.

We are only modifying the second step, but we need to describe the first one as well, because we are relying on the structure of the solution in the (YES) case. Before we are able to do that, let us give an overview of the reduction, to explain the origin of its various components.

---

[8] (A, B) is a bisection of X if $|A| = |B| = |X|/2$ and $A \cup B = X$.

3.1   Reducing from SSEH to UGC

In order to understand the reduction it makes sense to first see an outline of the proof that SSEH is strictly stronger than UGC, which was presented in [22].

We begin with a graph $G = (V, E, w)$—an instance of the Small Set Expansion problem. We will produce a bipartite graph $\mathcal{H}_1 = (\mathcal{U}_1 \uplus \mathcal{V}_1, \mathcal{E}_1, \{\pi_e\}_{e \in \mathcal{E}_1})$ which shall constitute an instance of Unique Label Cover problem.[9]

Reduction 5. Set $R = \frac{1}{\delta}$ (δ is a parameter from SSE). Let $\mathcal{U}_1 = \mathcal{V}_1 = V^R$, so every vertex in $\mathcal{H}_1$ corresponds to a vector of R vertices from G.

For $\mathbf{u} = \begin{bmatrix} u_1 \\ \vdots \\ u_R \end{bmatrix}$ and $\mathbf{v} = \begin{bmatrix} v_1 \\ \vdots \\ v_R \end{bmatrix}$ we add $(\mathbf{u}, \mathbf{v})$ to $\mathcal{E}_1$ if there is a permutation π such that $\pi(\mathbf{u})$ and $\mathbf{v}$ are neighbours in $G^{\otimes R}$.[9] We set $\pi_{(\mathbf{u},\mathbf{v})} = \pi$.

The idea behind this reduction is, that if there is a well-separated set $S \subset V$ of size $\delta|V|$ in G, then a vector of $\frac{1}{\delta}$ vertices should in expectation have exactly one vertex in S. We call a vector $\mathbf{v} \in V^R$ S-singular if $|\{v_1, \ldots, v_R\} \cap S| = 1$. It is clear that if $\mathbf{u} \in \mathcal{U}_1$ and $\mathbf{v} \in \mathcal{V}_1$ are both S-singular and $(\mathbf{u}, \mathbf{v}) \in \mathcal{E}_1$ then the constraint $\pi_{(\mathbf{u},\mathbf{v})}$ is satisfied. Unfortunately, only $\frac{1}{e}$-fraction of all the vectors will be S-singular, so Reduction 5 is clearly a failed attempt, as the easy part—completeness—is not working.

There is a simple way to overcome this problem. If we pick a large constant k and sample k distinct R-element vectors (blocks) of vertices, it is very likely that at least one of these vectors will be S-singular. This idea stands behind the cSSEH to UGC[10].

Reduction 6. Let $R = \frac{1}{\delta}$ and k be a constant depending on ϵ we wish to achieve in our UGC. Let $\mathcal{H}_2 = (\mathcal{U}_2 \uplus \mathcal{V}_2, \mathcal{E}_2, \{\pi_e\}_{e \in \mathcal{E}_2})$ where $\mathcal{U}_2 = \mathcal{V}_2 = V^{R \times k}$ (every vertex in $\mathcal{H}_2$ corresponds to k length-R vectors of vertices of G).

For $\mathbf{u} = \begin{bmatrix} u_1^1 & \cdots & u_1^k \\ & \cdots & \\ u_R^1 & \cdots & u_R^k \end{bmatrix}$ and $\mathbf{v} = \begin{bmatrix} v_1^1 & \cdots & v_1^k \\ & \cdots & \\ v_R^1 & \cdots & v_R^k \end{bmatrix}$ we add $(\mathbf{u}, \mathbf{v})$ to $\mathcal{E}_2$ if there are k permutations $\pi^1, \ldots, \pi^k$ such that $\pi^1(\mathbf{u}^1)$ and $\mathbf{v}^1$ are neighbours in $G^{\otimes R}$, and $\pi^2(\mathbf{u}^2)$ and $\mathbf{v}^2$ are neighbours in $G^{\otimes R}$ and so on[10]. We will set $\pi_{(\mathbf{u},\mathbf{v})} = \pi^{1,\ldots,k} = \pi^1|\pi^2|\ldots|\pi^k$ (concatenation of k permutations).

The reader may verify that the completeness of this reduction is satisfied. However, it turns out that the reduction presented above does not yet provide the desired soundness, and certain "noises" have to be added. We will not discuss these noises in this context, but they will be present in the next subsection (Reduction 8).

---

[9] The vectors $\mathbf{v}$ and $\mathbf{w}$ are neighbours in $G^{\otimes R}$ iff v_i is connected to w_i in G for every $i \in [R]$.

[10] Equivalently we will write that $\pi^{1,\ldots,k}(\mathbf{u})$ and $\mathbf{v}$ are neighbours in $G^{\otimes(R \times k)}$.

3.2 Max UnCut Hypergraph Bisection

The first phase of the reduction was described in [20]. It begins with a graph $G = (V, E)$ and produces a weighted hypergraph $H = (V_H, E_H, w_H)$, an instance of MUCHB. We will present two equivalent formulations of this reduction—one will be combinatorial (easier to work with), the other (original) is probabilistic and we present it for the sake of completeness (as it defines the weight function and may be more intuitive).[11]

Reduction 7 (combinatorial). *Let $R = \Theta(\frac{1}{\delta})$, k, l be constants. Let also $\Omega = \{0, 1, \perp\}$. The hypergraph is built as follows:*
- *The set of hypervertices $V_H = V^{R \times k} \times \Omega^{R \times k}$.*
- *Every $e \in E_H$ is associated with a tuple $\langle B^1, \ldots, B^l, x, D \rangle$ where $B^1, \ldots, B^l \in V^{R \times k}$, $x \in \Omega^{R \times k}$, $D \subseteq [R] \times [k]$.*
- *A hypervertex $v = (A_v, x_v)$ belongs to the hyperedge $e = \langle B_e^1, \ldots, B_e^l, x_e, D_e \rangle$ if there is a permutation $\pi \in (\Pi_R)^k$ such that:*
  (i) $\exists p$ s.t. $A_v(i) = B_e^p(\pi(i))$ for all i where $x_v(i) \neq \perp$.
  (ii) $x_v(i) = x_e(\pi(i))$ for all i such that $\pi(i) \notin D_e$.
We will denote as $\pi_{v,e}$ the permutation that is used to show that $v \in e$.[11] The weight $w_H(e)$ of an edge is defined by a probabilistic process described below.

To describe the probabilistic construction we need to introduce some notation.

- Let ε_V be a small constant. For each $A \in V^{R \times k}$, $T_V(A)$ denotes a distribution on $V^{R \times k}$ where (i, j)-th coordinate is set to $A_{i,j}$ with probability $1 - \varepsilon_V$ and a uniformly random vertex of V otherwise.
- For $A \in V^{R \times k}$, $G^{\otimes(R \times k)}(A)$ denotes a distribution on $V^{R \times k}$ where (i, j)-th coordinate is a neighbour of $A_{i,j}$ in G sampled proportionally to the edge weights.
- Let β be a small constant. $\Omega_\beta = \{0, 1, \perp\}_\beta$ denotes a distribution over $\Omega = \{0, 1, \perp\}$ where 0 and 1 are sampled with probability $\beta/2$ and $\perp$ is sampled with probability $1 - \beta$.
- Let ε_T be a small constant. $S_{\varepsilon_T}(R)$ is a distribution over subsets of $[R] = \{1, \ldots, R\}$ where each number is picked with probability ε_T.
- For $x \in \Omega^{R \times k}$ and $A \in V^{R \times k}$ let $M_x(A) = \Big\{ A' \in V^{R \times k} \mid \forall i \in [R \times k].\ x_i =\perp \vee A_i = A'_i \Big\}$.
- For a set $D \subseteq [R] \times [k]$ and $x \in \Omega_\beta^{R \times k}$ let $C_D(x) = \Big\{ x' \mid x'_{([R] \times [k]) \setminus D} = x_{([R] \times [k]) \setminus D} \Big\}$.
- Π_R is a set of all permutations of $(1, \ldots, R)$.

---

[11] This may be not unique. The choice of a permutation is arbitrary.

Reduction 8 (probabilistic)

1. *Sample $A \sim V^{R \times k}$ and $\tilde{A}^1, \ldots, \tilde{A}^l \sim T_V(A)$.*
2. *Sample $B^1 \sim G^{\otimes(R \times k)}(\tilde{A}^1), \ldots, B^l \sim G^{\otimes(R \times k)}(\tilde{A}^l)$ and $\tilde{B}^1 \sim T_V(B^1), \ldots, \tilde{B}^l \sim T_V(B^l)$.*
3. *Sample $x \sim \Omega_\beta^{R \times k}$ and $D \sim S_{\varepsilon_T}(Rk)$.*
4. *Output a hyperedge $\left\{ (\pi(B'), \pi(x')) \mid \exists p \in [l], \pi = \pi^1 | \ldots | \pi^k \in (\Pi_R)^k, \right.$*
   $\left. x' \in C_D(x), B' \in M_{x'}(\tilde{B}^p) \right\}$.

Remark 1. Reductions 7 and 8 produce a weighted hypergraph (with weights assigned to hyperedges). By copying the hyperedges with quantities proportional to their weights we can make the instance unweighted as defined in Problem 4.

Manurangsi [20] shows that, for every $\varepsilon > 0$, there exist constants R, k, l, β, ε_V, ε_T such that the following theorems hold:

Theorem 9 ([20] **[Lemma 9]**). *If G came from a (YES) case of SSEH, then there is a bisection (V_H^0, V_H^1) of V_H such that $|E_H(V_H^0)|, |E_H(V_H^1)| \geqslant (1/2 - \epsilon)|E_H|$.*

Theorem 10 ([20] **[Lemma 14]**). *If G came from a (NO) case of SSEH, then every subset $T \subset V_H$ with $|T| \leqslant \frac{|V_H|}{2}$ has $|E_H(T)| \leqslant \epsilon|E_H|$.*

Henceforth, we will assume that we use the same parameters R, k, l, β, ε_V, ε_T as in [20].

Completeness—The Bisection. We present the construction of sets V_H^0 and V_H^1 from Theorem 9 (in the *(YES)* case). They will be important in our further proceedings. The following additional notation will be used:

- For a vector $A \in V^{R \times k}$ let $A^1, \ldots, A^k \in V^R$ be its *blocks*. Let $W(A, x, j)$ denote the set of all coordinates i in j-th block such that $x_i \neq \perp$ and $A_i^j \in S$, i.e., $W(A, x, j) = \{i \in [R] \mid A_i^j \in S \wedge x_i \neq \perp \}$.
- Let $j^*(A, x) \in [k]$ denote the first *x-filtered S-singular* block, i.e., smallest j with $|W(A, x, j)| = 1$. Note that if such block does not exist, we set $j^*(A, x) = -1$.
- Let $i^*(A, x) \in [R] \times [k]$ be the only element in $W(A, x, j^*(A, x))$. If $j^*(A, x) = -1$, let $i^*(A, x) = -1$. For a hypervertex $v = (A_v, x_v)$ we will sometimes write $i^*(v)$ instead of $i^*(A_v, x_v)$.

We will split all hypervertices into three sets: $T_0, T_1, T_\perp$ [12]. Let $T_\perp$ be the set of hypervertices (A, x), such that $j^*(A, x) = -1$. Then, for every hypervertex (A, x) not in $T_\perp$ we look at the value $b = x_{i^*(A, x)}$ and include it in T_0 if $b = 0$ and in T_1 if $b = 1$. It is shown that $|E_H(T_0)|, |E_H(T_1)| \geqslant (\frac{1}{2} - \epsilon)|E_H|$.

---

[12] T_0 and T_1 are called T_0' and T_1' respectively in [20].

Not-disjointness Relation Between Hypergraph Edges. Following from the definition of the hypergraph (as in Reduction 7), we have the following property.

Observation 11. *Two hyperedges* $e_1 = \langle B_1^1, \ldots, B_1^l, x_1, D_1 \rangle$ *and* $e_2 = \langle B_2^1, \ldots, B_2^l, x_2, D_2 \rangle$ *of the hypergraph are* <u>*not disjoint*</u> *if two permutations* $\pi_1, \pi_2 \in (\Pi_R)^k$ *exist satisfying the following conditions*

(1) $x_1(\pi_1(i)) = x_2(\pi_2(i))$ *for i such that* $\pi_1(i) \notin D_1 \wedge \pi_2(i) \notin D_2$*;*
(2) $\exists p_1, p_2 \in [l]$ *such that* $B_1^{p_1}(\pi_1(i)) = B_2^{p_2}(\pi_2(i))$ *for all i satisfying*
    (i) $x_j(\pi_j(i)) = \perp \Rightarrow \pi_j(i) \in D_j$ *for $j \in \{1, 2\}$, and*
    (ii) $\pi_1(i) \notin D_1 \vee \pi_2(i) \notin D_2$.

We will denote the *not-disjointness* relation by Λ. So $(e_1, e_2) \in \Lambda$ iff there is a hypergraph vertex $v \in e_1 \cap e_2$. Similarly, we define our relation $\Delta \subset E_H \times E_H$. Its definition is similar to the conditions of not-disjointness, but it is not symmetrical. It is also a superset of Λ.

Definition 2. *Two hyperedges* $e_1 = \langle B_1^1, \ldots, B_1^l, x_1, D_1 \rangle$ *and* $e_2 = \langle B_2^1, \ldots, B_2^l, x_2, D_2 \rangle$ *of the hypergraph are in relation* Δ *if two permutations* $\pi_1, \pi_2 \in (\Pi_R)^k$ *exist satisfying the following conditions*

(1) $x_1(\pi_1(i)) \preccurlyeq x_2(\pi_2(i))$ *for i such that* $\pi_1(i) \notin D_1 \wedge \pi_2(i) \notin D_2$*;*
(2) $\exists p_1, p_2 \in [l]$ *such that* $B_1^{p_1}(\pi_1(i)) = B_2^{p_2}(\pi_2(i))$ *for all i satisfying*
    (i) $x_j(\pi_j(i)) = \perp \Rightarrow \pi_j(i) \in D_j$ *for $j \in \{1, 2\}$, and*
    (ii) $\pi_1(i) \notin D_1 \vee \pi_2(i) \notin D_2$.

$x_1 \preccurlyeq x_2$ is a point-wise inequality on vectors of Ω, where $a \preccurlyeq a$ for all $a \in \Omega$ and $0 \preccurlyeq 1$.

3.3   From MUCHB to Bipartite MMM

In [20] the bipartite graph $\mathcal{G}^{\mathcal{BIS}} = (\mathcal{V}_L \cup \mathcal{V}_R, \mathcal{E}^{\mathcal{BIS}})$ is created the following way.

– $\mathcal{V}_L, \mathcal{V}_R = E_H$.
– $e_l \in \mathcal{V}_L$ and $e_r \in \mathcal{V}_R$ are connected if they are not disjoint (i.e. $(e_l, e_r) \in \Lambda$).

We are applying a modification to this construction. Instead of Λ we are basing the graph on the relation Δ which will result in a superset of edges.

> **Reduction 12. (MUCHB to MMM)** *The bipartite graph* $\mathcal{G}^{\mathcal{MMM}} = \langle \mathcal{V}_L \cup \mathcal{V}_R, \mathcal{E}^{\mathcal{MMM}} \rangle$ *is built as follows.*
>
> – $\mathcal{V}_L, \mathcal{V}_R = E_H$.
> – $e_l \in \mathcal{V}_L$ and $e_r \in \mathcal{V}_R$ are connected in $\mathcal{E}^{\mathcal{MMM}}$ iff $(e_l, e_r) \in \Delta$.

Completeness. In the (YES) case we need to find a maximal matching that leaves roughly half of the vertices of $\mathcal{G}^{\mathcal{MMM}}$ unmatched. The next few pages will be devoted to constructing it, as promised in the following lemma.

Lemma 3 (Completeness). *If G came from the (YES) case of SSEH, then in $\mathcal{G}^{\mathcal{MMM}}$ there is a maximal matching of size at most $(\frac{1}{2} + \epsilon)|\mathcal{V}_L|$.*

Our matching M will be composed of two parts: In M_1, $\mathcal{V}_L(T_0)$ will be matched to $\mathcal{V}_R(T_1)$. In M_2, $\mathcal{V}_L \setminus (EH(T_0) \cup EH(T_1))$ will be matched to $\mathcal{V}_R \setminus (EH(T_0) \cup EH(T_1))$. The remaining part—$\mathcal{V}_L(T_1)$ and $\mathcal{V}_R(T_0)$—forms a bi-independent set (thanks to the graph being built on an assymetric relation).

Constructing M_1. We will find a bijection ϕ between $E_H(T_0)$ and $E_H(T_1)$ such that $(x, \phi(x)) \in \Delta$ for all x. For a hyperedge $t = \langle B_0^1, \ldots, B_0^l, x_0, D_0 \rangle$ define its *witness indices* $W(t) = \{i \in [R] \times [k] \mid \exists p \in [l], x' \in C_D(x). i = i^*(B^p, x')\}$. Now let $t_0 = \langle B_0^1, \ldots, B_0^l, x_0, D_0 \rangle$ be any hyperedge in $E_H(T_0)$. We obtain x_1 by modifying x_0 as follows.

$$x_1(i) = \begin{cases} 1, & \text{if } i \in W(t_0). \\ x_0(i), & \text{otherwise,} \end{cases} \tag{1}$$

and set $\phi(t_0) = t_1 = \langle B_0^1, \ldots, B_0^l, x_1, D_0 \rangle$. We need to prove the correctness of this definition: In Lemma 7 we show that $\phi(E_H(T_0)) \subseteq E_H(T_1)$. In Lemma 8 we show that ϕ indeed is a bijection. Before we are able to prove them, we need to gather some knowledge about our witness indices $W(t)$.

Lemma 4. *Let $W'(t)$ be constructed in the following way. Consider any hypervertex v in t. For every such v we include $\pi_{v,t}(i^*(v))$ in $W'(t)$. Then $W(t) = W'(t)$.*

Proof. We will first show that $W(t) \subseteq W'(t)$. Consider any $i \in W(t)$. Let $p \in [l]$, $x' \in C_D(x)$ be such that $i = i^*(B^p, x')$. Then $(B^p, x') \in t$, so $i \in W'(t)$.

Next, we will show that $W'(t) \subseteq W(t)$. Let $v = (A_v, x_v)$ be any vertex in t. Let p be such that $A_v(i) = B_e^p(\pi_{v,t}(i))$ for all i where $x_v(i) \neq \perp$. We consider another vertex $v' = (\pi_{v,t}(A_v), \pi_{v,t}(x_v))$, which also is in t. π permutes each block in A_v separately, so it preserves value i^*, that is $i^*(v') = \pi_{v,t}(i^*(v))$. Now we claim that $i^*(v') = i^*(B^p, x)$. It is so, because if $A_v(\pi_{v,t}(j)) \neq B^p(j)$ then $x(\pi_{v,t}(j)) = \perp$. $\square$

Lemma 5. *Let t_0 be any hyperedge in $E_H(T_0)$. Then $D_0 \cap W(t_0) = \emptyset$.*

Proof. Assume the contrary. This means that there is $i \in D_0 \cap W(t_0)$. Let $v = (A_v, x_v) \in t_0$ be such that $i^*(A, x) = i$. We set $x(i)$ to be 1. After this change v is still in t_0, because $i \in D_0$. v is, however, also in T_1 because $x_v(i^*(A, x)) = 1$, which contradicts $t_0 \in E_H(t_0)$. $\square$

Lemma 6. *For any $t_0 \in E_H(T_0)$, $\forall i \in W(t_0). x_0(i) = 0$. Similarly, $\forall t_1 \in E_H(T_1). \forall i \in W(t_1). x_1(i) = 1$.*

Proof. Consider any $p \in [l]$ and $x' \in C_{D_0}(x_0)$. Then vertex $v = (B^p, x')$ is in t_0, and, in consequence, in T_0. Thus, $x'(i^*(B^p, x')) = 0$. As $i^*(v) \notin D_0$, $x(i^*(B^p, x')) = 0$ as well.   □

We are ready to see that ϕ maps $E_H(T_0)$ to $E_H(T_1)$.

Lemma 7. *Let t_0 be any hyperedge in $E_H(T_0)$ and $t_1 = \phi(t_0)$. Let $v = (A, x)$ be any hypervertex in t_1. Then $v \in T_1$.*

Proof. By Lemma 4, $\pi_{v,t_1}(i^*(A, x)) \in W(t_1)$, so from (1) we have $x_1(\pi_{v,t_1}(i^*(A, x))) = 1$. $\pi_{v,t_1}(i^*(A, x)) \notin D_0$, so $x(i^*(A, x)) = x_1(\pi_{v,t_1}(i^*(A, x))) = 1$. Thus, $v \in T_1$.   □

Lemma 8. *Function ϕ is a bijection.*

Proof. We start with an easy observation:

Observation 13. *Let t_0 be any hyperedge in $E_H(T_0)$. Then $W(t_0) = W(\phi(t_0))$.*

This observation follows from the fact that $i^*(\cdot, x)$ does not distinguish between zeros and ones in x. Now, it is easy to construct the inverse function to ϕ. ϕ^{-1} will be almost the same as ϕ, except that in the vector x we will set $x(i)$ to be 0 instead of 1. Due to Observation 13, ϕ^{-1} changes the same coordinates in x as ϕ, so it is an inverse of ϕ.   □

We will use the bijection ϕ to define the perfect matching between $\mathcal{V}_L \cap E_H(T_0)$ and $\mathcal{V}_R \cap E_H(T_1)$ in a natural way: $M_1 = \{(t_0, \phi(t_0)) | t_0 \in \mathcal{V}_L \cap E_H(T_0)\}$. We need to show that all the edges of this matching are in graph $\mathcal{G}^{\mathcal{MMM}}$.

Lemma 9. *Let t_0, t_1 be any pair such that $\phi(t_0) = t_1$. Then $(t_0, t_1) \in \Delta$.*

Proof. We choose both π_1, π_2 to be identity permutation. Both conditions in Definition 2 (of the relation Δ) are satisfied:

(1) is satisfied, as for all i either $x_0(i) = x_1(i)$ or $x_0(i) = 0$ and $x_1(i) = 1$, which is exactly the definition of $x_0(i) \preccurlyeq x_1(i)$.
(2) is satisfied, because $B_0^1 = B_1^1$.   □

Constructing M_2. As we said earlier, the vertices in $\mathcal{V}_L \setminus (E_H(T_0) \cup E_H(T_1))$ and $\mathcal{V}_R \setminus (E_H(T_0) \cup E_H(T_1))$ need to be taken care of. We are going to match them: $M_2 = \{(e_l, e_r) | e_l \in \mathcal{V}_L \setminus (E_H(T_0) \cup E_H(T_1)), e_R \in \mathcal{V}_R \setminus (E_H(T_0) \cup E_H(T_1))$ st. $e_l = e_r\}$ (every vertex is matched with its own copy). These pairs are in $\mathcal{E}^{\mathcal{MMM}}$ as the relation Δ is reflexive. We define our matching $M = M_1 \cup M_2$.

Maximality of M. Now, all vertices left unmatched by M are either in $\mathcal{V}_L \cap E_H(T_1)$ or $\mathcal{V}_R \cap E_H(T_0)$. In the following lemma we prove that they form a bi-independent set, and therefore our matching is maximal.

Lemma 10. *For every $t_1 \in E_H(T_1)$ and $t_0 \in E_H(T_0)$, $(t_1, t_0) \notin \Delta$.*

Proof. Assume by contraposition, that there is $t_1 = \langle B_1^1, \ldots, B_1^l, x_1, D_1 \rangle \in E_H(T_1)$ and $t_0 = \langle B_0^1, \ldots, B_0^l, x_0, D_0 \rangle$ such that $(t_1, t_0) \in \Delta$. We will show that there exists a hypervertex $v \in t_0 \cap T_1$. It will clearly follow that $t_0 \notin E_H(T_0)$.

The relations Δ and Λ are very similar. If $(t_1, t_0) \in \Lambda$ (t_1 and t_0 are not disjoint), we have that $t_0 \cap T_1 \neq \emptyset$ right away. Otherwise, from the definition of Δ we have that there are two permutations π_1, π_0 (which by permuting the indices in t_1 and t_0 we may assume to be identity permutations) such that $\forall i \in ([R] \times [k]) \setminus (D_1 \cup D_0)$. $x_1(\pi_1(i)) \preccurlyeq x_0(\pi_0(i))$ and the inequality is strict for at least one i (since $(t_1, t_0) \notin \Lambda$). Let us consider another vertex $t_0' = \langle B_0^1, \ldots, B_0^l, x_0', D_0 \rangle$, where x_0' is defined as follows.

$$x_0'(i) = \begin{cases} 0, & \text{if } x_0(i) = 1 \text{ and } x_1(i) = 0. \\ x_0(i), & \text{otherwise.} \end{cases} \tag{2}$$

Clearly $(t_1, t_0') \in \Lambda$, which means that $t_0' \notin E_H(T_0)$. Let $v' = (A_{v'}, x_{v'})$ be such that $v' \in t_0' \cap T_1$ (with the witness permutation $\pi_{v', t_0'}$, which we may again assume to be the identity permutation). We will change v' slightly, to "reverse" changes made in (2). x_v will be defined as follows:

$$x_v(i) = \begin{cases} 1, & \text{if } x_0(i) = 1 \text{ and } x_{v'}(i) = 0. \\ x_{v'}(i), & \text{otherwise.} \end{cases} \tag{3}$$

Now, a hypervertex $v = (A_{v'}, x_v) \in t_0$. To finish the proof we need to show that $v \notin T_0$. Since $v' \in T_1$, $x_{v'}(i^*(A_v, x_v)) = 1$. Then, from (3), $x_v'(i^*(A_v, x_v')) = x_v(i^*(A_v, x_v))$, so v is also in T_1. $\qquad\square$

To finish the proof of Lemma 3 we bound the size of matching M.

$$|M| = |M_1| + |M_2| = |\mathscr{V}_L| - |E_H(T_1)| \leqslant \left(\frac{1}{2} + \epsilon \right) |\mathscr{V}_L|.$$

3.4   Soundness

We recall Lemma from [20] about graph $\mathcal{G}^{\mathcal{BIS}}$ in the (NO) case of MUCHB.

Lemma 11 ([20] [Appendix A]). *If for every $T \subset V_H$ of size at most $\frac{|V_H|}{2}$ we have $|E_H(T)| \leqslant \epsilon|E_H|$, then in $\mathcal{G}^{\mathcal{BIS}}$ there is no bi-independent set of size $2 \cdot (\epsilon|E_H| + 1)$.*

As the set of edges of graph $\mathcal{G}^{\mathcal{MMM}}$ is a superset of the set of edges of $\mathcal{G}^{\mathcal{BIS}}$, analogous statement is also true for $\mathcal{G}^{\mathcal{MMM}}$.

Lemma 12 (Soundness). *If H came from the (NO) case of Problem 4, then in $\mathcal{G}^{\mathcal{MMM}}$ there is no bi-independent set of size $2 \cdot (\epsilon|E_H| + 1)$.*

Altogether Lemmas 3 and 12 give us the following, from which Theorem 2 immediately follows.

Corollary 2. *Assuming Small Set Expansion Hypothesis, for any $\epsilon > 0$ it is NP-hard given a bipartite graph $G = (U \uplus V, E)$ (with $|U| = |V|$) to distinguish between two cases.*

- *(YES) There is a maximal matching in G of size $(\frac{1}{2} + \epsilon)|U|$.*
- *(NO) There is no bi-independent set of size $\epsilon|U \uplus V|$, so every maximal matching must have at least $(1 - \epsilon)|U|$ edges.*

4 Conclusion and Open Questions

In this work, we prove tight hardness of approximation for Minimum Maximal Matching (and equivalently Minimum Edge Dominating Set) on bipartite graphs, assuming the Strong Unique Games Conjecture or the Small Set Expansion Hypothesis. As a consequence, we also obtain tight inapproximability of Maximum Stable Matching with Ties and Incomplete Lists under the same assumptions.

An obvious open question from our work is to show the tight hardness for MMM under assumptions that are (possibly) weaker than those used in our work, such as the (vanilla) Unique Games Conjecture. Another open question is to obtain a polynomial time $\left(2 - \frac{1}{f(n)}\right)$-approximation algorithm for some slow growing function f. As discussed in the introduction, the best known result of this kind [12] has $f(n) = O(n/\log n)$. Can we improve this to, e.g., $f(n) = \text{poly} \log(n)$?

References

1. Austrin, P., Pitassi, T., Wu, Yu.: Inapproximability of treewidth, one-shot pebbling, and related layout problems. In: Gupta, A., Jansen, K., Rolim, J., Servedio, R. (eds.) APPROX/RANDOM -2012. LNCS, vol. 7408, pp. 13–24. Springer, Heidelberg (2012). https://doi.org/10.1007/978-3-642-32512-0_2
2. Bansal, N., Khot, S.: Optimal long code test with one free bit. In: 50th Annual IEEE Symposium on Foundations of Computer Science, FOCS 2009, Atlanta, Georgia, USA, 25–27 October 2009, pp. 453–462. IEEE Computer Society (2009)
3. Bhangale, A., Gandhi, R., Hajiaghayi, M.T., Khandekar, R., Kortsarz, G.: Bicovering: covering edges with two small subsets of vertices. In: Chatzigiannakis, I., Mitzenmacher, M., Rabani, Y., Sangiorgi, D. (eds.) 43rd International Colloquium on Automata, Languages, and Programming, ICALP 2016, Rome, Italy, 11–15 July 2016, vol. 55 of LIPIcs, pp. 6:1–6:12. Schloss Dagstuhl - Leibniz-Zentrum für Informatik (2016)
4. Carr, R.D., Fujito, T., Konjevod, G., Parekh, O.: A 2 1/10-approximation algorithm for a generalization of the weighted edge-dominating set problem. In: Paterson, M.S. (ed.) ESA 2000. LNCS, vol. 1879, pp. 132–142. Springer, Heidelberg (2000). https://doi.org/10.1007/3-540-45253-2_13
5. Charikar, M., Chatziafratis, V.: Approximate hierarchical clustering via sparsest cut and spreading metrics. In: Klein, P.N. (ed.) Proceedings of the Twenty-Eighth Annual ACM-SIAM Symposium on Discrete Algorithms, SODA 2017, Barcelona, Spain, Hotel Porta Fira, 16–19 January 2017, pp. 841–854. SIAM (2017)

6. Chlebík, M., Chlebíková, J.: Approximation hardness of edge dominating set problems. J. Comb. Optim. **11**(3), 279–290 (2006)
7. Dinur, I., Khot, S., Kindler, G., Minzer, D., Safra, M.: On non-optimally expanding sets in grassmann graphs. In: Diakonikolas, I., Kempe, D., Henzinger, M. (eds.) Proceedings of the 50th Annual ACM SIGACT Symposium on Theory of Computing, STOC 2018, Los Angeles, CA, USA, 25–29 June 2018, pp. 940–951. ACM (2018)
8. Dinur, I., Khot, S., Kindler, G., Minzer, D., Safra, M.: Towards a proof of the 2-to-1 games conjecture? In: Diakonikolas, I., Kempe, D., Henzinger, M. (eds.) Proceedings of the 50th Annual ACM SIGACT Symposium on Theory of Computing, STOC 2018, Los Angeles, CA, USA, 25–29 June 2018, pp. 376–389. ACM (2018)
9. Dudycz, S., Lewandowski, M., Marcinkowski, J.: Tight approximation ratio for minimum maximal matching. In: Lodi, A., Nagarajan, V. (eds.) IPCO 2019. LNCS, vol. 11480, pp. 181–193. Springer, Cham (2019). https://doi.org/10.1007/978-3-030-17953-3_14
10. Escoffier, B., Monnot, J., Paschos, V.T., Xiao, M.: New results on polynomial inapproximability and fixed parameter approximability of EDGE DOMINATING SET. In: Thilikos, D.M., Woeginger, G.J. (eds.) IPEC 2012. LNCS, vol. 7535, pp. 25–36. Springer, Heidelberg (2012). https://doi.org/10.1007/978-3-642-33293-7_5
11. Fujito, T., Nagamochi, H.: A 2-approximation algorithm for the minimum weight edge dominating set problem. Disc. Appl. Math. **118**(3), 199–207 (2002)
12. Gotthilf, Z., Lewenstein, M., Rainshmidt, E.: A $(2 - c\frac{\log n}{n})$ approximation algorithm for the minimum maximal matching problem. In: Bampis, E., Skutella, M. (eds.) WAOA 2008. LNCS, vol. 5426, pp. 267–278. Springer, Heidelberg (2009). https://doi.org/10.1007/978-3-540-93980-1_21
13. Hardt, M., Moitra, A.: Algorithms and hardness for robust subspace recovery. In: Shalev-Shwartz, S., Steinwart, I. (eds.) COLT 2013 - The 26th Annual Conference on Learning Theory, Princeton University, NJ, USA, 12–14 June 2013, vol. 30 of JMLR Workshop and Conference Proceedings, pp. 354–375. JMLR.org (2013)
14. Huang, C.-C., Iwama, K., Miyazaki, S., Yanagisawa, H.: A tight approximation bound for the stable marriage problem with restricted ties. In: Garg, N., Jansen, K., Rao, A., Rolim, J.D.M. (eds.) Approximation, Randomization, and Combinatorial Optimization. Algorithms and Techniques, APPROX/RANDOM 2015, Princeton, NJ, USA, 24–26 August 2015, vol. 40 of LIPIcs, pp. 361–380. Schloss Dagstuhl - Leibniz-Zentrum für Informatik (2015)
15. Khot, S.: On the power of unique 2-prover 1-round games. In: Reif, J.H. (ed.) Proceedings on 34th Annual ACM Symposium on Theory of Computing, Montréal, Québec, Canada, 19–21 May 2002, pp. 767–775. ACM (2002)
16. Khot, S., Minzer, D., Safra, M.: Pseudorandom sets in grassmann graph have near-perfect expansion. In: Thorup, M. (ed.) 59th IEEE Annual Symposium on Foundations of Computer Science, FOCS 2018, Paris, France, 7–9 October 2018, pp. 592–601. IEEE Computer Society (2018)
17. Khot, S., Regev, O.: Vertex cover might be hard to approximate to within $2 - \epsilon$. J. Comput. Syst. Sci. **74**(3), 335–349 (2008)
18. Khot, S., Vishnoi, N.K.: The unique games conjecture, integrality gap for cut problems and embeddability of negative-type metrics into ℓ_1. J. ACM **62**(1), 8:1–8:39 (2015)

19. Louis, A., Raghavendra, P., Vempala, S.S.: The complexity of approximating vertex expansion. In: 54th Annual IEEE Symposium on Foundations of Computer Science, FOCS 2013, Berkeley, CA, USA, 26–29 October 2013, pp. 360–369. IEEE Computer Society (2013)

20. Manurangsi, P.: Inapproximability of maximum edge biclique, maximum balanced biclique and minimum k-cut from the small set expansion hypothesis. In: 44th International Colloquium on Automata, Languages, and Programming, ICALP 2017, Warsaw, Poland, 10–14 July 2017, pp. 79:1–79:14 (2017)

21. Parekh, O.: Edge dominating and hypomatchable sets. In: Eppstein, D. (ed.) Proceedings of the Thirteenth Annual ACM-SIAM Symposium on Discrete Algorithms, San Francisco, CA, USA, 6–8 January 2002, pp. 287–291. ACM/SIAM (2002)

22. Raghavendra, P., Steurer, D.: Graph expansion and the unique games conjecture. In: Schulman, L.J. (ed.) Proceedings of the 42nd ACM Symposium on Theory of Computing, STOC 2010, Cambridge, Massachusetts, USA, 5–8 June 2010, pp. 755–764. ACM (2010)

23. Raghavendra, P., Steurer, D., Tulsiani, M.: Reductions between expansion problems. In: Proceedings of the 27th Conference on Computational Complexity, CCC 2012, Porto, Portugal, 26–29 June 2012, pp. 64–73 (2012)

24. Schmied, R., Viehmann, C.: Approximating edge dominating set in dense graphs. Theor. Comput. Sci. **414**(1), 92–99 (2012)

25. Yannakakis, M., Gavril, F.: Edge dominating sets in graphs. SIAM J. Appl. Math. **38**(3), 364–372 (1980)

On b-Matchings and b-Edge Dominating Sets: A 2-Approximation Algorithm for the 4-Edge Dominating Set Problem

Toshihiro Fujito$^{(\boxtimes)}$ and Takumi Tatematsu

Department of Computer Science and Engineering,
Toyohashi University of Technology, Toyohashi 441-8580, Japan
{fujito,tatematsu.takumi.ft}@tut.jp

Abstract. We consider a multiple domination version of the edge dominating set problem, called b-*EDS*. Here, an edge set $D \subseteq E$ of minimum cardinality is sought in a given graph $G = (V, E)$ with a demand vector $b \in \mathbb{Z}^E$ such that each edge $e \in E$ is dominated by b_e edges of D. When a solution D is not allowed to be a multi-set, it is called *simple* b-EDS. We present a 2-approximation algorithm for simple b-EDS for the case of $\max_{e \in E} b_e \leq 4$. The 2-approximation on general graphs has been known when $\max_{e \in E} b_e \leq 3$, and in the most general cases 8/3 is the best approximation ratio known attainable in polynomial time [2]. Our algorithms are designed based on the most natural LP relaxation of b-EDS and maximal b-matchings (or its generalization).

Keywords: b-edge dominating set problem · b-matchings · LP relaxation

1 Introduction

In an undirected graph an edge is said to *dominate* itself and all the edges adjacent to it, and a set of edges is an *edge dominating set* (abbreviated to *eds*) if every edge in the graph is dominated by some in the set. EDGE DOMINATING SET problem (EDS) asks to find an eds of minimum cardinality (cardinality case) or of minimum total cost (cost case). It was shown by Yannakakis and Gavril that, although EDS has important applications in areas such as telephone switching networking, it is NP-complete to tell if an eds of a specified size exists in a given graph G even when G is planar or bipartite of maximum degree 3 [16]. The classes of graphs for which its NP-completeness holds were later refined and extended by Horton and Kilakos to planar bipartite graphs, line and total graphs, perfect claw-free graphs, and planar cubic graphs [9], although EDS admits a PTAS (polynomial time approximation scheme) for planar [1] or λ-precision unit disk graphs [10]. Meanwhile, some polynomially solvable special cases have been

This work is supported in part by JSPS KAKENHI under Grant Numbers 20K11676 and 17K00013.

also discovered for trees [12], claw-free chordal graphs, locally connected claw-free graphs, the line graphs of total graphs, the line graphs of chordal graphs [9], bipartite permutation graphs, cotriangulated graphs [15], and so on. Chlebík and Chlebíková showed that finding a better than (7/6)-approximation for EDS is NP-hard [6] while it was shown UGC-hard by Schmied and Viehmann to approximate EDS to within any constant better than 3/2 [14].

There are various variants of the basic EDS problem, and the most general one among them was introduced by Berger et al. [2] in the form of *capacitated b-Edge Dominating Set*, (b, c)-EDS, where an instance consists of a graph $G = (V, E)$, a demand vector $b \in \mathbb{Z}_+^E$, a capacity vector $c \in \mathbb{Z}_+^E$ and a cost vector $w \in \mathbb{Q}_+^E$. A set D of edges in G is called a (b, c)-eds if each $e \in E$ is adjacent to at least b_e edges in D, where we allow D to contain at most c_e multiple copies of edge e. The problem asks to find a minimum cost (b, c)-eds. The (b, c)-EDS problem generalizes the EDS problem in much the same way that the set multicover problem generalizes the set cover problem. In the special case when all the capacities c_e are set to $+\infty$, we call the resulting problem *uncapacitated b-EDS*, whereas it is called *simple b-EDS* when $c_e = 1$ for all $e \in E$. If b_e's are set to a same value for all $e \in E$, it is called *uniform* (b, c)-EDS.

Let $b_{\max}$ denote $\max_{e \in E} b_e$. We mainly focus on simple b-EDS (i.e., $(b, 1)$-EDS), and b-EDS in what follows usually means the simple one unless otherwise stated explicitly. It should be noted, however, that this does not impose serious restrictions in problem solving, as long as $b_{\max}$ is bounded by some constant, since general (b, c)-EDS can be reduced to $(b, 1)$-EDS by introducing $(\min\{b_{\max}, c_e\} - 1)$ many copies of e, all of them parallel to e, for each $e \in E$.

1.1   Previous Work

It was shown by Yannakakis and Gavril that the minimum EDS can be efficiently approximated to within a factor of 2 by computing any maximal matching [16]. More recently, a 2.1-approximation algorithm first [5], and then 2-approximation algorithms [8,13] have been successfully obtained for the cost case of EDS problem via polyhedral approaches.

Among the approximation results obtained in [2] those relevant to ours are summarized in the following list (note: their results hold even for the cost case of (b, c)-EDS problems):

- The (b, c)-EDS problem can be approximated within a factor of 8/3.
- The uniform and uncapacitated b-EDS problem can be approximated within a factor of 2.1 if $b = 1$ or a factor of 2 if $b \geq 2$.
- The integrality gap of the LP relaxation they used for (b, c)-EDS, more complex than ours with additional valid inequalities, is at most 8/3 and it is tight even when $b_e \in \{0, 1\}$, $\forall e \in E$.

In case of *uncapacitated b-EDS* a linear-time 2-approximation algorithm was obtained by Berger and Parekh [3]. Interestingly it is *not* yet known whether uncapacitated b-EDS on trees is in P or NP-hard although it is polynomial when either $c(e) \equiv 1$ or $b(e) \equiv 1$, $\forall e \in E$ [4], and FPTAS and a fixed-parameter algorithm were obtained for uncapacitated b-EDS on trees by Ito et al. [11].

The work [7] of Fujito is an immediate ancestor of the current work, and the main focus therein was to investigate how good a locally optimal matching can approximate b-EDS. To simplify matters let us consider the case of uniform and simple b-EDS, and let $\gamma_b(G)$ denote the cardinality of any smallest $(b,1)$-eds for $G = (V, E)$ where $b_e \equiv b$, $\forall e \in E$. Following the way locally optimal solutions are often termed in the local search optimization, we say that a matching M in G is k-opt if, for any matching N in G larger than M, $|M \setminus N| \geq k$ (So, a *maximal* matching is 1-opt). It was observed in [7] that

- For any graph G and any matching M in G, $b|M|/2 \leq \gamma_b(G)$ $\forall b \in \mathbb{N}$.
- For any graph G and any b-opt matching M_b in G, $\gamma_b(G) \leq b|M_b|$ (and an eds of size $\leq b|M_b|$ can be constructed efficiently from M_b) for $b \in \{1, 2\}$.
- For $b \geq 3$, $\gamma_b(G) \leq b|M_b|$ does not necessarily hold.

The last item implies that 2-approximation for the case of $b \equiv 3$ (and $b_e \in \{1, 2, 3\}$) demands a stronger lower bound on the optimal size. For that purpose, starting with the most natural IP formulation for simple b-EDS:

$$\min \{x(E) \mid x(\delta(e)) \geq b_e \text{ and } x_e \in \{0, 1\}, \forall e \in E\},$$

where $x(F) = \sum_{e \in F} x_e$ for $F \subseteq E$, and $\delta(e) = \{e\} \cup \{e' \in E \mid e' \text{ is adjacent to } e\}$ for $e \in E$, the following LP relaxation for b-EDS and its dual LP are considered in [7]:

LP: (P') $\min \sum_{e \in E} x_e$ LP: (D') $\max \sum_{e \in E} b_e y_e$

subject to: $x(\delta(e)) \geq b_e$, $\forall e \in E$ subject to: $y(\delta(e)) \leq 1$, $\forall e \in E$

$x_e \geq 0$, $\forall e \in E$ $y_e \geq 0$, $\forall e \in E$

Notice here that (P') coincides with the LP relaxation of *uncapacitated* b-EDS rather than simple one. The main result of [7] is a 2-approximation algorithm for 3-EDS (or b-EDS with $b \in \{1, 2, 3\}^E$), in which a solution is constructed by building it upon a 3-opt matching such that its size never exceeds twice the optimal value of (D').

1.2 Our Work and LP Relaxations

The current work is motivated by the results as well as approaches presented in [7], and it is quite natural to wonder if it is possible to approximate b-EDS within a factor of 2 even if $b_e \geq 4$ for some $e \in E$. We shot that this is so for $b_{\max} = 4$. To design a primal-dual based algorithm one needs to choose a suitable LP relaxation of the original combinatorial optimization problem. A corollary deducible from the main result of [7] is that the integrality gap of LP:(P') is bounded by 2 when $b_{\max} \leq 3$. It turns out, however, that the integrality gap of LP:(P') can be larger than 2 when $b_e \equiv 4$, $\forall e \in E$ (See Sect. 4). This fact suggests us to use a more "legitimate" LP relaxation, namely, the one for *simple* b-EDS as given below along with its dual LP:

$$\text{LP:(P)} \quad \min \sum_{e \in E} x_e \qquad\qquad \text{LP:(D)} \quad \max \sum_{e \in E} (b_e y_e - z_e)$$

$$\begin{aligned} \text{subject to:} \quad & x(\delta(e)) \geq b_e, \ \forall e \in E & \text{subject to:} \quad & y(\delta(e)) - z_e \leq 1, \ \forall e \in E \\ & x_e \geq 0, \ \forall e \in E & & y_e \geq 0, \ \forall e \in E \\ & x_e \leq 1, \ \forall e \in E & & z_e \geq 0, \ \forall e \in E \end{aligned}$$

Besides using a different LP relaxation, we also abandon the idea of constructing a solution from a b-opt matching. Instead, we consider building a solution on top of a maximal b-matching in case of $b_e \equiv b$, $\forall e \in E$, or a maximal β-matching in the general case where $\beta : V \to \mathbb{Z}_+$ is determined by $b \in \{1, 2, 3, 4\}^E$. It will be shown that our algorithm, building a solution upon a maximal β-matching, approximates (*unweighted*) *simple* b-EDS within a factor of 2 and that the integrality gap of LP:(P) is bounded by 2 when $b_{\max} \leq 4$. While the algorithm itself is relatively simple and purely combinatorial, it contains some subtleties in the orders in which edges are processed for possible inclusion into a solution, to be explained later in details.

1.3   Notations and Definitions

In this paper only graphs with no loops are considered. For a vertex set $S \subseteq V$ let $\delta(S)$ denote the set of edges incident to a vertex in S. When S is an edge set, we let $\delta(S) = \delta(\cup_{e \in S} e)$ where edge e is a set of two vertices; then, $\delta(S)$ also denotes the set of edges dominated by S. When S is a singleton set $\{s\}$, $\delta(\{s\})$ is abbreviated to $\delta(s)$. The degree of a vertex u is denoted by $d(u)$. When $\delta(S)$ and $d(u)$ are considered only within a subgraph H of G (or when restricted to within a vertex subset or edge subset T), they are denoted by $\delta_H(S)$ and $d_H(u)$ (or $\delta_T(S)$ and $d_T(u)$), respectively.

When an edge e is dominated by b_e edges (or more), it is said to be *fully dominated*.

2   Algorithm for 4-EDS

We describe in this section how to compute a 4-EDS solution given a graph $G = (V, E)$ with $b \in \{1, 2, 3, 4\}^E$ (See a pseudocode of this algorithm in Algorithm 1 for more details). First of all we may assume that (G, b) is a feasible instance, meaning that it has at least one solution, because (G, b) is feasible if and only if $|\delta(e)| \geq b_e$, $\forall e \in E$, which can be easily verified.

Our algorithm for 4-EDS computes one feasible solution $D \subseteq E$ in two stages. Let E be partitioned into $E_1, E_2, E_3,$ and E_4 according to b_e values such that $E_k = \{e \in E \mid b_e = k\}$ for $k \in \{1, 2, 3, 4\}$ and $E = \bigcup_{1 \leq k \leq 4} E_k$.

First stage (corresponding to the first for-loop in Algorithm 1):

Recall that an optimal EDS solution can be approximated by a maximal matching M within a factor of 2 in any graph. One way to account for 2-approximation of M here is to notice that $y \in \mathbb{R}^E$ with $y_e = 1/2$ for $e \in M$ ($y_e = 0, \forall e \notin M$, and $z_e = 0$, $\forall e \in E$) is feasible in LP (D).

Let us now consider the case of 4-EDS, first assuming for the simplicity that $b_e \equiv 4$, $\forall e \in E$, and let M be a maximal 4-matching in G. Then, $y \in \mathbb{R}^E$ with $y_e = 1/8$ for $e \in M$ ($y_e = 0, \forall e \notin M$, and $z_e = 0$, $\forall e \in E$) is feasible in LP (D) as $|\delta_M(e)| \leq 8$, $\forall e \in E$, and the dual objective value is $\sum_{e \in E}(b_e y_e - z_e) = |M|/2$. So, if M were a 4-EDS solution, it would provide a 2-approximate solution. It is not in general, however, and M needs to be augmented with more edges to become a 4-EDS solution.

We also need to deal with the case of nonuniform b_e's in $\{1, 2, 3, 4\}$, and an edge set $M \subseteq E$ is constructed by repeatedly computing a maximal β-matching $M_k \subseteq E_k$ within $G_k = (V, E_k)$, in the *descending* order of k's, and taking the disjoint union of M_k's (i.e., $M = \bigcup_{1 < k \leq 4} M_k$). Here, $\beta : V \to \mathbb{Z}_+$ is chosen appropriately such that no edge in M becomes overwhelmingly dominated by M by setting $\beta(v) = \max\{0, k - d_{M_{>k}}(v)\}$, $\forall v \in V$, where $M_{>k} = \bigcup_{i>k} M_i$, and M_k is maximal in G_k in the sense that no other β-matching in G_k properly contains it.

By this choice of β, $e = \{u, v\} \in E_k$ can be added to M_k iff $k > d_{M_{>k}}(u)$ and $k > d_{M_{>k}}(v)$. Thus, when $e = \{u, v\}$ enters M, the number of incident M-edges is bounded above by b_e at each of u and v, and it never exceeds b_e afterwards since the bound on the number of admissible M-edges is monotonically decreasing as we compute M_k's in the descending order of k's. Hence, M thus computed has the following property:

Observation 1. *At any $v \in V$, $b_e \geq d_M(v)$ for each $e \in \delta_M(v)$.*

Meanwhile, any edge not included in M becomes fully dominated by M as $e = \{u, v\} \in E_k$ cannot be added to M_k iff $k \leq d_{M_{>k}}(u)$ or $k \leq d_{M_{>k}}(v)$.

On the other hand, any $e \in M$ is not necessarily fully dominated by M itself, and the second stage is initiated to amend such shortcomings.

Second stage (corresponding to the second for-loop in Algorithm 1):

Let $D \subseteq E$ be a solution to be constructed, initialized at the beginning of the second stage to be equal to M computed in the first stage. The algorithm checks edges in M_k in the *ascending* order of k, and if any $e \in M_k$ is found not yet dominated by k edges, it chooses $(k - |\delta_D(e)|)$ many edges and adds them to D. The order of edges e to be upgraded (to fully dominated edges) among those in M_k is in the ascending order of the number of M-edges dominating e, and edges are chosen from $\delta(e) \setminus M$ (then added to D), one at a time, until e becomes fully dominated. When choosing an edge within $\delta(e) \setminus M$, it is checked which is more effective in decreasing the number of under-dominated edges, to choose from $\delta(u) \setminus M$ or from $\delta(v) \setminus M$ for $e = \{u, v\}$. We pick an edge from the side (u or v) with which more under-dominated edges are incident and adds it to D (if it exists, and if not, we pick an edge from the other side).

Algorithm 1. Algorithm for 4-EDS

Partition E into $E_1 \cup E_2 \cup E_3 \cup E_4$ such that $E_k = \{e \in E \mid b_e = k\}$
Set $M \leftarrow \emptyset$
for $k = 4$ downto 1 **do**
    Compute a maximal β-matching $M_k \subseteq E_k$ in $G = (V, E)$
        where $\beta(v) = \max\{0, k - d_M(v)\}$
    $M \leftarrow M \cup M_k$
end for
$D \leftarrow M$
$F \leftarrow \{e \in E \mid |\delta_D(e)| \geq b_e\}$ ▷ Set of fully dominated edges
for $k = 1$ to 4 **do**
    while $E_k \nsubseteq F$ **do**
        Let $e = \{u, v\} \leftarrow \operatorname{argmin}\{|\delta_M(e')| \mid e' \in M_k \setminus F\}$
                                        ▷ Pick an edge in M_k with largest deficit
        while $|\delta_D(e)| < k$ **do**   ▷ **Upgrading** e: Make e fully k-dominated
                       ▷ Decide which of u and v has more edges not yet fully dominated
            if $|\delta_M(u) \setminus F| > |\delta_M(v) \setminus F|$ **then**
                if $\delta(u) \setminus \delta_D(u) \neq \emptyset$ **then**
                    Pick any $e' \in \delta(u) \setminus \delta_D(u)$
                else
                    Pick any $e' \in \delta(v) \setminus \delta_D(v)$
                end if
            else   /* $|\delta_M(v) \setminus F| \geq |\delta_M(u) \setminus F|$ */
                if $\delta(v) \setminus \delta_D(v) \neq \emptyset$ **then**
                    Pick any $e' \in \delta(v) \setminus \delta_D(v)$
                else
                    Pick any $e' \in \delta(u) \setminus \delta_D(u)$
                end if
            end if
            Add e' to D and update F.
                          ▷ Let owner$(e') \leftarrow e$ to be used in analysis only.
        end while
    end while
end for

---

3   Dual Assignments and Analysis

We describe here how to assign dual values to (y, z), and we do so in two stages.

1st (Initialization) stage: In the first stage the values of y_e's are initialized as follows:

$$y_e = \begin{cases} \frac{1}{2b_e} & \text{if } e \in M \\ 0 & \text{otherwise} \end{cases}$$

and $z_e = 0$ for each $e \in E$. Corresponding to the first **for**-loop, the value of y thus determined can account for one half of $|M|$ only; the objective function value incremented by $e \in M$ is $b_e \times y_e = 1/2$. We say this as paying $1/(2b_e)$ at y_e to account for e.

Moreover, because of Observation 1,

$$y(\delta(v)) = y(\delta_M(v)) \le d_M(v) \cdot \frac{1}{2d_M(v)} = \frac{1}{2}$$

for any $v \in V$;

Observation 2. *At this point (i.e., before the following upgrading phase)* $y(\delta(v)) \le 1/2$ *for all* $v \in V$.

2nd (Upgrading) stage: In the 2nd **for**-loop of the algorithm more edges are selected and added into a solution D; namely, $|D \setminus M|$ many edges. So in the 2nd stage we modify the value of (y, z) so that they can account for $|D|$, and we adjust the value of (y, z) every time one edge e not yet fully dominated is chosen from M_k at the beginning of the outer **while**-loop of Algorithm 1 and some number of edges are added to D to make e fully dominated.

Let $e \in M_k \setminus F$ be such an edge chosen in an iteration of the outer **while**-loop of Algorithm 1 by

$$e = \{u, v\} \leftarrow \operatorname{argmin}\{|\delta_M(e')| \mid e' \in E_k \setminus F\}$$

for some $k \in \{1, \ldots, 4\}$. Let $D(e)$ denote D at the time when e is picked but before any edge is added to D by the subsequent **while**-loop as a result. Thus, e is initially dominated by $|\delta_{D(e)}(e)|$ many edges, and hence, the subsequent inner **while**-loop would add $(k - |\delta_{D(e)}(e)|) = (b_e - |\delta_{D(e)}(e)|)$ many edges of $\delta(e)$ to D so that e becomes fully dominated within the same iteration of the outer **while**-loop (**Upgrading** phase). For such $e \in E_k$ the value of y_e is upgraded as follows:

1. In the 1st (Initialization) stage, the value of y_e is set $= 1/(2b_e) = 1/2k$ to account just for e itself.
2. To account for those edges consequently added to D, y_e is increased by $(k - |\delta_{D(e)}(e)|)/2k$.

Besides those additional edges selected into D after e, we also consider paying, under the name of e, for some edges already existing in $\delta_{D(e)}(e)$.

3. For any $e' \in \delta_{D(e)}(e) \cap M$, if $e' \in M_j$ with $j < k$ and e' is being paid by e' itself (i.e., if $y_{e'} > 0$), then we let e pay for e' by increasing y_e by $1/(2k)$ while decreasing $y_{e'}$ by $1/(2j)$.
4. Moreover, any $e' \in \delta_{D(e)}(e) \setminus M$ is included because it was paid by some $e'' \in M$ (i.e., the owner of e'), and if $e'' \in \delta_M(e)$ and $e'' \in M_j$ with $j < k$, then we transfer the ownership of e' from e'' to e by increasing y_e by $1/(2k)$ while decreasing $y_{e''}$ by $1/(2j)$.

   We don't do this if the owner of $e' \in \delta_{D(e)}(e) \setminus M$ is not in $\delta_{D(e)}(e)$ or if e' is not in $\delta_{D(e)}(e) \setminus M$ even if its owner is in $\delta_{D(e)}(e)$.

Note: The ownership can be stolen only from edges of lower b-value.

Clearly, y_e may have a positive value only when $e \in M$. Although the value of y_e can be increased from its initial value when e is upgraded, it never exceeds $b_e \times (1/2b_e)$ (e never becomes an owner of more than b_e edges including e itself), and hence,

Observation 3. *1. $y_e > 0$ implies that $e \in M$.*
2. $y_e \le 1/2$ for each $e \in M$ before and after e is upgraded.

The procedure above for dual assignments does not necessarily produce a dual feasible solution (y, z) (In fact z does not so far play any role). It can be identified, however, where (y, z) produced this way violates the dual constraints. Two types of edges are distinguished as special ones:

Type 1: $\bar{e} = \{\bar{x}, \bar{v}\} \notin M$ such that
- $d(\bar{v}) = 3$ and $d_M(\bar{v}) = 2$ (let $\delta_M(\bar{v}) = \{e_1, e_2\}$),
- $\delta_M(e_1) = \delta_M(e_2) = \{e_1, e_2\}$,
- $\{e_1, e_2\} \subseteq E_4$.

Type 2: $\bar{e} = \{\bar{u}, \bar{v}\} \in M$ such that
- $\delta_M(\bar{u}) = \delta(\bar{u}) = \{\bar{e}, e_1, e_2\} = \delta_M(e_1) = \delta_M(e_2)$,
- $\delta_M(\bar{v}) = \delta(\bar{v}) = \{\bar{e}, e_3, e_4\} = \delta_M(e_3) = \delta_M(e_4)$,
- $\{e_1, e_2, e_3, e_4\} \subseteq E_4$ and $\bar{e} \in E_3 \cup E_4$.

Lemma 1. *The procedure above constructs y such that $y(\delta(u)) \le 1/2$ and $y(\delta(v)) \le 1/2$ for each $e = \{u, v\} \notin M$ except for e of Type 1.*

Proof. See Sect. A.   □

Lemma 2. *The procedure above constructs y such that $y(\delta(e)) \le 1$ for each $e = \{u, v\} \in M$ except for e of Type 2.*

Proof. See Sect. B.   □

We now describe how to deal with edges of Type 1 and Type 2.

3rd (Fixing) Stage: Let $\bar{e} = \{\bar{x}, \bar{v}\} \notin M$ be an edge of Type 1; i.e.,
- $d(\bar{v}) = 3$ and $d_M(\bar{v}) = 2$ (let $\delta_M(\bar{v}) = \{e_1, e_2\}$),
- $\delta_M(e_1) = \delta_M(e_2) = \{e_1, e_2\}$,
- $\{e_1, e_2\} \subseteq E_4$.

By the y-assignment procedure given above, we know that $y_{e_1} + y_{e_2} \le (3 + 2)/8 = 1 + 1/8$. For such $\bar{e}$ we modify the dual assignments on $\bar{e}, e_1$, and e_2 as follows:
- $y_{e_1} \leftarrow y_{e_1} + 1/24$ (or $y_{e_2} \leftarrow y_{e_2} + 1/24$), and
- $z_{\bar{e}} \leftarrow 4/24 = 1/6$.

Namely, the y-value on e_1 and e_2 is increased by $1/3$ of the overflow $1/8$ while the z-value on $\bar{e}$ is set to 4 times of it.

Lemma 3. *After applying the fixing procedure to each edge of Type 1,*

1. *each of them becomes to satisfy the dual constraint on it, and*
2. *no dual constraint violation newly occurs anywhere else.*

Proof. 1. Before the fixing procedure is applied to $\bar{e} = \{\bar{x}, \bar{v}\}$, $y(\delta(\bar{x})) = 1/2$ and $y(\delta(\bar{v})) \leq 5/8$. Hence, after the fixing procedure is applied,

$$y(\delta(\bar{e})) - z_{\bar{e}} \leq \left(\frac{1}{2} + \frac{5}{8}\right) + \frac{1}{24} - \frac{4}{24} = 1.$$

2. Let $e_1 = \{u_1, \bar{v}\}$ and $e_2 = \{u_2, \bar{v}\}$. Since $y_{e_1} \leq 3/8 + 1/24$ (or $y_{e_2} \leq 3/8 + 1/24$) and $y(\delta(e_1)) = y(\delta(e_2)) \leq 5/8 + 1/24$, $y(\delta(u_1))$ or $y(\delta(u_2)) \leq 1/2$, and hence, the dual constraints on e_1, e_2, and any edge ($\notin M$) adjacent to them remain satisfied.

□

3rd (Fixing) Stage (continued): Let $\bar{e} = \{\bar{u}, \bar{v}\} \in M$ now be an edge of Type 2; i.e.,

- $\delta_M(\bar{u}) = \delta(\bar{u}) = \{\bar{e}, e_1, e_2\} = \delta_M(e_1) = \delta_M(e_2)$,
- $\delta_M(\bar{v}) = \delta(\bar{v}) = \{\bar{e}, e_3, e_4\} = \delta_M(e_3) = \delta_M(e_4)$,
- $\{e_1, e_2, e_3, e_4\} \subseteq E_4$ and $\bar{e} \in E_3 \cup E_4$.

By the y-assignment procedure given above, we know that either

- $y_{e_1}, y_{e_2}, y_{e_3}, y_{e_4} \leq 2/8$ while $y_{\bar{e}} = 1/8$ if $\bar{e} \in E_4$, or
- $y_{e_1} \leq 3/8$, $y_{e_2}, y_{e_3}, y_{e_4} \leq 2/8$ while $y_{\bar{e}} = 0$ if $\bar{e} \in E_3$ (assuming that e_1 is the first to be upgraded among e_i's),

and $y(\delta(\bar{e})) \leq 9/8$ in either case.

For such $\bar{e}$ we modify the dual assignments on $\bar{e}$ as follows:

- $y_{\bar{e}} \leftarrow y_{\bar{e}} + 1/24$, and
- $z_{\bar{e}} \leftarrow 4/24 = 1/6$.

Namely, the y-value on $\bar{e}$ is increased by $1/3$ of the overflow $1/8$ while at the same time the z-value on $\bar{e}$ is set to 4 times of it.

Lemma 4. *After applying the fixing procedure to each edge of Type 2,*

1. *each of them becomes to satisfy the dual constraint on it, and*
2. *no dual constraint violation newly occurs anywhere else.*

Proof. 1. Before the fixing procedure is applied to $\bar{e}$, $y(\delta(\bar{e})) \leq 1 + 1/8$. Hence, after the fixing procedure is applied,

$$y(\delta(\bar{e})) - z_{\bar{e}} \leq \frac{9}{8} + \frac{1}{24} - \frac{4}{24} = 1.$$

2. The dual constraints affected by application of the fixing procedure are, besides the one on $\bar{e}$, only those on e_1, e_2, e_3, and e_4. Since $y(\delta(e_i)) \leq (2 + 2 + 1)/8 + 1/24$ (or $\leq (3 + 2 + 0)/8 + 1/24$) for $i = 1, \ldots, 4$, all of them remain to satisfy ≤ 1.

□

Theorem 1. *Algorithm 1 is a 2-approximation algorithm for b-EDS for $b_{\max} \leq$ 4.*

Proof. The correctness is verified by observing that 1) any edge $\notin M$ is fully dominated by M due to maximality of M, and 2) any edge $\in M$ is fully dominated by D because adjacent edges are added to D until it becomes fully dominated.

For the approximation performance, consider y constructed by dual assignment procedure, and observe that $y(E) = y(M) = |D|/8$ since the dual value is increased by $1/8$ every time an edge is picked and added to D. Although (y, z) at this point is not necessarily dual feasible, notice that the objective function value of (D) is

$$\sum_{e \in E}(4y_e - z_e) = \frac{|D|}{2}.$$

By the fixing procedures we obtain a dual feasible (y, z) (due to Lemmas 3 and 4), and in each of them, every time y-value is increased, so z-value is increased 4 times of it. Therefore, the objective function value remains unchanged no matter how many times the fixing procedure is applied. It shows that (y, z) finally obtained is dual feasible and its value is $|D|/2$. We may thus conclude that Algorithm 1 outputs D of size bounded by twice the optimal value.   □

4   Integrality Gaps and Additional Remarks

As stated in Sect. 1 the integrality gap of LP:(P') is known to be bounded by 2 when $b \in \{1, 2, 3\}^E$ [7]. Here we show that it exceeds 2 when $b_e \equiv 4$, $\forall e \in E$. Let $C = (V_C = \{v_0, v_1, \ldots, v_7\}, E_C = \{\{v_0, v_1\}, \{v_1, v_2\}, \ldots, \{v_7, v_0\}\})$ denote a cycle of length 8, and $\tilde{C} = (V_C, E_C \cup \{\{v_2, v_6\}\})$ be C with one chordal edge $\{v_2, v_6\}$ attached to it. Let $\tilde{C}_i = (V_{C_i} = \{v_{0,i}, v_{1,i}, \ldots, v_{7,i}\}, E_{C_i} = \{\{v_{0,i}, v_{1,i}\}, \{v_{1,i}, v_{2,i}\}, \ldots, \{v_{7,i}, v_{0,i}\}, \{v_{2,i}, v_{6,i}\}\})$ denote a copy of $\tilde{C}$ for each $i = 1, 2, \ldots, k$, and construct a connected graph G' from all of $\tilde{C}_i$'s by superimposing $v_{0,1}, v_{0,2}, \ldots, v_{0,k}$ into a single node named $\bar{v}_0$ and $v_{4,1}, v_{4,2}, \ldots, v_{4,k}$ into another node named $\bar{v}_4$. By attaching 2 new edges, $\{u_1, \bar{v}_0\}, \{u_2, \bar{v}_0\}$ and another 2, $\{w_1, \bar{v}_4\}, \{w_2, \bar{v}_4\}$ to G', construction of a connected graph G is completed. Consider $x \in \mathbb{R}^E$ with the following primal values:

$$x_e = \begin{cases} 4 & \text{if } e = \{v_{2,i}, v_{6,i}\} \text{ (i.e., a chordal edge) for some } 1 \leq i \leq k, \\ 4 & \text{if } e = \{u_1, \bar{v}_0\} \text{ or } = \{w_1, \bar{v}_4\}, \\ 0 & \text{otherwise.} \end{cases}$$

It is easy to see that x is feasible for LP:(P') since $\delta(e)$ contains either $\{u_1, \bar{v}_0\}, \{w_1, \bar{v}_4\}$, or a chordal edge $\{v_{2,i}, v_{6,i}\}$ for some i for any edge e in G. Thus, the objective value of LP:(P') can be as low as $4k + 8$. To consider next the size of minimum simple 4-eds for G, it suffices just to notice that no edge of G can be missing from any simple 4-eds, and G has $9k + 4$ edges in total. Therefore, the integrality gap of LP:(P') is at least $(9k+4)/(4k+8)$ which grows arbitrarily close to $9/4$.

It follows immediately from Theorem 1 that the integrality gap of the LP relaxation LP:(P) for b-EDS is no larger than 2 when $b_e \leq 4$ for all the edges $e \in E$:

Corollary 1. *The integrality gap of the following natural LP relaxation for 4-EDS (i.e., $b_e \leq 4, \forall e \in E$):*

$$LP:(P) \quad \min \left\{ \sum_{e \in E} x_e \,\middle|\, x(\delta(e)) \geq b_e, 0 \leq x_e \leq 1, \forall e \in E \right\}$$

is bounded by 2.

Meanwhile, it was shown in [2] that improving the integrality gap of a natural LP relaxation for b-EDS beyond 8/3, even if augmented with some valid inequalities, is hard. These two results on the integrality gap are not contradictory as the latter hardness result applies to the case of "weighted" b-EDS while ours is only to the unweighted case. It is interesting to see and left open, together with the 2-approximability of (unweighted) simple b-EDS, if the integrality gap of the LP relaxation above remains bounded by 2 or not, when $b_e \geq 5$ for some $e \in E$.

A   Proof of Lemma 1

Proof. We consider only $y(\delta(v))$ for $e = \{u, v\} \notin M$ (as the same arguments apply to $y(\delta(u))$). Moreover, if no edge in $\delta_M(v)$ is upgraded, $y(\delta(v)) \leq 1/2$ (recall Observation 2), and we thus consider the case when at least one in $\delta_M(v)$ is upgraded.

Let $\delta_M(v) = \{e_1, e_2, \ldots, e_s\}$ for $s = d_M(v)$ in what follows, sorted in the ascending order of b_e (i.e., $i < j$ implies that $b_{e_i} \leq b_{e_j}$).

Case $d_M(v) = 0$. Clearly $y(\delta(v)) = 0$ since $y_e = 0, \forall e \in \delta(v)$.
Case $d_M(v) = 1$. Since $y_e \leq 1/2$ for any $e \in M$, $y(\delta(v)) \leq 1/2$.
Case $d_M(v) = 4$. For each $e \in \delta_M(v)$, $b_e = 4$, and none in $\delta_M(v)$ can be upgraded as it is fully dominated by M.
Case $d_M(v) = 3$. For each $e \in \delta_M(v)$, $b_e \in \{3, 4\}$.
    Subcase $|\delta_M(v) \cap E_3| = 3$. Then, none in $\delta_M(v)$ can be upgraded as it is fully dominated by M.
    Subcase $|\delta_M(v) \cap E_3| = 2$. The only edge that can be possibly upgraded is e_3 with $b_{e_3} = 4$. Such an upgrade on e_3 results in $y_{e_3} = 1/2$ while the initial values assigned on e_1, e_2 being nullified at the same time, and hence, $y(\delta(v)) = 1/2$.
    Subcase $|\delta_M(v) \cap E_3| = 1$. The edges in $\delta_M(v)$ that can be possibly upgraded are only e_2 and e_3 since e_1 with $b_{e_1} = 3$ is fully dominated by M. Suppose e_2 is upgraded before e_3 is. This means that $|\delta_M(e_2)| = 3$, and when e_2 is upgraded, an edge to be added to D must be chosen form $\delta(v) \setminus F$ if e_3 is not yet fully dominated, by addition of such an edge, e_3 becomes fully dominated. Therefore, e_3 already is or becomes fully dominated when e_2 is

upgraded, and hence, it never happens to upgrade e_3 itself. The upgrading stage also shifts the values of y such that, while resetting y_{e_1} to 0, y_{e_2} is increased from initial $1/8$ by $1/8$ for e_1 and another $1/8$ for an edge newly added to D, thus y_{e_2} totaling to $3/8$. Meanwhile, y_{e_3} remains at the initial value of $1/8$, and hence, $y(\delta(v)) = \sum_{i=1}^{3} y_{e_i} = 0 + 3/8 + 1/8 = 1/2$.

Subcase $|\delta_M(v) \cap E_3| = 0$. So, $\delta_M(v) \subseteq E_4$, and if $e_i \in \delta_M(v)$ is to be upgraded, it must be the case that $|\delta_M(e_i)| = 3$, which means that $\delta_M(e_i) = \delta_M(v)$. If only one of e_1, e_2, e_3 is to be upgraded, then $y(\delta(v)) = \sum_{i=1}^{3} y_{e_i} = 2/8 + 1/8 + 1/8 = 1/2$ after the upgrade. If two of them are possibly upgradable, then such an upgrade would hinder any subsequent upgrade for the following reason; the first upgrade on e would cause an addition of one edge to D, and that edge must come from $\delta(v) \setminus F$ because $|\delta_M(e)| = 3$ and $\delta(v)$ contains other edges not yet fully dominated. So the first upgrade makes all edges in $\delta_M(v)$ fully dominated. Therefore, at most one out of three is possibly upgraded again in this case, and we have $y(\delta(v)) = \sum_{i=1}^{3} y_{e_i} = 2/8 + 1/8 + 1/8 = 1/2$.

Case $d_M(v) = 2$. Let $\delta_M(v) = \{e_1, e_2\}$ and we know that $2 \leq b_{e_1} \leq b_{e_2} \leq 4$.

Subcase $b_{e_1} = 2$: So, e_1 is fully dominated by M (cannot be upgraded), and e_2 must be upgraded, implying that $b_{e_2} \geq 3$. Such an upgrade would reset y_{e_1} to 0 while $y_{e_2} \leq 1/2$, and hence, $y(\delta(v)) \leq 1/2$.

Subcase $b_{e_1} = 3$: We first consider the case when e_1 is upgraded, which implies that $\delta_M(e_1) = \delta_M(v)$. So, this upgrade chooses an edge from $\delta(v) \setminus F$ and adds it to D. If $e_2 \in E_3$, e_2 becomes fully dominated as a result of upgrading e_1, and hence, $y_{e_1} = 2/6$ and $y_{e_2} = 1/6$. Suppose now that $e_2 \in E_4$, and after the upgrade on e_1, either e_2 can be fully dominated or not. In the former case, $y(\delta(v))$ ends up with $y_{e_1} = 2/6$ and $y_{e_2} = 1/8$. In the latter case, e_2 is also upgraded and it causes resetting $2/6$ of y_{e_1} to 0 while increasing y_{e_2} to no larger than $1/2$, and hence, $y(\delta(v))$ becomes $\leq 1/2$.

Let us consider next the case when e_1 is *not* upgraded, implying that upgraded is e_2. In this case again, $y_{e_1} = 1/6$ and $y_{e_2} = 2/6$ if $e_2 \in E_3$, or $y_{e_1} = 0$ and $y_{e_2} \leq 1/2$ if $e_2 \in E_4$, and hence, $y(\delta(v)) \leq 1/2$ in either case.

Subcase $b_{e_1} = 4$: In this case $\delta_M(v) \subseteq E_4$. If only one of e_1, e_2 is upgraded, say e_2, then $y_{e_1} = 1/8$ and y_{e_2} becomes no larger than $3/8$, and hence, $y(\delta(v)) \leq 1/2$. It remains to consider the case both e_1, e_2 are upgraded, and assume w.l.o.g. that $|\delta_M(e_1)| \leq |\delta_M(e_2)|$ and that e_1 is upgraded before e_2 is.

Subsubcase $|\delta_M(e_1)| = 3$: So, $|\delta_M(e_2)| = 3$, and y_{e_1} becomes $2/8$ while y_{e_2} does at most $2/8$, and hence, $y(\delta(v)) \leq 1/2$.

Subsubcase $|\delta_M(e_1)| = 2$ (or $\delta_M(e_1) = \delta_M(v)$): When e_1 is upgraded, two edges are added to D. The algorithm would choose the 1st of those two edges from $\delta(v) \setminus D$ since e_2, not yet fully dominated edge, is incident with v. Since e_2 is supposed to be still subsequently upgraded, it must be the case that $\delta_M(e_2) = \delta_M(v)$, and the algorithm would try to choose the 2nd of those two again from $\delta(v) \setminus D$, if it exists, and if it does, e_2 becomes fully dominated, unnecessitating an upgrade on

e_2. Therefore, it is possible for both e_1 and e_2 to be upgraded only when $\delta_M(e_1) = \delta_M(e_2) = \delta_M(v)$ and $d(v) = 3 = d_M(v) + 1$, but this is exactly the case when e is an edge of Type 1.

□

B   Proof of Lemma 2

Proof. First of all if no edge in $\delta(e)$ is upgraded, $y(\delta(e)) = y(\delta(u)) + y(\delta(v)) - y_e \leq 1$ because of Observation 2.

Because of Observation 3.2, clearly $y(\delta(e)) \leq 1$ if $|\delta_M(e)| \leq 2$. So we consider only cases when $|\delta_M(e)| \geq 3$.

Case $|\delta_M(e)| = 3$. If none in $\delta_M(e)$ is upgraded, it can be verified that

$$y(\delta(e)) \leq 2 \cdot \frac{1}{4} + \frac{1}{6} < 1.$$

So, we may assume that at least one in $\delta_M(e)$ is upgraded.

Subcase $d_M(u) = 3, d_M(v) = 1$ (or vice versa). For any $e' \in \delta_M(e)$ we have $b_{e'} \geq 3$. This means that $y_{e'} \leq 2/8$ even after those edges are upgraded, and hence, $y(\delta(e)) \leq 6/8$.

Subcase $d_M(u) = d_M(v) = 2$. Let $\delta_M(u) = \{e, e_1\}$ and $\delta_M(v) = \{e, e_2\}$. Then, $b_{e_1}, b_{e_2} \geq 2$ and $b_e \geq 3$. This means that $y_{e_1}, y_{e_2} \leq 3/8$ and $y_e \leq 2/8$ even after those edges are upgraded, and hence, $y(\delta(e)) \leq 1$.

Case $|\delta_M(e)| = 4$. If none in $\delta_M(e)$ is upgraded, it can be verified that

$$y(\delta(e)) \leq \frac{1}{4} + \frac{1}{8} + 2 \cdot \frac{1}{6} < 1.$$

So, we may assume that at least one in $\delta_M(e)$ is upgraded.

Subcase $d_M(u) = 4, d_M(v) = 1$ (or vice versa). In this case, $\delta_M(e) \subseteq E_4$, and none can be upgraded.

Subcase $d_M(u) = 3, d_M(v) = 2$ (or vice versa). Let $\delta_M(u) = \{e, e_1, e_2\}$ and $\delta_M(v) = \{e, e_3\}$. If $e \in E_4$, then it cannot be upgraded while $y_{e_1}, y_{e_2} \leq 2/8$ and $y_{e_3} \leq 3/8$. Hence,

$$y(\delta(e)) \leq \frac{1 + 2 + 2 + 3}{8} = 1$$

if $e \in E_4$.

Consider next the case when $e \in E_3$, which cannot be upgraded. If $e_3 \in E_3 \cup E_2$, then $y_{e_3} \leq 2/6$, and by the same reasoning as above,

$$y(\delta(e)) \leq \frac{2 + 2}{8} + \frac{1 + 2}{6} = 1.$$

So, let us assume that $e_3 \in E_4$. If e_3 is not upgraded, easily $y(\delta(e)) \leq 1$, and so suppose it is upgraded. But then, such an operation would absorb $1/6$ of y_e while $y_{e_3} \leq 4/8$. Thus again,

$$y(\delta(e)) \leq \frac{2 + 2}{8} + 0 + \frac{4}{8} = 1.$$

Case $|\delta_M(e)| = 6$. So, $d_M(u) = 4$ and $d_M(v) = 3$ (or vice versa). Then, $\delta_M(u) \subseteq E_4$, no edge in it can be upgraded, and $y(\delta_M(u)) = 4/8$. Meanwhile, $y_{e'}$ cannot become larger than $2/8$ for each edge $e' \in \delta_M(v) \setminus \{e\}$. Therefore, $y(\delta(e)) \leq (4 + 2 + 2)/8 = 1$.

Case $|\delta_M(e)| = 7$. So, $d_M(u) = d_M(v) = 4$, $\delta_M(e) \subseteq E_4$, and none in $\delta_M(e)$ can be upgraded.

Case $|\delta_M(e)| = 5$. If $d_M(u) = 4$ and $d_M(v) = 2$ (or conversely, $d_M(u) = 2$ and $d_M(v) = 5$), since every edge in $\delta_M(u)$ belongs to E_4 and is not upgradable,

$$y(\delta(e)) \leq \frac{4}{8} + \frac{3}{8} = \frac{7}{8}.$$

It remains to consider the case when $d_M(u) = d_M(v) = 3$. Observe first that $\delta_M(e) \subseteq E_3 \cup E_4$, and e as well as any edge in E_3 is not upgradable. So, only such an edge $e' \in \delta_M(e) \cap E_4$ with $|\delta_M(e')| = 3$ is upgradable. Let $\delta_M(u) = \{e, e_1, e_2\}$ and suppose e_1 is such an edge and the first edge upgraded within $\delta_M(e)$. Observe here that, if $e_2 \in E_3$ or if $d(u) > d_M(u)$, e_2 cannot be upgraded after e_1 is, and when e_1 is upgraded, $y(\delta(u))$ becomes $4/8$ whether e, e_2 are in E_3 or in E_4. So, in this case, $y(\delta(e)) \leq y(\delta(u)) + 2/8 + 2/8 = 1$. Similarly, if any in $\delta_M(v) \setminus \{e\}$ is not upgraded, $y(\delta(e)) \leq 1$.

Thus, the only remaining case is when all the edges in $\delta_M(e)$ except for e are upgraded, and this is possible if and only if $d(u) = d(v) = 3$, $e' \in E_4$, and $|\delta_M(e')| = 3$ for all edges $e' \in \delta_M(e) \setminus \{e\}$, and this is exactly when e is an edge of Type 2.

Thus, we may conclude that $y(\delta(e)) > 1$ for $e \in M$ only if e is an edge of Type 2.

$\square$

References

1. Baker, B.S.: Approximation algorithms for NP-complete problems on planar graphs. J. Assoc. Comput. Mach. **41**(1), 153–180 (1994)
2. Berger, A., Fukunaga, T., Nagamochi, H., Parekh, O.: Approximability of the capacitated b-edge dominating set problem. Theoret. Comput. Sci. **385**(1–3), 202–213 (2007)
3. Berger, A., Parekh, O.: Linear time algorithms for generalized edge dominating set problems. Algorithmica **50**(2), 244–254 (2008)
4. Berger, A., Parekh, O.: Erratum to: linear time algorithms for generalized edge dominating set problems. Algorithmica **62**(1–2), 633–634 (2012)
5. Carr, R., Fujito, T., Konjevod, G., Parekh, O.: A $2\frac{1}{10}$-approximation algorithm for a generalization of the weighted edge-dominating set problem. J. Comb. Optim. **5**(3), 317–326 (2001)
6. Chlebík, M., Chlebíková, J.: Approximation hardness of edge dominating set problems. J. Comb. Optim. **11**(3), 279–290 (2006)
7. Fujito, T.: On Matchings and b-edge dominating sets: a 2-approximation algorithm for the 3-edge dominating set problem. In: Ravi, R., Gørtz, I.L. (eds.) SWAT 2014. LNCS, vol. 8503, pp. 206–216. Springer, Cham (2014). https://doi.org/10.1007/978-3-319-08404-6_18

8. Fujito, T., Nagamochi, H.: A 2-approximation algorithm for the minimum weight edge dominating set problem. Disc. Appl. Math. **118**(3), 199–207 (2002)
9. Horton, J.D., Kilakos, K.: Minimum edge dominating sets. SIAM J. Disc. Math. **6**(3), 375–387 (1993)
10. Hunt, H.B., Ravi, S.S., Marathe, M.V., Rosenkrantz, D.J., Radhakrishnan, V., Stearns, R.E.: A unified approach to approximation schemes for NP- and PSPACE-hard problems for geometric graphs. In: van Leeuwen, J. (ed.) ESA 1994. LNCS, vol. 855, pp. 424–435. Springer, Heidelberg (1994). https://doi.org/10.1007/BFb0049428
11. Ito, T., Kakimura, N., Kamiyama, N., Kobayashi, Y., Okamoto, Y.: Minimum-cost b-edge dominating sets on trees. Algorithmica **81**(1), 343–366 (2019)
12. Mitchell, S., Hedetniemi, S.: Edge domination in trees. In: Proceedings of the Eighth Southeastern Conference on Combinatorics, Graph Theory and Computing, Louisiana State University, Baton Rouge, LA, 1977, pp. 489–509. Congressus Numerantium, No. XIX (1977)
13. Parekh, O.: Edge dominating and hypomatchable sets. In: Proceedings of 13th SODA, pp. 287–291 (2002)
14. Schmied, R., Viehmann, C.: Approximating edge dominating set in dense graphs. Theoret. Comput. Sci. **414**, 92–99 (2012)
15. Srinivasan, A., Madhukar, K., Nagavamsi, P., Rangan, C.P., Chang, M.-S.: Edge domination on bipartite permutation graphs and cotriangulated graphs. Inf. Process. Lett. **56**(3), 165–171 (1995)
16. Yannakakis, M., Gavril, F.: Edge dominating sets in graphs. SIAM J. Appl. Math. **38**(3), 364–372 (1980)

The Traveling k-Median Problem: Approximating Optimal Network Coverage

Dylan Huizing[1]([🖂])[ID] and Guido Schäfer[1,2][ID]

[1] Centrum Wiskunde and Informatica, Amsterdam, The Netherlands
{dylan.huizing,g.schaefer}@cwi.nl
[2] University of Amsterdam, Amsterdam, The Netherlands

Abstract. We introduce the *Traveling k-Median Problem (TkMP)* as a natural extension of the k-Median Problem, where k agents (medians) can move through a graph of n nodes over a discrete time horizon of ω steps. The agents start and end at designated nodes, and in each step can hop to an adjacent node to improve coverage. At each time step, we evaluate the coverage cost as the total connection cost of each node to its closest median. Our goal is to minimize the sum of the coverage costs over the entire time horizon.

In this paper, we initiate the study of this problem by focusing on the uniform case, i.e., when all edge costs are uniform and all agents share the same start and end locations. We show that this problem is NP-hard in general and can be solved optimally in time $\mathcal{O}(\omega^2 n^{2k})$. We obtain a 5-approximation algorithm if the number of agents is large (i.e., $k \geq n/2$). The more challenging case emerges if the number of agents is small (i.e., $k < n/2$). Our main contribution is a novel rounding scheme that allows us to round an (approximate) solution to the 'continuous movement' relaxation of the problem to a discrete one (incurring a bounded loss). Using our scheme, we derive constant-factor approximation algorithms on path and cycle graphs. For general graphs, we use a different (more direct) approach and derive an $\mathcal{O}(\min\{\sqrt{\omega}, n\})$-approximation algorithm if $d(s,t) \leq 2\sqrt{\omega}$, and an $\mathcal{O}(d(s,t) + \sqrt{\omega})$-approximation algorithm if $d(s,t) > 2\sqrt{\omega}$, where $d(s,t)$ is the distance between the start and end point.

Keywords: k-median problem · Network coverage · Routing over time · Approximation algorithms

1 Introduction

Background and Motivation. The *k-Median Problem (k-MED)* is a classic problem in combinatorial optimization that has been studied extensively, both because of its theoretical appeal as well as its practical relevance. In this problem, we are given a finite metric space (V, d) and a parameter k. Our goal is to select

© Springer Nature Switzerland AG 2021
J. Koenemann and B. Peis (Eds.): WAOA 2021, LNCS 12982, pp. 80–98, 2021.
https://doi.org/10.1007/978-3-030-92702-8_6

k points (called *medians*) such that the sum of the connection costs with respect to d from each node $u \in V$ to its nearest median is minimized.

In this paper, we consider the following natural extension of the k-Median Problem, which we term the *Traveling k-Median Problem (TkMP)*: Suppose we are given a fixed time horizon $T = \{0, \ldots, \omega\}$ and a set $A = \{1, \ldots, k\}$ of k medians (called *agents* in this context) which can move through the graph $G = (V, E)$ in discrete steps. Each agent $a \in A$ starts at time $\tau = 0$ at a designated start node $s_a \in V$, and ends at time $\tau = \omega$ at a designated end node $t_a \in V$. In each time step, an agent may either hop from her current location $u \in V$ to an adjacent node $v \in V$, or simply stay at u. Our goal is to determine a set of k walks (one for each agent) such that the total coverage costs (summed over all time steps) is minimized. Here, the *coverage cost* at time τ is defined the same way as in the k-Median Problem (with respect to the locations of the agents at time τ).

Intuitively, in the Traveling k-Median Problem one wants to distribute the agents over the graph such that they guarantee a low coverage cost, ideally positioning them at the optimal k-median locations. However, in many real-life applications we cannot simply assume that the agents can be 'teleported' to their ideal locations. Instead, the agents might have to move there over time, which in turn influences the achievable coverage. Assuming that the time horizon is fixed adds some new characteristics to the problem which are relevant to several applications in emergency logistics. For example, the agents might correspond to an 'emergency fleet' (e.g., police patrol force, emergency medical coverages, railway emergency responders) that one needs to schedule to provide a good emergency response time over a pre-specified time horizon. In fact, we arrived at the Traveling k-Median Problem introduced in this paper through our (more applied) investigations in the domain of emergency logistics; more specifically, the Traveling k-Median Problem considered here is a stylized variant of the so-called *Median Routing Problem* studied in [12].

Our Contributions. In this paper, we initiate the study of the Traveling k-Median Problem (TkMP) by focusing on the uniform case (referred to as U-TkMP), i.e., when all edge costs are uniform and all agents start from the same start location and end at the same end location. Our main contributions are as follows:

(1) We first argue that even U-TkMP is NP-hard in general by a simple reduction from the Set Cover Problem. We then show that the problem can be solved optimally in $\mathcal{O}(\omega^2 n^{2k})$ time. (These results are presented in Sect. 3.)

(2) We derive a 5-approximation algorithm for U-TkMP if the number of agents is large, i.e., $k \geq n/2$. This allows us to assume $k \leq n/2$ in the remainder. Basically, the idea here is to let each agent take care of an edge in a maximum matching of the underlying graph. (These results are given in Sect. 3).

(3) We introduce a novel way of approximating U-TkMP by rounding an (approximate) solution to a 'continuous movement' relaxation of the problem to a discrete solution, thereby incurring a bounded loss in the approximation factor. (This Rounding Theorem is presented in Sect. 4).

(4) We use this Rounding Theorem to derive 10-approximation algorithms for U-TkMP on path graphs and cycle graphs. (These results are given in Sect. 5).

(5) Finally, for U-TkMP on general graphs we obtain the following results: If the shortest path distance $d(s,t)$ between s and t satisfies $d(s,t) \leq 2\sqrt{\omega}$, then we obtain an $\mathcal{O}(\min\{\sqrt{\omega}, n\})$-approximation algorithm. If $d(s,t) > 2\sqrt{\omega}$, we instead have an $\mathcal{O}(d(s,t) + \sqrt{\omega})$-approximation algorithm. This is useful for those graph topologies for which we do not yet know how to solve or approximate their continuous counterparts. (These results are given in Sect. 6).

Related Work. To the best of our knowledge, the Traveling k-Median Problem introduced in this paper has not been studied in the literature before. The following are the key distinguishing features of TkMP with respect to other problems considered in the literature: (i) We consider a fixed time horizon and all agents have to start and end at designated nodes; in particular, this restricts the walks that can be chosen by each median. (ii) The coverage cost accounts for the total connection cost of each node to its closest median (k-MED-objective). However, there are several other optimization problems that are related to TkMP. Below is a partial list of related problems.

At the core of TkMP lies the k-Median Problem which has been studied extensively in the literature. The metric problem introduced above was first shown to be NP-hard by Kariv and Hakimi [15]. The more general variant of the problem where the connection cost is non-negative and symmetric (but does not necessarily satisfy the triangle inequality) was first shown to be APX-hard by Lin and Vitter [18]. On the positive side, the problem is polynomial time solvable on trees [15,21] and there exists a polynomial-time approximation scheme if (V, d) constitutes a Euclidean space [1]. The current best approximation algorithm for the problem achieves an approximation ratio of $\alpha = 2.675 + \varepsilon$ [4]. In terms of negative results, the current best lower bound is the $(1 + 2/e)$-inapproximability result by Guha and Khuller [9].

TkMP is also related to classic online problems such as the k-*Server Problem* [17] or, more generally, the *Metrical Task Systems* [3]. In the k-Server Problem [17], there are k servers that can move through a graph and have to serve a sequence of requests. Also here, the servers incur some movement costs (given by a metric). In each step a node is requested and a server has to move there. The main difference with respect to TkMP is that the time horizon is not limited and that there is no coverage cost. In Metrical Task Systems [3], an online algorithm starts at a designated node of a given graph and has to cover a sequence of tasks. Also here, the algorithm incurs some coverage cost and some movement cost (given by a metric) to serve a sequence of tasks. However, the coverage cost is given in terms of a non-negative cost incurred at the node in which the algorithm resides; this is different from TkMP, where the coverage cost accounts for the total connection cost of the nodes to the medians.

Dynamic facility location models are often on a strategic level [5]. Wesolowsky [23] extends several static facility location models, including k-MED, to a problem where facility locations have to be chosen for several periods. If the facility locations differ from one period to the next, then a period-dependent cost is incurred, independent of how many facilities are relocated and how far.

Of the many dynamic facility location problems reviewed and classified by Farahani et al. in 2012 [5], none fits our setting of one configuration determining the decision space in the next time-step. Distance-dependent relocation costs appear to be less commonly discussed than time-dependent costs, but they are studied by Huang et al. [11] and Gendreau et al. [7]. Galvão and Santibanez-Gonzalez, instead, charge period-dependent costs for having a facility open [6]. Melo et al. consider a combination of facility opening, closing and maintenance costs, and costs of reallocating capacity between facilities [19].

On a more operational level, ambulance relocation models seek to deploy an ambulance to a revealed emergency and relocate the rest for optimal response to the next emergency [8,13]. Even these operational models assume 'teleportation' from one steady-state configuration to the next. The same holds on the online level for the classical k-Server Problem [16]. Bertsimas and Van Ryzin coordinate vehicles over the Euclidean plane to minimize the completion time of dynamically revealed requests [2].

From a game-theoretical standpoint, Hotelling [10] discusses the *Ice-Cream Man Problem*, in which two competing ice cream vendors position themselves on the line to capture the largest portion for themselves. When moving to increase the portion of the line they cover, they surprisingly reach an equilibrium when their distance approaches zero. However, an equilibrium is no longer reached for three vendors [22].

2 Preliminaries

In the *Traveling k-Median Problem* (TkMP), we are given a connected graph $G = (V, E)$ with n nodes, a set $A := \{1, \dots, k\}$ of k agents, and a discrete time horizon $T := \{0, \dots, \omega\}$. Each agent $a \in A$ has their own start location $s_a \in V$, where they must be at time 0, and an end location $t_a \in V$, where they must be at time ω. Between each time-step and the next, agents may stay where they are or 'hop' to adjacent nodes.

At each time-step, we evaluate how well 'spread' the agents currently are. A *configuration* $\sigma = (\sigma_1, \dots, \sigma_k) \in V^A$ specifies for each agent $a \in A$ the node $\sigma_a \in V$ at which a resides. Let $d(u, v)$ denote the cost function for covering $v \in V$ from $u \in V$. We assume that every node will be covered from its nearest median in σ, i.e., the cost of node v is $D(\sigma, v) := \min_{a \in A} d(\sigma_a, v)$, and the total cost of this configuration is $D(\sigma) := \sum_{v \in V} D(\sigma, v)$. We say that a configuration σ is *optimal* if it minimizes the total cost $D(\sigma)$ (over all possible configurations).

A *feasible solution* $\mu \in V^{A \times T} = (\mu^0, \dots, \mu^\omega)$ is a sequence of $(T + 1)$ configurations, with μ^0 and μ^ω given by the start and end locations, respectively, and each configuration μ^τ being *adjacent* to the previous configuration $\mu^{\tau-1}$ for every $\tau \in \{1, \dots, \omega\}$. We say that μ^τ is adjacent to configuration $\mu^{\tau-1}$ if for each agent $a \in A$, either $\mu_a^{\tau-1} = \mu_a^\tau = u$ (agent a remains at node u), or $\mu_a^{\tau-1} = u$,

$\mu_a^\tau = v$ and $(u, v) \in E$ (agent a moves along edge (u, v) from u to v). For any feasible solution μ, we define the cost as $D(\mu) := \sum_{\tau=0}^{\omega} D(\mu^\tau)$. TkMP is the problem of finding a solution μ with minimal $D(\mu)$.

Given a feasible solution μ, we call the sequence $(\mu_a^0, \ldots, \mu_a^\omega)$ of nodes visited by agent a also the *walk* of a. The cost of the walk of agent $a \in A$ with respect to some fixed node v is defined as $\sum_{\tau=0}^{\omega} D(\mu_a^\tau, v)$.

In the *Uniform Traveling k-Median Problem* (U-TkMP), which is a special case of TkMP, we make the following three assumptions: (i) d is defined as the shortest path distance (with respect to the number of edges) induced by the underlying graph G, (ii) all agents start at the same node s, and (iii) all agents end at the same node t.

3   Hardness and Bounds

In this section, we will derive some lemmas and observations that will serve as instrumental building blocks for the approximation algorithms in this paper. As a first result, it deserves mentioning that even U-TkMP is NP-hard.

Theorem 1. *U-TkMP and its generalization TkMP are NP-hard.*

Proof. This follows from a reduction from the Set Cover Problem. See Appendix A.   □

Theorem 2. *TkMP is solvable in $\mathcal{O}(\omega^2 n^{2k})$.*

Proof. This follows from formulating a shortest path problem on a digraph with one node per agent configuration and time-step. See Appendix A.   □

However, if we want to determine a minimum cost walk of one agent with respect to a given node v, we can do this quite easily.

Lemma 1. *In U-TkMP, given any node v and an agent $a \in A$, we can find a minimum cost (s, t)-walk $\pi = (\pi^0, \ldots, \pi^\omega)$ of agent a with respect to v in $\mathcal{O}(n^4)$ time.*

Proof. See Appendix A.

We now provide several bounds on performance and parameters.

Lemma 2. *Consider U-TkMP. Then an optimal configuration σ^* has cost at least $D(\sigma^*) \geq (n - k)$. Further, any feasible solution μ has cost at least $D(\mu) \geq (\omega + 1)D(\sigma^*) \geq (\omega + 1)(n - k)$.*

Proof. In any configuration σ, at most k nodes can be occupied. All other nodes account for at least a cost of 1, implying $D(\sigma^*) \geq (n - k)$. Because any solution μ decomposes into $(\omega + 1)$ feasible configurations, $D(\mu) \geq (\omega + 1)D(\sigma^*)$.   □

Theorem 3. *In U-TkMP, if $k \geq n/2$, we can find a 5-approximation in $\mathcal{O}(n^5)$ time.*

Proof. We construct the approximate solution μ as follows. Using a greedy $\mathcal{O}(n^2)$ algorithm, find any (inclusion-wise) maximal matching on G. For every edge in this matching, assign an agent to one of its end points: we have enough agents to do this, because no matching can consist of more than $n/2$ edges. Assign the remaining agents to t. For every agent, determine a minimum cost walk with respect to the assigned node by using Lemma 1 (which requires at most $\mathcal{O}(n^4)$ time). We need to determine at most $n/2 + 1$ such walks, resulting in an algorithm that needs at most $\mathcal{O}(n^5)$ time.

We can compare μ to the hypothetical solution μ^n where the number of available agents is n, and we assign each node its own agent. In μ^n, we again send each agent over a walk that minimizes the total distance to their picked node. Note that $D(\mu^n)$ is minimal, because every node that gets its own agent is served optimally.[1] If node v is 'picked' by μ, v contributes this same optimal amount to $D(\mu)$. If v is not 'picked', we show here that the cost increases by at most 4 per time-step.

Take any node v and time-step τ, and let us examine the contribution of (v, τ) to the cost function. Let a be the agent in μ^n assigned to v. Let $u \in V$ be the location of a at time τ. In μ, there may or may not be someone at u at time τ. If there is, the request (v, τ) has the same cost contribution to $D(\mu)$ as to $D(\mu^n)$, remarking again that nodes with their own agent are served optimally. If not, there is a node $u' \in V$ that was picked, at most 2 hops away from u; otherwise, the underlying matching would not be maximal. If the agent in μ assigned to u' is still headed for it, that agent is at most 2 hops away from u', because $d(s, u') \leq d(s, u) + d(u, u') \leq \tau + 2$. Similarly, if the agent is moving away from u' towards t, the agent is at most 2 hops away from u' because $d(u', t) \leq d(u', u) + d(u, t) \leq (\omega - \tau) + 2$. That is, there are only $\omega - \tau$ hops left to get to t, and while this is no problem when $d(u', t)$ is small, the agent will be pulled away by at most 2 hops if $d(u', t)$ is large. So indeed, by having less agents then in μ^n, each unpicked node v (at most $n - k$) contributes at most 4 extra per time-step. Thus, if μ^* is some optimal solution, then

$$D(\mu) \leq D(\mu^n) + 4(\omega + 1)(n - k) \leq D(\mu^*) + 4(\omega + 1)(n - k) \leq 5D(\mu^*). \quad \square$$

As a result of Theorem 3, we will often assume in the remainder that $k \leq n/2$, which yields the following useful refinement of Lemma 2.

Corollary 1. *Consider U-TkMP and assume that $k \leq n/2$. Then an optimal configuration σ^* has cost at least $D(\sigma^*) \geq \frac{1}{2}n$. Further, any feasible solution μ has cost at least $D(\mu) \geq (\omega + 1)D(\sigma^*) \geq \frac{1}{2}(\omega + 1)n$.*

---

[1] That is, for every node v, there exists an agent who, through Lemma 1, is always as nearby as possible, so far as the start and end conditions allow. While it may seem nontrivial which nodes this agent could give coverage to on the way to v, in fact those intermediate nodes have agents of their own serving them optimally, and the agent never has to cover anything else but v.

Lemma 3. *In U-TkMP, for any feasible solution μ and time-steps τ_1 and τ_2, we have $D(\mu^{\tau_2}) \leq D(\mu^{\tau_1}) + |\tau_2 - \tau_1|n$.*

Proof. For any $v \in V$, let a be the agent nearest to it in μ^{τ_1}. The contribution of v to $D(\mu^{\tau_2})$ cannot exceed the distance of v to a, who cannot have moved more than $|\tau_2 - \tau_1|$ hops away from their position in μ^{τ_1}.   □

Theorem 4. *Suppose we have an α-approximation algorithm for k-MED that runs in $\mathcal{O}(kmed(n))$ time. In U-TkMP, if $k \leq n/2$ and $\omega \geq 4n^2 + 2n$, we can find an $(\alpha + 1)$-approximation in $\mathcal{O}(kmed(n) + n^5)$ time.*

The proof of Theorem 4 is given in Appendix A. Note that this theorem helps us in the following way: instances with $\omega \geq 4n^2 + 2n$ are 'easy', and we can assume in the remainder that $\omega \leq 4n^2 + 2n$. This, in turn, means that algorithms with runtime depending on ω (polynomially) can still be considered polynomial-time.

We also remark that by combining Theorem 2 and Theorem 4, U-TkMP can be solved in $\mathcal{O}(n^6)$ if there is only one agent.

4 Rounding Continuous Graph Movement

A related problem to U-TkMP is the *Continuous Uniform Traveling k-Median Problem* (CU-TkMP), which we can use to approximate U-TkMP. In CU-TkMP, the allowed positions for agents are not only the nodes, but also any point on the edges.

Technically, we define CU-TkMP as follows. Given an unweighted, undirected graph $G = (V, E)$, we construct its 'continuous version' as the following metric space $(\tilde{E}, c)$. Assuming an arbitrary orientation on the edges of E, we build $\tilde{E}$ out of V and all points 'strictly on' edges: $\tilde{E} := V \cup E \times (0, 1)$, where a point $((u, v), \theta) \in E \times (0, 1)$ represents the point that is on the oriented edge (u, v), at a fraction θ from u to v.

As for the distance metric c, we base it on a continuous version of the shortest path distance as follows. Take any $x \in \tilde{E}$ and $y \in \tilde{E}$. If both are in V, then $c(x, y) = d(x, y)$, where d is the shortest path distance on G. If $x \in V$ but $y \notin V$, let $y = ((u, v), \theta)$. Then $c(x, y) = \min\{\theta + d(u, x), 1 - \theta + d(v, x)\}$. We uphold an analogous definition if $y \in V$ but $x \notin V$. Finally, if both points are not in V, let again $y = ((u, v), \theta)$. Then $c(x, y) = \min\{\theta + c(u, x), 1 - \theta + c(v, x)\}$.

Now, in CU-TkMP, we are again given k agents and a discrete time horizon $\{0, \ldots, \omega\}$. The agents must start at $s \in V$ at time $\tau = 0$ and must be at their end location $t \in V$ at time $\tau = \omega$. If an agent is at position $x \in \tilde{E}$ at time τ, then at time $\tau + 1$, they may only be at a position $y \in \tilde{E}$ with $c(x, y) \leq 1$. The continuous cost of a configuration $\sigma \in \tilde{E}^A$ is not the discrete sum of distances to nodes, but the integrated distance to all points in $\tilde{E}$, i.e., $C(\sigma) := \int_{\tilde{E}} \min_{x \in C} c(x, y) dy$. The continuous cost of a solution $\nu \in \tilde{E}^{A \times T}$ is again the sum over time, i.e., $C(\nu) := \sum_{\tau \in T} C(\nu^\tau)$.

Suppose G has maximum degree Δ, and $k \leq n/2$. Suppose also we can find a β-approximate solution ν to CU-TkMP. We prove that we can 'round' ν to a $(4\beta\Delta + 2)$-approximate solution $\tilde{\nu}$ for the original, 'discrete' problem.

We first show that the cost functions D and C are related. Remark that, if ρ is a continuous configuration, we mean by $D(\rho)$ that we sum distances to V rather than integrating over $\tilde{E}$: we define $D(\rho) := \sum_{v \in V} \min_{u \in \rho} c(u, v)$. If ν is a continuous solution, we define $D(\nu)$ similarly.

Lemma 4. *Let $\rho \in \tilde{E}^A$ be a continuous configuration. Then $D(\rho) \leq 2C(\rho) + k$.*

Proof. Given ρ, we split E into three classes. Denote by $E_0 \subset E$ all edges $\{u, v\}$ that have at least one agent 'strictly' on them, meaning we disregard agents on u and v. Note that $|E_0| \leq k$. Let $E_1 \subset E \setminus E_0$ denote all unoccupied edges $\{u, v\}$ with $d(\rho, u) \neq d(\rho, v)$, oriented as (u, v) if $d(\rho, u) < d(\rho, v)$ and as (v, u) otherwise. Let $E_2 \subset E \setminus E_0$ refer to all unoccupied edges $\{u, v\}$ for which $d(\rho, u) = d(\rho, v)$. We orient E_0 and E_2 arbitrarily.

For any edge $(u, v) \in E_0$, we have $d(\rho, u) + d(\rho, v) \leq 1$, as any one agent on (u, v) is $0 < \theta < 1$ away from u and $1 - \theta$ away from v. Notice that $\{u : (u, v) \in E_0\} \cup \{v : (u, v) \in E_0\} \cup \{v : (u, v) \in E_1\} = V$, as any node $v \in V$ is either incident to E_0, or has a neighbor u on the shortest (v, ρ)-path, thus $(u, v) \in E_1$. Thus summing over these three node sets is sufficient, though we may be over-counting:

$$D(\rho) \leq \sum_{(u,v) \in E_0} (d(\rho, u) + d(\rho, v)) + \sum_{(u,v) \in E_1} d(\rho, v) \leq |E_0| + \sum_{(u,v) \in E_1} (d(\rho, u) + 1)$$

Each $(u, v) \in E_1$ has $\int_u^v d(\rho, w)\mathrm{d}w = d(\rho, u) + \int_0^1 w\mathrm{d}w = d(\rho, u) + \frac{1}{2}$. So

$$C(\rho) = \sum_{(u,v) \in E} \int_u^v d(\rho, w)\mathrm{d}w \geq \sum_{(u,v) \in E_1} (d(\rho, u) + \tfrac{1}{2}) \geq \tfrac{1}{2}|E_1|$$

Putting this together, we find

$$D(\rho) \leq \sum_{(u,v) \in E_1} (d(\rho, u) + \tfrac{1}{2}) + \tfrac{1}{2}|E_1| + |E_0| \leq 2C(\rho) + k. \qquad \square$$

Lemma 5. *Let $\sigma \in V^A$ be a discrete configuration and $n \geq 2k$. Then $C(\sigma) \leq 2\Delta D(\sigma)$.*

Proof. Because σ is a discrete configuration, $E_0 = \emptyset$. Each $(u, v) \in E_1$ has $\int_u^v d(\sigma, w)\mathrm{d}w = d(\sigma, u) + \int_0^1 w\mathrm{d}w = d(\sigma, u) + \frac{1}{2}$. Each $(u, v) \in E_2$ has $\int_u^v d(\sigma, w)\mathrm{d}w = d(\sigma, u) + 2\int_0^{\frac{1}{2}} w\mathrm{d}w = d(\sigma, u) + \frac{1}{4}$.

Observe some oriented version $\vec{E}$ of E, such that if $(v, u) \in \vec{E}$, then $d(\sigma, u) \leq d(\sigma, v)$. For any $u \in V$, let $\delta^-(u)$ be the edges in E such that, in $\vec{E}$, they point at u. Then $|\delta^-(u)| \leq \Delta$ for any $u \in V$.

$$C(\sigma) = \sum_{(u,v)\in E_1} (d(\sigma,u)+\tfrac{1}{2}) + \sum_{(u,v)\in E_2}(d(\sigma,u)+\tfrac{1}{4}) \leq \sum_{u\in V}|\delta^-(u)|(d(\sigma,u)+\tfrac{1}{2})$$

$$\leq \Delta \sum_{u\in V} d(\sigma,u) + \tfrac{1}{2}\Delta n \leq \Delta D(\sigma) + \Delta\cdot(n-k) \leq 2\Delta D(\sigma) \qquad \square$$

Suppose we have some continuous solution $\nu = (\nu_0,\ldots,\nu_\omega)$, which has $C(\nu) \leq \beta C^*$, with C^* the optimal coverage solution value for CU-TkMP. We can for each agent a find a continuous movement line across $(\nu_0)_a,\ldots,(\nu_\omega)_a$ by taking shortest walks through $\tilde{E}$ from each $(\nu_\tau)_a$ to $(\nu_{\tau+1})_a$, waiting at $(\nu_{\tau+1})_a$ until $\tau+1$ if necessary. We can obtain a 'rounded solution' $\tilde{\nu}$ by setting $(\tilde{\nu}_\tau)_a$ equal to $(\nu_\tau)_a$ if that is a node. If instead it is a point on an edge (u,v), because the agent is moving from u to v in the continuous movement line, then we set $(\tilde{\nu}_\tau)_a$ to the closest of these two to $(\nu_\tau)_a$. If the point is exactly halfway between u and v, we set $(\tilde{\nu}_\tau)_a$ to the destination v. Assuming again that $k \leq n/2$ and $\omega \leq 4n^2$, we can perform this rounding in $\mathcal{O}(k\omega) = \mathcal{O}(n^3)$.

Theorem 5 (Rounding Theorem). *If $n \geq 2k$, then rounding ν to the solution $\tilde{\nu}$ yields a $(4\beta\Delta + 2)$-approximation with respect to $D(\mu^*)$, where μ^* is an optimal solution.*

Proof. Take some instance of U-TkMP with $n \geq 2k$. Let μ^* be an optimal solution for this instance, and ν^* an optimal solution for its CU-TkMP-equivalent. Let ν be a β-approximate solution to the CU-TkMP-instance, and let $\tilde{\nu}$ be the result of rounding ν in the way described above.

Solution $\tilde{\nu}$ is feasible for the discrete problem. That is, each agent still starts at their start point and ends at their end point; we show that they also only move from nodes to neighboring nodes. Take any agent a and any time-step $\tau < \omega$. Define $q := (\nu_\tau)_a$ and $r := (\nu_{\tau+1})_a$. By feasibility of ν in CU-TkMP, $d(q,r) \leq 1$. If q and r are on the same edge (u,v), including the end points, then they are rounded to u and/or v, which are neighboring. If q is on (u,v) and r is not, then r is on some (v,w). If q or r is rounded to v, then q and r are rounded to neighboring nodes. Suppose not, so q is rounded to u and r to w. Then $d(q,v) > \tfrac{1}{2}$ and $d(v,r) \geq \tfrac{1}{2}$, so $d(q,r) = d(q,v) + d(v,r) > 1$, but this contradicts feasibility of ν in CU-TkMP.

It remains to show the approximation factor. Using Corollary 1 for OPT, we have

$$OPT := D(\mu^*) \geq (\omega+1)(n-k) \geq (\omega+1)\tfrac{n}{2} \geq (\omega+1)k$$

$$D(\tilde{\nu}) \leq D(\nu) + (\omega+1)n\cdot\tfrac{1}{2} \leq D(\nu) + OPT \qquad (1)$$

$$D(\nu) \leq \sum_{\tau=0}^{\omega}(2C(\nu_\tau)+k) \leq 2C(\nu) + OPT \qquad (2)$$

$$C(\nu) \leq \beta C(\nu^*) \leq \beta C(\mu^*) \qquad (3)$$

$$C(\mu^*) \leq 2\Delta D(\mu^*) = 2\Delta\cdot OPT \qquad (4)$$

(1) follows from the fact that every demand point (v, τ) has its median shifted away at most $\frac{1}{2}$ in rounding. (2) follows from Lemma 4. (3) follows from the fact that ν is β-approximate for CU-TkMP, while μ^* is feasible. (4) follows from Lemma 5. $\square$

5   Topologies with Constant Factor Guarantees

In this section, we show how to compute an optimal solution to the continuous problem CU-TkMP on two simple graph topologies, namely on a path and on a cycle. In both cases, we need to solve a linear program with a convex minimization objective, which can be done in polynomial time (see, e.g., [20]). Throughout this section, we use $convex(k_1, k_2)$ to refer to the time that is needed to solve such a program consisting of k_1 variables and k_2 constraints. Recall that in light of Theorem 4, we can assume that $k \leq n/2$ and $\omega \leq 4n^2 + 2n$; in particular, this implies that $\mathcal{O}(k\omega) = O(n^3)$

5.1   Path Case

We show that one can compute an optimal solution to the continuous problem CU-TkMP if the underlying topology is a line of length L, i.e., $\tilde{E} = [0, L]$. By applying our Rounding Theorem (Theorem 5) to the optimal solution, we obtain the result stated in Theorem 6 below (with $\beta = 1$ and $\Delta = 2$).

Theorem 6. *Let $n \geq 2k$. Then there is a 10-approximation algorithm for U-TkMP on the path graph that runs in $\mathcal{O}(convex(k_1, k_2) + n^3)$ time, where $k_1 = k_2 = \mathcal{O}(k\omega)$.*

A key observation that we can exploit in the path case is that we can impose an order on the positions of the agents. More precisely, let $\sigma \in \tilde{E}^A$ be a configuration. Assume that the agents in $A = \{1, \ldots, k\}$ are indexed such that $\sigma_1 \leq \sigma_2 \leq \cdots \leq \sigma_k$ (break ties arbitrarily). For notational convenience, we introduce two dummy agents 0 and $k + 1$ which are fixed at $\sigma_0 = 0$ and $\sigma_{k+1} = L$, respectively. Further, let λ_a be a scalar that is defined as $\lambda_a = \frac{1}{4}$ for all $a \in \{1, \ldots, k - 1\}$ and $\lambda_0 = \lambda_k = \frac{1}{2}$.

Note that we do not have to worry about agents 'switching' their index: because everyone has the same start and end location, there obviously exists an optimal solution in which they do not switch their index. This is because for any feasible solution with a switch, there also exists a feasible solution with identical cost in which the switch is not made, namely the solution in which the agents approach each other for the switch but go back again. Therefore, we are free to enforce in our convex program that agents will always keep their index.

The following lemma shows that the coverage cost reduces to a function that depends on the distances between consecutive agents.

Lemma 6. *Consider a continuous configuration* $\sigma \in \tilde{E}^A$ *and suppose that the agents are indexed as stated above. Then the coverage cost is*

$$C(\sigma) = \sum_{a=0}^{k} \lambda_a (\sigma_{a+1} - \sigma_a)^2.$$

Proof. Consider the cost of an interior line segment $[\sigma_a, \sigma_{a+1}]$ between two consecutive agents σ_a and σ_{a+1} with $a \in \{1, \ldots, k-1\}$. In this case, each point of the segment is covered by the closest agent among σ_a and σ_{a+1}. Thus, the coverage cost is

$$\int_{\sigma_a}^{\sigma_{a+1}} \min_{u \in \{\sigma_a, \sigma_{a+1}\}} c(u, v) dv = 2 \int_{0}^{\frac{1}{2}(\sigma_{a+1} - \sigma_a)} v dv = \frac{1}{4}(\sigma_{a+1} - \sigma_a)^2.$$

Next, consider the first line segment $[0, \sigma_1]$. In this case, each point of the segment is covered by σ_1. Thus, the coverage cost is

$$\int_{0}^{\sigma_1} c(\sigma_1, v) dv = \int_{\sigma_0}^{\sigma_1} v dv = \frac{1}{2}(\sigma_1 - \sigma_0)^2.$$

A similar argument proves that the coverage cost of the last line segment $[\sigma_k, \sigma_{k+1}]$ is $\frac{1}{2}(\sigma_{k+1} - \sigma_k)^2$. Summing over all line segments proves the claim. $\square$

Exploiting the above observation, we can solve CU-TkMP for the path case simply by formulating the problem as a linear program with a convex (in fact quadratic) objective function.

We introduce a variable $x_a^\tau \in [0, L]$ for each $a \in \{0, \ldots, k+1\}$ and time step $\tau \in T$. Our interpretation is that x_a^τ represents the position of the a-th leftmost agent (breaking ties arbitrarily). The following convex program then captures the continuous version of the problem:

$$\min \quad \sum_{\tau \in T} \sum_{a=0}^{k} \lambda_a (\sigma_{a+1} - \sigma_a)^2 \tag{QP1}$$

$$\begin{aligned}
\text{s.t.} \quad & 0 = x_0^\tau \leq x_1^\tau \leq \cdots \leq x_k^\tau \leq x_{k+1}^\tau = L && \forall \tau \in T \\
& x_a^\tau - x_a^{\tau-1} \leq 1 && \forall a \in A \;\; \forall \tau \in T \\
& -x_a^\tau + x_a^{\tau-1} \leq 1 && \forall a \in A \;\; \forall \tau \in T \\
& x_a^\tau \in [0, L] && \forall a \in A \;\; \forall \tau \in T
\end{aligned}$$

Note that the second and third set of constraints together ensure that the distance that each agent $a \in A$ (more precisely, the a-th leftmost agent) moves in each time step $\tau \in T$ is at most 1, i.e., $|x_a^\tau - x_a^{\tau-1}| \leq 1$.

(QP1) is a linear program with a convex minimization objective and can thus be solved in polynomial time, e.g., by using standard interior point methods (see [20]). Let $\nu \in \tilde{E}^{A \times T}$ be the continuous solution that we obtain by solving (QP1). If we then round ν according to our Rounding Theorem, we obtain a discrete solution which is a 10-approximation for the corresponding U-TkMP problem on the path graph. This completes the proof of Theorem 6.

5.2 Cycle Case

We show that ideas similar to the ones used for the path case above, can be used to compute an optimal solution for the continuous problem CU-TkMP if the underlying topology is a cycle of length L, i.e., $\tilde{E} = ([0, L] \mod L)$.[2]

We obtain the following result from our Rounding Theorem (with $\beta = 1$ and $\Delta = 2$):

Theorem 7. *Let $n \geq 2k$. Then there is a 10-approximation algorithm for U-TkMP on the cycle graph $\mathcal{O}(convex(k_1, k_2) + n^3)$ time, where $k_1 = k_2 = \mathcal{O}(k\omega)$.*

As before, given a configuration $\sigma \in \tilde{E}^A$ we assume that the agents in $A = \{1, \ldots, k\}$ are indexed such that $\sigma_1 \leq \sigma_2 \leq \cdots \leq \sigma_k$. We introduce a single dummy agent $k + 1$ whose position is defined as $\sigma_{k+1} = L + \sigma_1$. We can again assume that the agents will never want to switch index, because there exists an optimal solution in which they never switch. Note that we have $(\sigma_{k+1} \mod L) = \sigma_1$.

Lemma 7. *Consider a continuous configuration $\sigma \in \tilde{E}^A$ and suppose that the agents are indexed as stated above. Then the coverage cost is*

$$C(\sigma) = \frac{1}{4} \sum_{a=1}^{k} (\sigma_{a+1} - \sigma_a)^2.$$

Proof. As in the proof of Lemma 6, we subdivide the cycle into k line segments and argue for each segment separately. Using the same arguments as in the proof of Lemma 6, the cost of each line segment $[\sigma_a, \sigma_{a+1}]$ with $a \in \{1, \ldots, k-1\}$ is $\frac{1}{4}(\sigma_{a+1} - \sigma_a)^2$.

It remains to consider the (cyclic) segment $([\sigma_k, \sigma_{k+1}] \mod L)$ covered by agents k and 1. It is not hard to see that the coverage cost of this segment is

$$\int_{\sigma_k}^{\sigma_{k+1}} \min_{u \in \{\sigma_k, \sigma_{k+1}\}} |v - u| dv = 2 \int_0^{\frac{1}{2}(\sigma_{k+1} - \sigma_k)} v \, dv = \frac{1}{4}(\sigma_{k+1} - \sigma_k)^2.$$

Summing over all line segments proves the claim. $\square$

In light of the above lemma, we can now adapt the convex program for the path case to the cycle case as follows:

$$\min \quad \frac{1}{4} \sum_{\tau \in T} \sum_{a=1}^{k} (\sigma_{a+1} - \sigma_a)^2 \qquad \text{(QP2)}$$

$$\begin{array}{rcll}
\text{s.t.} & 0 \leq x_1^\tau \leq \cdots \leq x_k^\tau \leq L & \forall \tau \in T \\
& x_{k+1}^\tau = L + x_1^\tau & \forall \tau \in T \\
x_a^\tau - x_a^{\tau-1} & \leq & 1 & \forall a \in A \ \forall \tau \in T \\
-x_a^\tau + x_a^{\tau-1} & \leq & 1 & \forall a \in A \ \forall \tau \in T \\
x_a^\tau & \in & [0, L] & \forall a \in A \ \forall \tau \in T
\end{array}$$

As before, we can solve this program efficiently and use our Rounding Theorem to obtain a 10-approximation for U-TkMP in the cycle case.

---

[2] Note that the modulo operator ensures that 0 and L correspond to the same point.

Algorithm 1: A 'mediate at $\sqrt{\omega}$'-algorithm for U-TkMP if $d(s,t) \leq 2\sqrt{\omega}$.

Input: U-TkMP-instance, α-approximation algorithm for k-MED
Output: Solution μ

1 Find the reachable set $B := \{v \in V : d(s,v) \leq \theta \wedge d(v,t) \leq \theta\}$, with $\theta := \lfloor \sqrt{\omega} \rfloor$.;
2 Find medians σ by approximating k-MED with facilities V and clients V.;
3 Take any bijection between agents and medians in σ, then form solution μ by minimizing each agent's summed distance to their median, using Lemma 1.;
4 Find medians $\sigma^{[B]}$ by approximating k-MED with facilities B and clients B.;
5 Take any bijection between agents to medians in $\sigma^{[B]}$, then form solution $\mu^{[B]}$ by minimizing each agent's summed distance to their median, using Lemma 1.;
6 **return** $\mu \leftarrow \arg\min\{D(\mu), D(\mu^{[B]})\}$;

---

6 Approximation Algorithms for General Graphs

Our Rounding Theorem can be used whenever we can solve (or approximate) the corresponding continuous version CU-TkMP of the problem. In this section, we derive approximation algorithms that do not rely on this rounding scheme. We first derive an $\mathcal{O}(\min\{\sqrt{\omega}, n\})$-approximation algorithm if the distance between s and t satisfies $d(s,t) \leq 2\sqrt{\omega}$. We then show that we can obtain an $\mathcal{O}(d(s,t) + \sqrt{\omega})$-approximation algorithm if $d(s,t) > 2\sqrt{\omega}$.

Our first algorithm is described in Algorithm 1. The idea of this algorithm is as follows: We first find approximate medians, and then try to stay as close as possible to these medians. We do not only try to find good medians for all of V, but also for a smaller subset B that we can cover well without moving more than $\sqrt{\omega}$ hops from s and t. If the 'optimal medians' are indeed within $\sqrt{\omega}$ hops from s and t, then the V-medians will give us a constant-factor approximation; otherwise, the B-medians will give us an $\mathcal{O}(\sqrt{\omega})$-approximation.

We assume we have access to an α-approximation algorithm that runs in $\mathcal{O}(kmed(n))$ time for k-MED for some given facility and client sets; e.g., like the 6-approximation by Jain and Vazirani [14].

Lemma 8. *Algorithm 1 runs in $\mathcal{O}(kmed(n) + n^5)$ time.*

Proof. Line 1 costs $\mathcal{O}(n)$ time. Lines 2 and 4 cost $\mathcal{O}(kmed(n))$ time. In lines 3 and 5, we seek $k \leq n/2$ walks, which according to Lemma 1, each cost $\mathcal{O}(n^4)$ time to find.

Theorem 8. *If there exists an optimal configuration σ^* that is within $\lfloor \sqrt{\omega} \rfloor$ hops from the starting configuration and the ending configuration, then Algorithm 1 is an $(\alpha + 2 + \sqrt{2})$-approximation that runs in $\mathcal{O}(kmed(n) + n^5)$ time.*

Proof. Let OPT be the optimal solution value, and ALG the value of the solution from Algorithm 1. In μ, the agents are at some configuration σ (as determined by the algorithm) throughout $\tau = \theta, \theta + 1, \ldots, \omega + 1 - \theta$. We assumed that there is some optimal configuration $\sigma^* \subseteq B$, so $D(\sigma) \leq \alpha D(\sigma^*)$. With Lemma 3 and Corollary 1, we obtain

$$D(\mu) \leq (\omega + 1)D(\sigma) + n \cdot \sum_{\tau=1}^{\theta} \tau + n \cdot \sum_{\tau=1}^{\theta} \tau = (\omega + 1)D(\sigma) + (\theta + 1)\theta n$$

$$\leq \alpha \cdot (\omega + 1)D(\sigma^*) + \omega n + \sqrt{\omega}n \leq (\alpha + 2 + \sqrt{2})OPT$$

where the last inequality holds because $\frac{\sqrt{\omega}n}{\frac{1}{2}(\omega+1)n} \leq \frac{2}{\sqrt{\omega}} \leq \sqrt{2}$ for $\omega \geq 2$ (which we can assume). We conclude that $ALG \leq D(\mu) \leq (\alpha + 2 + \sqrt{2})OPT$. $\square$

Theorem 9. *Assuming $d(s,t) \leq 2\sqrt{\omega}$, Algorithm 1 is an $\mathcal{O}(\sqrt{\omega})$-approximation algorithm that runs in $\mathcal{O}(kmed(n) + n^5)$ time.*

Proof. For any configuration σ and $U \subseteq V$, define $D(\sigma, U) := \sum_{u \in U} d(\sigma, u)$. Let $V' := V \setminus B$ be the nodes that are not reachable in 2θ hops, with B as defined in Algorithm 1. Also let $\vec{D} := \sum_{v \in V'} \min_{u \in B} d(u, v)$ be the summed distance of V' to B. Let σ^* be an optimal configuration, and σ_B^* be a minimizer of $D(\sigma, B)$.

We can split the value of some optimal solution μ^* over its contribution to B and V'. The former can be lower-bounded with a configuration σ_B^* that minimizes $D(\sigma, B)$ instead of $D(\sigma)$. The latter can be lower-bounded by the fact that there can be no agents in V' if $\tau \leq \sqrt{\omega} - 1$ or $\tau \geq \omega + 1 - (\sqrt{\omega} - 1)$, so in those time-steps, at least a cost $\vec{D}$ is incurred. If $\vec{D} = 0$ or $\theta = 0$, then Theorem 8 gives us a constant-factor approximation, so assume $\vec{D} > 0$. Then

$$OPT \geq (\omega + 1)D(\sigma^*, B) + D(\mu^*, V') \geq (\omega + 1)D(\sigma_B^*, B) + (\sqrt{\omega} + \sqrt{\omega})\vec{D}.$$

In $D(\mu^{[B]})$, the agents will be at $\sigma^{[B]}$ throughout $\tau = \theta, \theta + 1, \ldots, \omega + 1 - \theta$. By Lemma 3, we again know that the total cost cannot exceed $(\omega + 1)D(\sigma^{[B]}) + (\theta + 1)\theta n$. While we know that $\sigma^{[B]}$ is favorable for serving B, we need to upper-bound the cost of V' as well. For any $v \in V'$, if u is the nearest node in B to v, then we know that there is a median m at most 2θ hops away, because $d(u, m) \leq d(u, s) + d(s, m) \leq 2\theta$. So $D(\sigma^{[B]}) \leq D(\sigma^{[B]}, B) + \vec{D} + |V'| \cdot 2\theta$. We have

$$ALG \leq D(\mu^{[B]}) \leq (\omega + 1)D(\sigma^{[B]}) + (\theta + 1)\theta n \leq (\omega + 1)D(\sigma^{[B]}) + \omega n + \sqrt{\omega}n$$

$$\leq (\omega + 1)\left(D(\sigma^{[B]}, B) + \vec{D} + |V'| \cdot 2\sqrt{\omega}\right) + \omega n + \sqrt{\omega}n$$

$$\frac{ALG}{OPT} \leq \frac{(\omega+1)D(\sigma^{[B]}, B)}{(\omega+1)D(\sigma_B^*, B)} + \frac{(\omega+1)\vec{D}}{(2\sqrt{\omega})\vec{D}} + \frac{(\omega+1)|V'| \cdot 2\theta}{|V'| \cdot \frac{1}{2}(\theta+1)\theta} + \frac{\omega n + \sqrt{\omega}n}{\frac{1}{2}(\omega+1)n}$$

$$\leq \alpha + \frac{\omega+1}{2\sqrt{\omega}} + \frac{4(\omega+1)}{\sqrt{\omega}+1} + 2 + \sqrt{2} = \mathcal{O}(\sqrt{\omega})$$

$\square$

Though it is true that $\sqrt{\omega}$ can become arbitrarily large, we can always apply Theorem 8 when $\sqrt{\omega}$ exceeds n. Therefore, these results together make Algorithm 1 an $\mathcal{O}(\min\{\sqrt{\omega}, n\})$-approximation algorithm.

Algorithm 2: A 'mediate at $\sqrt{\omega}$'-algorithm for U-TkMP if $d(s,t) > 2\sqrt{\omega}$.

Input: U-TkMP-instance, α-approximation algorithm for k-MED
Output: Solution μ

1 Find the reachable set $H := \{v \in V : d(s,v) + d(v,t) \le d(s,t) + \theta\}$, with
   $\theta := \lfloor\sqrt{\omega}\rfloor$.;
2 Find medians σ by approximating k-MED with facilities V and clients V.;
3 form solution μ by minimizing each agent's summed distance to their median,
   using Lemma 1.;
4 Find medians $\sigma^{[H]}$ by approximating k-MED with facilities H and clients H.;
5 Take any bijection between agents to medians in $\sigma^{[H]}$, then form solution $\mu^{[H]}$
   by minimizing each agent's summed distance to their median, using Lemma 1.;
6 **return** $\mu \leftarrow \arg\min\{D(\mu), D(\mu^{[H]})\}$;

Next, we consider the case $d(s,t) > 2\sqrt{\omega}$. Then it may occur that B is empty. In that case, we resort to Algorithm 2, the performance of which is $\mathcal{O}(d(s,t) + \sqrt{\omega})$. Analogous statements of Lemma 8 and Theorem 8 clearly hold for Algorithm 2, but we explain the weaker performance bound in Theorem 10.

Theorem 10. *Assuming $d(s,t) > 2\sqrt{\omega}$, Algorithm 1 is an $\mathcal{O}(d(s,t) + \sqrt{\omega})$-approximation algorithm that runs in $\mathcal{O}(kmed(n) + n^5)$ time.*

Proof. We follow a similar argument as in Theorem 9. That is, we separately bound the cost that H incurs in $\mu^{[H]}$ and that $\tilde{V}' := V \backslash H$ incurs. Denote $\vec{H} := \sum_{v \in \tilde{V}'} \min_{u \in H} d(u,v)$ the summed distance of $\tilde{V}'$ to H. Note first that any $u \in H$ and $v \in H$ can have distance at most $d(u,v) \le d(s,t) + \sqrt{\omega}$:

$$2d(u,v) \le d(u,s) + d(s,v) + d(u,t) + d(t,v) \le 2(d(s,t) + \sqrt{\omega})$$

Let μ^* be an optimal solution, let $\sigma^{*[H]}$ be an optimal k-MED-configuration if the client set and facility set are both H, and let $\mu^{*[H]}$ be an optimal U-TkMP-solution for the induced subgraph $G[H]$. Note also that $OPT := D(\mu^*) \ge \sqrt{\omega}\vec{H}$, because for at least $\sqrt{\omega}$ time-steps, no agent can be in $\tilde{V}'$. Therefore:

$$ALG \le D(\mu^{[H]}) = D(\mu^{[H]}, H) + D(\mu^{[H]}, \tilde{V}') \tag{5}$$

$$D(\mu^{[H]}, H) \le (\omega + 1)D(\sigma^{[H]}, H) + \frac{1}{2}(d(s,t) + \sqrt{\omega})(d(s,t) + \sqrt{\omega} + 1)n \tag{6}$$

$$(\omega + 1)D(\sigma^{[H]}, H) \le (\omega + 1)\alpha D(\sigma^{*[H]}, H) \le \alpha D(\mu^*) \tag{7}$$

$$\frac{D(\mu^{[H]}, H)}{OPT} \le \frac{\alpha D(\mu^*)}{D(\mu^*)} + \frac{\frac{1}{2}n(d(s,t) + \sqrt{\omega})(d(s,t) + \sqrt{\omega} + 1)}{\frac{1}{2}n(\omega + 1)} \tag{8}$$

$$\le \alpha + \frac{d(s,t)^2 + 2d(s,t)\sqrt{\omega} + \omega + d(s,t) + \sqrt{\omega}}{\omega} \tag{9}$$

$$D(\mu^{[H]}, \tilde{V}') \le (\omega + 1)(\vec{H} + (d(s,t) + \sqrt{\omega})n) \tag{10}$$

$$\frac{D(\mu^{[H]}, \tilde{V}')}{OPT} \le \frac{(\omega + 1)\vec{H}}{\sqrt{\omega}\vec{H}} + \frac{(\omega + 1)n(d(s,t) + \sqrt{\omega})}{\frac{1}{2}(\omega + 1)n} \tag{11}$$

In (6), we use Lemma 3 repeatedly, knowing that movement occurs in at most $(d(s,t) + \sqrt{\omega})$ hops. In (10), we use the fact that the diameter of H does not exceed $d(s,t) + \sqrt{\omega}$. So if $v \in \tilde{V}'$ and $u = \min_{u \in H} d(u,v)$, then the cost contribution of v cannot exceed $d(u,v) + d(s,t) + \sqrt{\omega}$. So in every time-step, the nodes of $\tilde{V}'$ together cannot contribute more than $\vec{H} + n(d(s,t) + \sqrt{\omega})$ to the cost. Put together, this means that $\frac{ALG}{OPT} \leq \frac{D(\mu^{[H]}, H)}{OPT} + \frac{D(\mu^{[H]}, \tilde{V}')}{OPT}$ is an $\mathcal{O}(\sqrt{\omega} + d(s,t))$-approximation factor. $\square$

Studying the approximation factor in the proof of Theorem 10, it would seem that Algorithm 2 performs poorly for high $d(s,t)$ and low ω. However, we remark that the most extreme version of this is observed only in the path graph, which is again 'easy'. It could be interesting to see if this 'gradual transition to path graphs' can help strengthen the performance bound.

7   Conclusions

We introduced TkMP and its uniform variant, U-TkMP. We provided several approximation algorithms for U-TkMP, including a novel method in which we round back an optimal or β-approximate solution to a 'continuous graph movement relaxation'.

As venues for future research, we suggest the following. There are likely more topologies than just path graphs and cycle graphs for which we can solve or approximate CU-TkMP. Extending the results of this paper to heterogeneous start and end locations seems imaginable, once an equivalent to Lemma 3 can be found that allows us to again lower-bound $D(\sigma)$ by some function of only n. (It seems more difficult to let go of the edge lengths being uniform, because this makes rounding back non-trivial, and we can no longer bound distances to unoccupied nodes to be between 1 and n.)

Acknowledgements. We thank the three anonymous reviewers for their valuable suggestions.

A   Omitted proofs

A.1   Theorem 1

Proof. U-TkMP is obviously in NP. We will prove that U-TkMP is at least as hard as the Set Cover Problem, which is NP-hard. Let I_1 be the following instance of the decision version of the Set Cover Problem: given universe U and a collection S of subsets of U, can we find k (or less) members of S which have union U? Let I_2 be an instance of the decision version of U-TkMP that we construct as follows. Let the node set be $V := V_U \cup V_S \cup \{s\}$, where each element in U is represented by its own node in V_U, and each member in S is represented by its own node in V_S, and s is some additional node. In the graph $G = (V, E)$, we have $(u,v) \in E$ for nodes $u, v \in V$ exactly when either of these two conditions hold: $u, v \in V_S \cup \{s\}$, or $u \in V_U$ and $v \in V_S$ and the element that corresponds to u is contained in the subset that corresponds to v. In I_2, we set $\omega = 2$, and let all k agents start and end at s, and we ask: can we find a feasible solution with cost at most $2|S| + 4|U| + (n - k)$? I_2 can be constructed in polynomial time and space with respect to the input of I_1. In any feasible solution, all agents are at s at times $\tau = 0$ and $\tau = 2$, so we incur a cost of $|S| + 2|U|$. So the answer to I_2 is 'yes' if and only if we can find a configuration at $\tau = 1$ with cost $(n - k)$, meaning every node is adjacent to at least one agent position. The agents can only be in $V_S \cup \{s\}$, which is a clique. Adjacency to all nodes in V_U is achieved if and only if the chosen positions in V_S correspond to members of S which have union U. (If agents stay at s, this corresponds to choosing less than k subsets in I_1.) So any 'yes'-answer to I_1 supplies positions that will give a 'yes'-answer to I_2, and any 'yes'-answer to I_2 gives a 'yes'-answer to I_1 by reading out the positions at $\tau = 1$. So I_1 and I_2 are equivalent, implying U-TkMP is NP-hard. □

A.2 Theorem 2

Proof. Let $G = (V, E)$ be the original graph of our TkMP-instance. We construct a weighted, expanded graph $G^+ = (V^+, E^+)$ as follows. First, make a 'directed looping version' of G: take V, and for every $u, v \in V$, include arc (u, v) if edge $\{u, v\} \in E$ or $u = v$. Let this digraph be denoted by G'. Next, let $\vec{T}$ be the digraph with node set T, and arc set $\cup_{\tau \in T \setminus \{\omega\}} \{(\tau, \tau+1)\}$, so $\vec{T}$ is a path digraph representing the time horizon.

Then, let G^+ be the tensor product $(G')^k \times \vec{T}$. So every node $(u_1, u_2, \ldots, u_k, \tau_1)$ in G^+ represents a position for each agent, and a time-step. There is an arc in G^+ from node $(u_1, u_2, \ldots, u_k, \tau_1)$ to node $(v_1, v_2, \ldots, v_k, \tau_2)$ if and only if the arc (u_a, v_a) is in G' for each agent a, and $\tau_2 = \tau_1 + 1$.

Finally, let each such arc in G^+ have weight $D(\sigma_2)$, where σ_2 is the visited configuration $(v_1, v_2, \ldots, v_k)$. This means that, walking through G^+, we incur at each time-step $\tau = 1, \ldots, \omega$ the cost of the configuration currently visited. The cost made at $\tau = 0$ is a constant that is the same for each feasible solution. Therefore, finding an optimal TkMP-solution is equivalent to finding a cheapest $(s_1, s_2, \ldots, s_k, 0) - (t_1, t_2, \ldots, t_k, \omega)$-path. Because G^+ has $\mathcal{O}(\omega n^k)$ nodes, finding this path with Dijkstra's algorithm has time complexity $\mathcal{O}(\omega^2 n^{2k})$. □

A.3 Lemma 1

Proof. Suppose first that $\omega \geq 2n$. Then, certainly, a walk exists from s to v and then t. We take a shortest (s, v)-walk and a shortest (v, t)-walk, and spend all remaining time 'looping' at v. This is obviously a walk π that minimizes $\sum_{\tau=0}^{\omega} d(\pi^\tau, v)$, because at every time-step, the agent is as near to v as the start and end conditions allow.

Suppose now that $\omega < 2n$. In this case, we construct G^+ exactly as in the proof of Theorem 2, setting $k = 1$. Each arc $((u, \tau), (u', \tau + 1))$ in G^+ now has weight $d(u', v)$. Because $\omega < 2n$, we find our walk in $\mathcal{O}(n^4)$ by applying Dijkstra's algorithm on G^+ from $(s, 0)$ to (t, ω). $\qquad\square$

A.4 Theorem 4

Proof. We construct our solution as follows. First, if σ^* is an optimal k-MED-configuration, we use the α-approximation algorithm to find a good configuration σ with $D(\sigma) \leq \alpha D(\sigma^*)$, requiring a running time of $\mathcal{O}(kmed(n))$. Next, we assign these medians to agents arbitrarily. Finally, for each of the $k \leq n/2$ agents, we find in $\mathcal{O}(n^4)$ time the (s, t)-walk that minimizes the total distance to that agent's median, using Lemma 1. This adds a running time of $\mathcal{O}(n/2 \cdot n^4)$, for a total of $\mathcal{O}(kmed(n) + n^5)$.

We now prove that this yields an $(\alpha + 1)$-approximation. Let μ be the found solution, and μ^* an optimal solution. Obviously, μ consists of shortest paths from s to points in σ, a lot of waiting in σ, then shortest paths to t.

Observe any time-step τ. If $n - 1 \leq \tau \leq (\omega + 1) - (n - 1)$, then clearly all agents have reached their position in σ and have not departed them yet. In each of those time-steps, we have $D(\mu^\tau) \leq \alpha D(\sigma^*)$. In the other time-steps τ, we at least know that $D(\mu^\tau) \leq n^2$, because each of the n nodes has an agent at most n hops away.

So, recalling Corollary 1, we find

$$\frac{D(\mu)}{D(\mu^*)} \leq \frac{2n \cdot n^2}{\frac{1}{2}\omega n} + \frac{(\omega + 1 - 2n) \cdot \alpha D(\sigma)}{(\omega + 1)D(\sigma^*)} \leq \frac{2n \cdot n^2}{\frac{1}{2} \cdot 4n^2 \cdot n} + \alpha = \alpha + 1$$

meaning μ is an $(\alpha + 1)$-approximation. $\qquad\square$

References

1. Arora, S., Raghavan, P., Rao, S.: Approximation schemes for Euclidean k-medians and related problems. In: Proceedings of the Thirtieth Annual ACM Symposium on Theory of Computing, pp. 106–113. Association for Computing Machinery (1998)
2. Bertsimas, D.J., Van Ryzin, G.: Stochastic and dynamic vehicle routing in the Euclidean plane with multiple capacitated vehicles. Oper. Res. **41**(1), 60–76 (1993)
3. Borodin, A., Linial, N., Saks, M.E.: An optimal on-line algorithm for metrical task system. J. ACM **39**(4), 745–763 (1992)
4. Byrka, J., Pensyl, T., Rybicki, B., Srinivasan, A., Trinh, K.: An improved approximation for k-median and positive correlation in budgeted optimization. ACM Trans. Algorithms **13**(2) (2017)
5. Farahani, R.Z., Abedian, M., Sharahi, S.: Dynamic facility location problem. In: Zanjirani Farahani, R., Hekmatfar, M. (eds.) Facility Location. CMS, pp. 347–372. Physica-Verlag HD, Heidelberg (2009). https://doi.org/10.1007/978-3-7908-2151-2_15
6. Galvão, R.D., Santibanez-Gonzalez, E.D.R.: A Lagrangean heuristic for the pk-median dynamic location problem. Eur. J. Oper. Res. **58**(2), 250–262 (1992)
7. Gendreau, M., Laporte, G., Semet, F.: A dynamic model and parallel tabu search heuristic for real-time ambulance relocation. Parallel computing **27**(12), 1641–1653 (2001)

8. Gendreau, M., Laporte, G., Semet, F.: The maximal expected coverage relocation problem for emergency vehicles. J. Oper. Res. Soc. **57**(1), 22–28 (2006)
9. Guha, S., Khuller, S.: Greedy strikes back: improved facility location algorithms. J. Algorithms **31**(1), 228–248 (1999)
10. Hotelling, H.: Stability in competition. In: The Collected Economics Articles of Harold Hotelling, pp. 50–63. Springer (1990). https://doi.org/10.1007/978-1-4613-8905-7_4
11. Huang, W.V., Batta, R., Babu, A.: Relocation-promotion problem with Euclidean distance. Eur. J. Oper. Res. **46**(1), 61–72 (1990)
12. Huizing, D., Schäfer, G., van der Mei, R.D., Bhulai, S.: The median routing problem for simultaneous planning of emergency response and non-emergency jobs. Eur. J. Oper. Res. **285**(2), 712–727 (2020)
13. Jagtenberg, C.J., Bhulai, S., van der Mei, R.D.: An efficient heuristic for real-time ambulance redeployment. Oper. Res. Health Care **4**, 27–35 (2015)
14. Jain, K., Vazirani, V.V.: Approximation algorithms for metric facility location and k-median problems using the primal-dual schema and Lagrangian relaxation. J. ACM (JACM) **48**(2), 274–296 (2001)
15. Kariv, O., Hakimi, S.L.: An algorithmic approach to network location problems. II: the p-medians. SIAM J. Appl. Math. **37**(3), 539–560 (1979)
16. Koutsoupias, E.: The k-server problem. Comput. Sci. Rev. **3**(2), 105–118 (2009)
17. Koutsoupias, E., Papadimitriou, C.H.: On the k-server conjecture. J. ACM **42**(5), 971–983 (1995)
18. Lin, J.H., Vitter, J.S.: ε-approximations with minimum packing constraint violation. In: Proceedings of the Twenty-Fourth Annual ACM Symposium on Theory of Computing, pp. 771–782. Association for Computing Machinery (1992)
19. Melo, M.T., Nickel, S., Da Gama, F.S.: Dynamic multi-commodity capacitated facility location: a mathematical modeling framework for strategic supply chain planning. Comput. Oper. Res. **33**(1), 181–208 (2006)
20. Nesterov, Y., Nemirovskii, A.: Interior-point Polynomial Algorithms in Convex Programming. Society for Industrial and Applied Mathematics (1994)
21. Tamir, A.: An $O(pn^2)$ algorithm for the p-median and related problems on tree graphs. Oper. Res. Lett. **19**(2), 59–64 (1996)
22. Teitz, M.B.: Locational strategies for competitive systems. J. Regional Sci. **8**(2), 135–148 (1968)
23. Wesolowsky, G.O.: Dynamic facility location. Manage. Sci. **19**(11), 1241–1248 (1973)

EPTAS for Load Balancing Problem on Parallel Machines with a Non-renewable Resource

G. Jaykrishnan and Asaf Levin[✉]

Faculty of Industrial Engineering and Management, The Technion, Haifa, Israel
levinas@ie.technion.ac.il

Abstract. The problem considered is the non-preemptive scheduling of independent jobs that consume a resource (which is non-renewable and replenished regularly) on parallel uniformly related machines. The input defines the speed of machines, size of jobs, the quantity of resource required by the jobs, the replenished quantities, and replenishment dates of the resource. Every job can start processing only after the required quantity of the resource is allocated to the job. The objective function is the minimization of the convex combination of the makespan and an objective that is equivalent to the l_p-norm of the vector of loads of the machines. We present an EPTAS for this problem. Prior to our work only a PTAS was known in this non-renewable resource settings and this PTAS was only for the special case of our problem of makespan minimization on identical machines.

Keywords: Scheduling · Approximation algorithms · EPTAS

1 Introduction

The problem ums considered is the non-preemptive scheduling of n independent jobs that consume a single resource (which is non-renewable and replenished regularly) on m uniformly related machines. In the setting of uniformly related machines we let $s_i > 0$ be the speed of machine i. Each job j has a size $p_j > 0$ associated with it and the processing time of job j on machine i is p_j/s_i. The speed of the fastest machine is without loss of generality 1, else the speed of the machines can be converted to relative speeds with respect to the speed of the fastest machine. In our problem, each job j consumes $d_j \geq 0$ quantity of resource to start its processing. That is, in order to start processing job j, the resource should be allocated to the job at the start of the processing of job j. Therefore, jobs may have to wait if sufficient quantity of the resource is not available and thus idle time is essential for this problem. The resource is supplied at e different time points. The resource replenishment time for time period k is denoted by u_k and the quantity of resource supplied at this time point is $q_k \geq 0$. Both u_k and

Supported by ISF - Israeli Science Foundation grant number 308/18.

J. Koenemann and B. Peis (Eds.): WAOA 2021, LNCS 12982, pp. 99–116, 2021.
https://doi.org/10.1007/978-3-030-92702-8_7

q_k (for all k) are given as part of the input. In the given instance of the problem we have $q_k > 0$ for all k, but we construct additional instances in which it is convenient to allow zero resource supplied in some periods.

A solution of the problem is feasible if one job is assigned to only one of the machines for the entirety of the continuous processing time of that job and no other job should be assigned to that machine during this time, so there is no overlap of time slots of jobs assigned to a common machine (this is the standard non-preemptive requirement). Also, the necessary quantity of the resource required by the jobs should be supplied, thus for every point of time t, the inequality

$$\sum_{j:j \text{ starts strictly before } t} d_j \leq \sum_{k:u_k < t} q_k,$$

should be also satisfied for feasibility, so the resource is common for all machines.

The parameters mentioned above define an input instance I for the problem. All these parameters are rational numbers.

The objective function considered is the minimization of the convex combination of the makespan and the sum of the ϕth powers of the load of the machines, where $\psi \in [0, 1]$ is the coefficient of the convex combination, and $\phi > 1$. Given the possibility of idle time, the load of a machine is the time of completion of the last job assigned to the machine. That is, the objective value of a schedule σ is defined as $obj(\sigma) = \psi \max_i \Lambda_i + (1 - \psi) \sum_i \Lambda_i^\phi$. where Λ_i is the load of machine i in σ, and obj denotes the objective function. The sum of the ϕth power of the completion times of the machines is equivalent to the l_ϕ-norm of the completion times of the machines that is a standard objective in the load balancing literature. Define *norm-cost* of a schedule σ as $\Phi(\sigma) = \sum_{i=1}^m \Lambda_i^\phi$. The objective considered in this work is a unified generalization of the makespan minimization and the l_ϕ-norm of the loads so we consider the two extreme cases $\psi = 0$ and $\psi = 1$ as the most interesting cases of this objective. For $\phi \leq 1$ and $\psi = 0$, the optimization problem of finding a minimum cost solution where there is a unique replenishment date at time zero is solved easily by allocating all jobs to one of the fastest machines and thus in this work we consider the problem assuming that $\phi > 1$. Problem ums is to find a job assignment function σ which assigns a job j to a machine i and a starting time on the machine that is a feasible solution so that obj is minimized. Our result is an EPTAS for ums.

Definitions of Approximation Algorithms. A ρ-approximation algorithm for a minimization problem is a polynomial time algorithm that always finds a feasible solution of cost at most ρ times the cost of an optimal solution. A polynomial time approximation scheme (PTAS) for a given problem is a family of approximation algorithms such that the family has a $(1+\varepsilon)$-approximation algorithm for any $\varepsilon > 0$. An efficient polynomial time approximation scheme (EPTAS) [5,6,9] is a PTAS whose time complexity is upper bounded by the form $f(\frac{1}{\varepsilon}) \cdot poly(n)$ where f is some computable (not necessarily polynomial) function and $poly(n)$ is a polynomial of the length of the (binary) encoding of the input. A fully polynomial time approximation scheme (FPTAS) is defined like an EPTAS, with the added restriction that f must be upper bounded by a polynomial in $\frac{1}{\varepsilon}$. Note that our problem generalizes the standard minimum makespan on identical machines that is known to be strongly NP-hard, and thus our problem does not admit an FPTAS (unless P = NP). Hence, an EPTAS is the fastest scheme that can be established for ums.

Notation and Preliminaries. In what follows machines are usually specified by the index $i, i = 1, 2, \ldots, m$, jobs are specified by the index $j, j = 1, 2, \ldots, n$, and time periods are specified by the index $k, k = 1, 2, \ldots, e$. Furthermore, let J be the set of all jobs, M be the set of all machines, and $m(s)$ denote the number of machines with speed s. Without loss of generality, $m \leq n$ because otherwise the $m - n$ slowest machines can be removed from the instance. Let $\varepsilon > 0$ be such that $1/\varepsilon$ is an integer, and our goal is to find an approximation algorithm of approximation ratio $1 + \varepsilon$ with the required time complexity bound. Based on standard scaling (of ε) it suffices to exhibit an algorithm that runs in the required time complexity and returns a feasible solution of cost at most $(1 + \varepsilon)^c$ times the optimal cost for an arbitrary constant $c > 0$. Note that if we modify a solution for ums so that the new load of every machine is at most $1 + \varepsilon$ times its old load then the makespan of the new solution is at most $1 + \varepsilon$ time the old makespan and the norm-cost of the new solution is at most $(1 + \varepsilon)^\phi$ times the old norm-cost.

Previous Studies of Scheduling Problems with Non-renewable Resources. In [2,17,28] there are studies motivating the setting of non-renewable resources in scheduling problems like problems arising in steel production or in order picking in a platform with a distribution company. Słowiński [29] Carlier and Kan [4] and Grigoriev et al. [10] were among the first to study such problems and considered minimizing the resource consumed as an objective but the former also added the makespan minimization objective. The settings of non-renewable resources was considered in other studies like [3,11,12,14–16,24]. Last and most relevant to our work, Györgyi [13] derived a PTAS for parallel machine scheduling with single non-renewable resource and makespan objective on identical machines. This PTAS is not an EPTAS, and thus our work both improves the time complexity of the known scheme for this problem to an EPTAS and furthermore it generalizes the machine settings to uniformly related machine and the objective to include also the sum of the ϕ-powers of machines' completion times that was not considered before in the settings of non-renewable resource. Furthermore, prior to our work there was no approximation algorithm in the literature for minimizing the makespan on uniform machines with non-renewable resource.

Earlier Approximation Schemes for Special Cases of the Problem Without Non-renewable Resources. A special case of ums is when there are no resource consuming jobs with parallel machines (this is the situation when the entire quantity of the resource is available to all jobs is released at time 0). For this machine setting Jansen [21] showed an EPTAS for the makespan minimization objective improving the seminal PTAS developed by Hochbaum and Shmoys [20]. Epstein and Levin [7] showed an EPTAS for ℓ_p-norm minimization problem (see also [8] for an alternative approach leading to an EPTAS for this problem), and Kones and Levin [25] established an EPTAS for the problem of load balancing with the objective that generalizes the one considered here.

The problem of scheduling on identical machines is a special case of uniformly related machines when speed of all machines are equal. Hochbaum and Shmoys [19] showed an EPTAS (see also [18, Chapter 9]) for the makespan minimization objective for this setting. Alon et al. [1] showed an EPTAS for ℓ_p-norm minimization. Jansen et al. [22] improved the time complexity of the approximation schemes of the earlier results for makespan minimization on identical machines and uniformly related machines and also for ℓ_p-norm of the vector of machines loads on identical machines. Lenstra et al. [26] showed that for the more general settings of unrelated machines unless P = NP there is no approximation algorithm with an approximation ratio less than 1.5 for the makespan minimization objective. They also presented a 2-approximation algorithm to this classical scheduling problem.

Outline of the Scheme. We apply geometric rounding to the parameters of the input (Sect. 2) followed by a guessing step (Sect. 3) to guess partial information on the optimal solution. This guessing step that is based on a careful characterization of a subset of solutions containing a near optimal solution. The so-called configuration mixed-integer linear program, MILP, (Sect. 4) is based on configurations in addition to the guessed information and we solve the MILP to find an optimal solution of this mathematical program. The optimal solution of the MILP is found in polynomial time using [23, 27] since the number of integer variables is a constant. For every possible value for the guessed information, we solve the MILP, and the least cost solution among all solutions to the MILP is converted to a feasible schedule for the scheduling problem whose cost is approximately the cost of the optimal solution of the MILP (see Sect. 5).

2   Rounding

Our scheme starts by applying standard geometric rounding of the input parameters stated below. We apply rounding of the job sizes, of the machine speeds (and as a result also the processing times), and of the replenishment dates. The rounding is carried out as follows. For every job j, the size of j is rounded up to the nearest integer power of $(1 + \varepsilon)$, the speed of every machine i is rounded down to the nearest integer power of $(1 + \varepsilon)$, and for the replenishment dates we first add $\varepsilon \cdot p_{\min}$ to every replenishment date and then we round up to the nearest integer power of $(1+\varepsilon)$ where $p_{\min} = \min_j p_j > 0$ is the minimum size of a job in the original instance. The speed of the fastest machine remains 1 after the rounding since 1 is an integer power of $1 + \varepsilon$. That is, $p'_j = (1+\varepsilon)^{\lceil \log_{(1+\varepsilon)} p_j \rceil}$, $s'_i = (1+\varepsilon)^{\lfloor \log_{1+\varepsilon} s_i \rfloor}$ and $u'_k = (1+\varepsilon)^{\lceil \log_{(1+\varepsilon)}(u_k + \varepsilon p_{\min}) \rceil}$. Note that as a result of the rounding the processing time of job j on machine i becomes p'_{ij} and satisfies that

$$p'_{ij} = \frac{p'_j}{s'_i} \in \left[\frac{p_j}{s_i}, (1+\varepsilon)^2 \frac{p_j}{s_i} \right). \tag{1}$$

We let I' be the rounded instance.

Let α be the smallest integer such that $(1 + \varepsilon)^\alpha \geq u'_1$, and β be the smallest integer such that $(1 + \varepsilon)^\beta \geq u'_e$, and let $\mu = \beta - \alpha$ be the number of time periods in I' where a period is between two consecutive replenishment dates even if the amount of resource supplied in some of these periods (i.e., at the starting time of a period) are perhaps zero. Then $\mu = \beta - \alpha = \lceil \log_{(1+\varepsilon)}(u_e + \varepsilon p_{\min}) \rceil - \lceil \log_{(1+\varepsilon)}(u_1 + \varepsilon p_{\min}) \rceil + 1 \leq \log_{(1+\varepsilon)}(u_e + \varepsilon p_{\min}) - \log_{(1+\varepsilon)} \varepsilon p_{\min} + 2 \leq \log_{(1+\varepsilon)} \frac{u_e}{\varepsilon p_{\min}} + 3$, which is upper bounded by a polynomial in the input encoding length when ε is fixed.

Last, we note that if there are different replenishment dates that are identical (in I') we combine identical replenishment dates into one date by summing up the replenished quantity of the dates that are combined. There are at most μ replenishment dates.

With a slight abuse of notation, let u'_k be the kth replenishment date after the above rounding. The resource supplied at time period k, in increasing order of k, is given by $q'_k = \sum_{v:u_v \leq u'_k} q_v - \sum_{v:u'_v < u'_k} q'_v$.

Next, we prove that we can assume without loss of generality that the input to our problem is rounded.

Proposition 1. *Any schedule σ feasible to instance I, of makespan C_{max} and norm-cost $\Phi(\sigma)$, can be used to generate a schedule σ' which is feasible to the rounded instance I' with makespan at most $(1 + \varepsilon)^3 C_{max}$ and $\Phi(\sigma') \leq (1 + \varepsilon)^{3\phi} \Phi(\sigma)$. Any schedule σ' feasible to the rounded instance I', of makespan C'_{max} and norm-cost $\Phi(\sigma')$, is feasible to the original instance I with makespan at most C'_{max} and norm-cost at most $\Phi(\sigma')$.*

Using the last proposition, we assume without loss of generality that the original instance is already rounded accordingly so with a slight abuse of notation we let p_j, s_i, p_{ij}, u_k and q_k be the (rounded) size of a job j, the (rounded) speed of a machine i, the (rounded) processing time of job j on machine i, the kth (rounded) replenishment date and resource supplied at u_k respectively and last we assume that I is the (rounded) input instance. In the next sections we provide an EPTAS for rounded instances of ums.

3   Characterization of Near-Optimal Solutions and the Guessing Step

In the known approximation schemes for the makespan minimization objective (for various machine models) one can apply the standard step of guessing the optimal makespan using binary search and then discard loads that are sufficiently smaller than this guessed value. Since our objective is different, we are not able to discard small loads and we are not able to use binary search for finding the approximate makespan. Thus we need another tool to characterize near optimal solution. In this section we provide the necessary characterization for our scheme.

Lemma 1. *There is a feasible solution whose objective function value is $(1+\varepsilon)^\phi$ times the optimal cost satisfying that the makespan is attained on a machine of speed at least ε^2.*

A machine is said to be active for a time interval consisting of some time periods if the machine is processing a job (or a part of a job) during that time interval and it is idle outside of this time interval. Let $\tilde{\mu} = \left\lceil \log_{(1+\varepsilon)} \frac{1}{\varepsilon} + 2 \right\rceil$ be a function of ε, and note that $\tilde{\mu}$ is a constant once ε is fixed. Our next goal is to show that we can restrict ourselves to schedules in which every machine is active only for a time interval consisting of a constant number of consecutive time periods. We prove the next lemma by applying time stretching and moving the new idle time of a machine to be at the beginning of the time horizon.

Lemma 2. *The feasible schedule σ to the instance I resulting from Lemma 1 can be converted into another schedule $\tilde{\sigma}$ in which first, every machine is active in a time interval consisting of at most $\tilde{\mu}$ consecutive time periods, second, $obj(\tilde{\sigma}) \leq (1+\varepsilon)^\phi obj(\sigma)$, and last, the makespan is attained on a machine of speed at least ε^2.*

Thus, for the succeeding discussions, it suffices to consider schedules that assign jobs to at most $\tilde{\mu}$ consecutive time periods on every machine and the makespan is attained on a machine of speed at least ε^2. Let opt be a least cost solution among all the near optimal solutions that satisfy Lemma 2, and such a feasible solution exists when the input instance has a feasible solution. Let the makespan value of opt rounded down to an integer power of $1+\varepsilon$ be C_{opt} and the speed of the machine on which the makespan is achieved be s_{opt}. Then the makespan of all near optimal solutions that we would like to consider lies in $[C_{opt}, (1+\varepsilon)C_{opt}]$. In the guessing step that we exhibit now we guess the pair (C_{opt}, s_{opt}).

We *guess* both pieces of information regarding opt. The term guess is used to mean that we perform the following steps of the algorithm for every possible value of this information and we output the feasible solution of minimal cost among all iterations that results in a feasible solution to ums. We note that our solution (for a given value of the guess) need not satisfy this guessed information but we will use the existence of a near optimal solution corresponding to this guess. In the analysis of the approximation ratio of our scheme it suffices to consider the iteration of this exhaustive search in which we used the correct information regarding opt. We next verify that the number of possibilities of this guessed information (i.e., the number of iterations of the exhaustive enumeration) is upper bounded by a polynomial in the input encoding length.

Lemma 3. *The number of possibilities of the guessed information (C_{opt}, s_{opt}) is at most $n\left(\log_{(1+\varepsilon)} n + 2\right) \cdot \left(1 - \log_{(1+\varepsilon)} \varepsilon^2\right)$.*

4   The Mixed-Integer Linear Program

Our scheme is based on solving a mixed-integer linear program (MILP) for every given value of the guessed information. The MILP is based on configurations and

assignment of jobs to configurations. We first define some preliminaries definitions, then our notion of configurations, and we conclude this section presenting
the MILP.

Huge, Big, and Small Jobs. We classify jobs into huge, big, and small jobs with
respect to the pair (s, w), where s is the speed of a machine and w is its load.
The formal definition applies for an arbitrary pair of positive reals. Let $\rho \geq 1$
be a constant that will be specified later. A job j is a *huge job* for a pair (s, w)
if $p_j/s > (1 + \varepsilon)w$, is a *big job* for a pair (s, w) if $(1 + \varepsilon)w \geq p_j/s \geq \varepsilon^\rho w$, and
is a *small job* for a pair (s, w) if $p_j/s < \varepsilon^\rho w$. Let $\Delta = \left\lceil \log_{(1+\varepsilon)}(1 + \varepsilon)w \right\rceil$ and
$\delta = \left\lceil \log_{(1+\varepsilon)} \varepsilon^\rho w \right\rceil$. Then, the number of distinct sizes for big jobs for a fixed
pair (s, w) is at most $\lambda = \Delta - \delta + 1$, and λ depends only on ε (for a constant ρ).

Large and Small Loads. Recall that C_{opt} is the guessed approximated value of
makespan of a near optimal solution that satisfies Lemma 2. A load w of a
machine is a *large load* if $w \geq \kappa C_{\text{opt}}$, else it is a *small load*, where κ is a function
of ε which will be specified later. Thus a large load lies in $[\kappa C_{\text{opt}}, (1 + \varepsilon)C_{\text{opt}}]$.
Therefore, given a value of C_{opt}, the number of distinct values of large loads is at
most $\log_{(1+\varepsilon)} \frac{(1+\varepsilon)C_{\text{opt}}}{\kappa C_{\text{opt}}} + 1 = \log_{(1+\varepsilon)} \frac{1}{\kappa} + 2$ which is a constant for a fixed value
of ε and κ.

Fast and Slow Speeds. Let S be the set of all values for the rounded speed of
machines. A speed s or a machine of speed s is *fast* if $s \geq \kappa$, else it is *slow*, where
κ is the same function of ε we used to define large loads. The number of distinct
values of fast speeds is at most $\log_{(1+\varepsilon)} 1 - \log_{(1+\varepsilon)} \kappa + 1 \leq \log_{(1+\varepsilon)} \frac{1}{\kappa} + 2$ which
is constant when ε is fixed.

Configuration. We use the term *configuration* as a compact and approximated
representation of a schedule of one machine. Each configuration c is a vector
with $\tilde{\mu} + 2$ components.

The first component stores the speed s_c of the machine assigned configuration
c. The second component, w_c, is the upper bound on the load (load bound) of
a machine to which configuration c is assigned, rounded up to an integer power
of $(1 + \varepsilon)$. Based on our characterization of near optimal solutions, a machine
assigned c will be active for at most $\tilde{\mu}$ time periods intersecting $[\varepsilon w_c, (1 + \varepsilon)w_c]$.
Let $\Delta_c = \left\lceil \log_{(1+\varepsilon)}(1 + \varepsilon)w_c \right\rceil$ and $\delta_c = \left\lceil \log_{(1+\varepsilon)} \varepsilon^\rho w_c \right\rceil$. The number of distinct
sizes for big jobs for the configuration c is at most $\lambda_c = \Delta_c - \delta_c + 1$.
Let $\nu_c^l = \left\lceil \log_{(1+\varepsilon)} \varepsilon w_c \right\rceil$ and let $\nu_c^u = \left\lceil \log_{(1+\varepsilon)}(1 + \varepsilon)w_c \right\rceil$. Each of the
remaining $\tilde{\mu}$ components of the configuration corresponds to time periods
$k = \nu_c^l, \nu_c^l + 1, \nu_c^l + 2, \ldots, \nu_c^u$. For each period k in this interval, there is an
associated subvector of $\lambda_c + 1$ components. Each of the first λ_c components of
the subvector associated to a period k stores the number of big jobs, for the
pair (s_c, w_c), of processing time $(1 + \varepsilon)^{\delta_c + r - 1}$ on machines with speed s_c for

$r = 1, 2, \ldots, \lambda_c$ that are started in that period on a machine assigned configuration c, and we denote this value of the component by $B_c(k, r)$. The last component of the subvector associated to a period k stores the floor of the ratio of the sum of the processing time of all the small jobs, for the pair (s_c, w_c), that are started in that period on a machine assigned configuration c and $\varepsilon^\rho w_c$, denoted by $S_c(k)$. That is, $S_c(k) = \left\lfloor \frac{\sum(p_j/s_c)}{\varepsilon^\rho w_c} \right\rfloor$ where the summation is over the set of the small jobs assigned to period k and configuration c on such machine.

Number of Configurations. Our next goal is to upper bound the number of configurations with a common given pair (s_c, w_c) by a constant depending on ε and to upper bound the number of configurations for all such pairs by a polynomial in the input encoding length. We will prove the following upper bounds.

Lemma 4. *For a fixed value for speed and load bound, the number of possible configurations is at most $\left(\frac{1+\varepsilon}{\varepsilon^\rho} + 1 \right)^{\tilde{\mu}(\lambda+1)}$. In total there are polynomial number of configurations.*

Generating a Schedule of One Machine from a Configuration c. We would like to restrict ourselves to configurations that represent a non-preemptive schedule. To do that, we define a partial schedule of one machine corresponding to a configuration c, and the possibility to create this partial schedule will mean that the configuration will be called *feasible*. Based on the configuration c with its speed s_c and the load bound w_c, we create a schedule on $\tilde{\mu}$ periods intersecting $[\varepsilon w_c, (1 + \varepsilon)w_c]$ on that machine. For every period k in this interval, in an increasing order of k, perform the steps below.

1. Initialize the starting time of the first job in period k denoted as $Start_k$ to be

    $\max\{u_k$, the finishing time of the last processed job in period at most $k - 1\}$.

    When k is the first period in this interval, finishing time of last processed job in period $k - 1$ is the starting time of period k.
2. If $S_c(k)$ is not zero, then assign a *virtual job* of size $S_c(k) \cdot \varepsilon^\rho w_c s_c$ to start at time $Start_k$ and increase the value of $Start_k$ by the processing time of this virtual job.
3. The remaining jobs assigned to period k are big jobs and these big jobs are scheduled sequentially without idle time in non-decreasing order of their size, the number of big jobs of each size are as described in c, and the first such job is scheduled to start at $Start_k$.

A configuration is *feasible* if for each period in the schedule generated, there is at most one job (that is not a virtual job) whose processing crosses into the successive period(s). Let C be the set of all feasible configurations and we compute C in advance with the required time complexity of the form $f(\frac{1}{\varepsilon}) \cdot$

$poly(n)$. We can indeed compute C in advance in this required time complexity as the algorithm for generating a schedule of one machine based on a configuration is a linear time algorithm and verifying feasibility is carried out during this generation process.

Notation for the MILP. A MILP is developed that uses the guessed information of opt to schedule the jobs to configurations for the rounded instance, I'. For each configuration $c \in C$, the MILP assigns the jobs to only a particular set of consecutive periods such that when that configuration is assigned to a machine, the machine will be active for only that particular set of consecutive periods. The set of periods is defined by the configuration. Let $\nu_c^l = \lfloor \log_{(1+\varepsilon)} \varepsilon w_c \rfloor$ and $\nu_c^u = \lceil \log_{(1+\varepsilon)} (1 + \varepsilon) w_c \rceil$. Then the set of consecutive time periods for which a machine assigned configuration c will be active (or more precisely it will not be active in other time periods) is denoted by $K_c = \{(1 + \varepsilon)^k : k = \nu_c^l, \nu_c^l + 1, \nu_c^l + 2, \dots, \nu_c^u\}$, $\forall c \in C$. Furthermore, let $K = \{(1 + \varepsilon)^i : i = \log_{1+\varepsilon} u_1, \log_{1+\varepsilon} u_1 + 1, \dots, \log_{1+\varepsilon} C_{\text{opt}} + 1\}$, denotes the set of universal time periods for any configuration and recall that $|K| = \mu$ is bounded by a polynomial in the input encoding length.

Since the classification of the jobs depend on both the speed of the configuration and the load of the configuration the distinct size of the big jobs change with respect to the configuration. Hence let R_c, $\forall c \in C$ denote the set of indexes for integer power of $1 + \varepsilon$ corresponding to sizes of big jobs for the pair (s_c, w_c) of configuration c.

For each speed s and load bound w that is an integer power of $1 + \varepsilon$, create the sets $HJ(s, w)$, $BJ(s, w)$ and $SJ(s, w)$ for huge jobs, big jobs and small jobs for the pair (s, w) respectively. Order the jobs in each list in non-decreasing order of the resource requirements.

The Decision Variables of the MILP. We have two families of decision variables.

Configuration Counters. Denoted by z_c. The variable z_c counts the number of machines assigned configuration $c \in C$. Each configuration counter for a configuration of fast speed and large load is required to be integer and configuration counters for remaining configurations are allowed to be fractional. Thus the number of integer configuration counters is at most $\left(\log_{(1+\varepsilon)} \frac{1}{\kappa} + 2\right)^2 \cdot$ $\left(\frac{1+\varepsilon}{\varepsilon^\rho} + 1\right)^{\tilde{u}(\lambda+1)}$, which is a constant when ε is fixed. The number of fractional configuration counters is polynomial in the input encoding length.

Assignment Variables. Denoted by x_{jkc}. In an integer solution, if x_{jkc} is 1, job j is assigned to configuration c and to start in period k. All these variables are allowed to be fractional. The number of assignment variables is at most $n \cdot |K| \cdot |C|$, which is upper bounded by a polynomial in the input encoding length.

The MILP. The objective function of the MILP is the minimization of

$$\psi C_{\text{opt}} + (1 - \psi) \sum_{c \in C} z_c \cdot w_c^\phi .$$

We next list the constraints of the MILP (together with their meaning).

A job can be assigned to a configuration and period combination only as a non-huge job.

$$\sum_{c \in C : j \notin HJ(s_c, w_c)} \sum_{k \in K_c} x_{jkc} = 1 , \forall j \in J . \tag{2}$$

For each size, the number of big jobs of that size assigned to a period k of a configuration c is equal to the number of big jobs of that size in period k of configuration c times the number of machines of this configuration.

$$\sum_{j : j \in BJ(s_c, w_c), p_j = (1+\varepsilon)^{\delta_c + r - 1}} x_{jkc} = z_c \cdot B_c(k, r) , \forall c \in C, k \in K_c, r \in R_c . \tag{3}$$

An upper bound on the sum of processing time of small jobs (for every configuration and every period), while allowing some additional space that is needed since $S_c(k)$ was obtained by rounding down.

$$\sum_{j \in SJ(s_c, w_c)} (x_{jkc} \cdot p_j) \le z_c \cdot s_c \cdot (S_c(k) + 1) \cdot \varepsilon^\rho w_c , \forall c \in C, k \in K_c . \tag{4}$$

The number of chosen configurations for a particular speed should be equal to the number of machines of that speed.

$$\sum_{c \in C : s_c = s} z_c = m(s) , \forall s \in S . \tag{5}$$

In each period, the resource requirement of a job is met before the job starts processing. Resource can be carried over to future periods, if available.

$$\sum_{c \in C} \sum_{k' : k' \le k} \left(\sum_j (x_{jk'c} \cdot d_j) \right) \le \sum_{k' : k' \le k} q_{k'} , \forall k \in K . \tag{6}$$

Constraint that enforces the guessing of the makespan and the machine on which makespan is achieved, once again we slightly relax the condition and allow a machine of that speed with slightly smaller load bound:

$$\sum_{c \in C : s_c = s_{\text{opt}}, (1+\varepsilon) w_c \ge C_{\text{opt}}} z_c \ge 1 . \tag{7}$$

The bounds for the assignment variable.

$$0 \le x_{jkc} \le 1 , \forall c \in C, k \in K, j \in J . \tag{8}$$

For configurations with fast speed and large load, the configuration counter variable is an integer and all configuration counters are non-negative.

$$z_c \ge 0 , \forall c \in C , \text{ and} \tag{9}$$

$$z_c \in \mathbb{Z} , \forall c \in C : s_c \ge \kappa, w_c \ge \kappa C_{\text{opt}} . \tag{10}$$

5   Using the MILP

The remaining parts of our scheme and its analysis are as follows. We first show in Theorem 1 that (for the correct guessed value) there is a solution to the MILP whose objective function value is at most $(1 + \varepsilon)^\phi$ times the objective value of the near optimal solution satisfying Lemma 2. That is, we prove that the MILP has a feasible solution of cost not significantly larger than the cost of the near optimal schedule of our problem. Then, our scheme continues by finding an optimal solution for the MILP, and then transform it into a feasible schedule (a feasible solution to our scheduling problem) of cost not much larger than the cost of the given optimal solution for the MILP. The use of the solution of the MILP for constructing our output is considered next.

Theorem 1. *The optimal objective function value of the MILP is at most $(1 + \varepsilon)^\phi$ times the objective value of the near optimal solution satisfying Lemma 2 (as a solution to the scheduling problem).*

The transformation of an optimal solution to the MILP to a near optimal solution for ums that we describe in the rest of this section proves the following theorem.

Theorem 2. *If there exists a solution to the MILP for I for the given value of the guessed information, then there exists a feasible schedule to instance I of cost at most $(1 + 6\varepsilon)^\phi$ times the cost of the solution of the MILP.*

Let the MILP solution be sol, the values of the decision variables in this MILP solution be $(\mathbf{x}, \mathbf{z})$, and the guessed information be C_{opt} and s_{opt}. $\mathbf{x} \in \mathbb{R}$ but the z value is integer for configurations with fast speed and large load. Let $\hat{z}_c = \lceil z_c \rceil$, $\forall c \in C$. This converts all fractional configuration counters to integers while the configuration counters for fast speed and large loads remain unchanged and integer. Thus for every speed s we add $\sum_{c \in C : s_c = s} \hat{z}_c - m(s)$ *virtual machines*. Virtual and real machines in this discussion do not refer to the machines in the instance I, and only act as terms to explain the last rounding phase of our scheme. Thus, now we have more machines than specified in the input instance.

Next, we assign configurations to machines, i.e., every machine i is now associated with a configuration c whose first component is the speed of i such that the number of machines that are assigned a configuration c is equal $\hat{z}_c$. We allocate the configurations to real and virtual machines in a way that for every configuration c, at least $\lfloor z_c \rfloor$ machines that were assigned configuration c are real machines (and at most one machine assigned configuration c is a virtual machine). This last requirement is possible as the z values satisfy constraint (5). For further discussions let C be the set of configurations c with $\hat{z}_c > 0$.

Fix a real machine of speed s_{opt} as the machine which attains the makespan by assigning the configuration of largest load of speed s_{opt} to this machine. From constraint (7), the load bound of this machine is at least $C_{\mathrm{opt}}/(1 + \varepsilon)$. Let that machine be denoted as $\rangle$.

We first show how to allocate the jobs to the machines (virtual or real) such that the load of every such machine is at most $1 + O(\varepsilon)$ times its load bound and so that the resource required by the jobs is supplied on time. Later on we will analyze the impact of not using virtual machines and changing the assignment of all jobs that were assigned to virtual machines to be processed on machine $\rangle$.

Allocating Most Jobs and Leaving Only Some Jobs that are Small for at Least One Machine. Recall that K denotes the set of universal time periods for any configuration. Let R' denote the set of integer powers of $(1 + \varepsilon)$ corresponding to distinct sizes of all jobs in the (rounded) instance I. Let $\alpha_{r'k}$ be the number of jobs of size $(1 + \varepsilon)^{r'}, r' \in R'$, assigned to start in period k based on the MILP solution, that is $\alpha_{r'k} = \left\lfloor \sum_{j \in J : p_j = (1+\varepsilon)^{r'}} \sum_{c \in C} x_{jkc} \right\rfloor$. Thus according to the MILP solution there is enough resource to start processing $\alpha_{r'k}$ jobs of size $(1 + \varepsilon)^{r'}$ in period k and furthermore there is enough available time to process so many jobs of each size on the machines of I. Intuitively, since $\alpha_{r'k}$ is rounded down, for every size for which there is a machine with speed and load bound for which the size is a size of a small job, there is at most one job of this (small) size left unassigned (with respect to the MILP solution) in each such period.

For every machine and every active period, in the first rounding step, we increase the total size available for small jobs in period k by $2\varepsilon^\rho s_c w_c$, where c is the configuration assigned to the machine. This is done to ensure that the process we present will be feasible. Now the total size available for small jobs on a machine in period k is $(S_c(k) + 2)\varepsilon^\rho s_c w_c$, where k is one of the active periods of that machine. We will allow adding the additional small jobs at the end of the schedule of this machine.

Next in the second rounding step that we describe below, we assign almost all jobs to the real or virtual machines where the load of every machine with configuration c will be approximately w_c. More precisely, for each period $k \in K$, in an increasing order of k, perform the following.

Create a list of jobs J_k for which there exists $c \in C$ such that $x_{jkc} > 0$ and j was not scheduled to start in an earlier period. Sort J_k in non-decreasing order of the resource requirement. The assignment of big jobs is detailed next. Go over each machine one by one and perform the following operations. Identify the set of sizes of jobs that are big for the machine assigned configuration c and let R_c denote the set of indexes for integer powers of $1 + \varepsilon$ corresponding to sizes of big jobs for the pair (s_c, w_c) for configuration c where c is the configuration of the current machine. For each $r \in R_c$, in an increasing order of r, perform the following operations. Identify the jobs of size $(1 + \varepsilon)^{\delta_c + r - 1}$ in J_k, in the order the jobs appear in J_k. Assign the jobs one by one in order starting with the first job as follows. A big job of size $(1 + \varepsilon)^{\delta_c + r - 1}$ can be assigned to the current machine, assigned configuration c, if after assigning the job to this machine and period the following two conditions hold. First, the total number of jobs of size $(1 + \varepsilon)^{\delta_c + r - 1}$ assigned to all the machines in period k is at most $\alpha_{(\delta_c + r - 1)k}$ and second the total number of big jobs of size $(1 + \varepsilon)^{\delta_c + r - 1}$ assigned to the current machine in period k is at most $B_c(k, r)$. If a job is assigned to a machine, remove it from J_k.

Once the above procedure is completed for the current value of k we move on to schedule jobs as small jobs (in this period) as follows. Go over the machines in non-decreasing order of the product of speed and load bound (of its assigned configuration) and perform the following operations for assigning jobs to the current machine. Identify the set of sizes of jobs that are small for the machine assigned configuration c and let R'_c denote the set of indexes for integer powers of $1+\varepsilon$ corresponding to sizes of small jobs for the pair (s_c, w_c) for configuration c. For each $r \in R'_c$, in an increasing order of r, perform the following operations. Identify the jobs of size $(1+\varepsilon)^{\delta_c + r - 1}$ in J_k, in the order the jobs appear in J_k. Assign the jobs one by one in order starting with the first job as follows. Assign the job to the machine if the total size of small jobs on that machine in period k, after assignment of the job, is at most $(S_c(k)+2)\varepsilon^\rho s_c w_c$, and the total number of jobs of size $(1+\varepsilon)^{\delta_c+r-1}$ assigned in period k on all machines, after assignment of the job, is at most $\alpha_{(\delta_c+r-1)k}$. Once we have considered all machines (for this period k), we increase k by one and go back to creating the lists J_k.

After completing the assignment of jobs using the above two rules for all k, the second rounding step is completed. The remaining unassigned jobs are small jobs that were left unassigned due to the rounding down of $\alpha_{r'k}$. In the last rounding step all non-assigned jobs are assigned to the end of the schedule on machine $\rangle$ and if necessary we add idle time before processing these jobs to ensure that they do not start before time C_{opt}. This assignment of jobs in the third rounding step to the end of the schedule on $\rangle$ is feasible because there exists enough resource to assign all these jobs by the time machine $\rangle$ finishes processing its scheduled jobs using the fact that C_{opt} is not smaller than the last time in which a resource is supplied (considering only the resource needed by the set of jobs in the instance).

In the last rounding step of our scheme all jobs that were assigned to virtual machines in the second rounding step are moved to be processed on machine $\rangle$ in an arbitrary order without idle time after the last job assigned to that machine (in the second or third step) is completed (once again starting not earlier than C_{opt}). Similarly to the argument regarding the feasibility of the assignment of jobs in the third rounding step, the resulting schedule is feasible if we are able to show that the assignment of jobs in the second rounding step is feasible with respect to the resource constraint. This completes the description of the procedure of transforming the MILP solution into a feasible schedule for ums. We next turn our attention to the analysis of this transformation procedure.

Proving that the Partial Assignment of Jobs to Virtual and Real Machines is Feasible. The feasibility of the partial schedule at the end of the second rounding step with respect to the resource constraint follows from the feasibility of the MILP solution and noting that in every prefix of periods (i.e., all periods k' with $k' \leq k$ for a fixed value of k) and every size of jobs $(1+\varepsilon)^r$ (for a fixed $r \in R$) we have the following. The partial schedule schedules no more than $\sum_{k':k'\leq k} \alpha_{rk'}$ jobs of size $(1+\varepsilon)^r$ to start in this prefix of periods and these jobs have the least resource requirement (among the jobs of that size). However, by the definition of the α values, the MILP solution schedules at least so many jobs of this size to start during this prefix, and the Constraint (6) guarantees that the total consumption of resource in this partial schedule in this prefix of period does not exceed the total amount of resource that is supplied in this prefix of periods.

Next, we would like to argue that at the end of the second rounding step for every period k we have the following. For every size of jobs x for which there is no machine active in this period where x is a size of a small job for that machine, the number of jobs of this size assigned to start in this period is exactly as in the MILP solution, and for every other size we are left with at most one unassigned job. Thus we will prove the following lemma.

We will let $\rho = 10$ and require the property that $\kappa \leq \varepsilon^{2\rho+3}$ so if a job cannot be assigned as a small job to machine $\rangle$ then it must be assigned as a large job to a fast machine with large load and all these jobs are assigned (integrally) by the MILP solution.

Lemma 5. *For every period k, let $S(k)$ be the set of sizes of jobs satisfying that for every $x \in S(k)$, there is at least one configuration c where a job of size x is not huge for (s_c, w_c) and x is small for machine $\rangle$. At the end of the second rounding step we have that the set of unassigned jobs have at most one job of each size for every period where this size belongs to $S(k)$ (and no such job of this size if the size does not belong to $S(k)$).*

Proof. Fix a size of jobs $(1+\varepsilon)^r$ (for $r \in R$), we want to show that the number of unassigned jobs of this size is one or zero. Consider a configuration $c \in C$ and a period $k \in K$. Notice that $B_c(k,r)\hat{z}_c$ is integer and it upper bounds the number of jobs of this size assigned to a machine of this configuration and period as large jobs in the MILP solution. We argue that the second rounding step assigns exactly α_{kr} jobs of this size to start in period k and this number is exactly $\sum_{j \in J: p_j = (1+\varepsilon)^r} \sum_{c \in C} x_{jkc}$ if $(1+\varepsilon)^r \notin S(k)$. First, consider the second part, namely if r is large enough so that there is no machine in which a job of size $(1+\varepsilon)^r$ is a non-huge job, then such jobs cannot be assigned to start at this period in the MILP solution. It suffices to show that we indeed assign α_{kr} jobs of this size for every period k and we would like to prove that for every $r \in R$.

Assume that this is not the case. Since the assignment of jobs have over-used the positions as large jobs, the assignment of jobs as small jobs does not guarantee the feasibility of this assignment for some of the sizes. Notice that the machines are ordered in the non-decreasing order of the product of speed and load bound for the assignment of the small jobs. Thus when a job is small for a machine then the job will be small for all succeeding machines in the ordering.

Identify the first machine i', in the ordering, such that each machine in the suffix of machines starting from i' and up to the end of the (ordered list of) machines, satisfies that the total size of small jobs is at least $(S_c(k) + 1)\varepsilon^\rho s_c w_c$ and is about to exceed $(S_c(k) + 2)\varepsilon^\rho s_c w_c$, for c that is the configuration of the corresponding machine. Let $\mathcal{M}$ be the machine set of this suffix of machines. If the claim does not holds, then $\mathcal{M} \neq \emptyset$. Denote by $\mathcal{R}$ the sizes of jobs that are small only for machines in $\mathcal{M}$. For the last machine in $M \setminus \mathcal{M}$ (if it exists), the total size of small jobs assigned to the machine, by the algorithm, is at most

$(S_c(k) + 1)\varepsilon^\rho s_c w_c$. This means that the total size of unassigned jobs that were small for this machine was less that $(S_c(k) + 1)\varepsilon^\rho s_c w_c$. Let the last small job assigned to such a machine be j'. Then the last job that was small for this machine was j' and all jobs of size greater than the size of j' are small only for the machines succeeding this machine in the ordering. Moreover, j' was the last job of size not larger than $p_{j'}$ (along J_k) or we have already scheduled the maximum number of jobs of these sizes to this period, else the algorithm would assign more jobs of these sizes to this machine by the choice of the suffix of machines.

The MILP solution was able to assign x_{jkc} fraction of all jobs of sizes in $\mathcal{R}$ from J_k to all configurations assigned to machines in $\mathcal{M}$ such that for each configuration the total size of small jobs on each configuration is at most $(S_c(k) + 1)\varepsilon^\rho s_c w_c$ times z_c, from constraint (4). Furthermore, the jobs with sizes in $\mathcal{R}$ could not be assigned to any machine in $M \backslash \mathcal{M}$ as small jobs, since these jobs were big jobs for all those machines. The total (fractions) of configuration counters selected by the MILP solution for this assignment is no more than the number of machines in $\mathcal{M}$, and we get a contradiction to our assumption that $\mathcal{M} \neq \emptyset$ and the claim follows.   □

Observe that at the end of the second rounding step a machine with configuration c has load of at most $(1 + 2\varepsilon)w_c$. To see this fact observe that the first rounding step increases the load of a machine with configuration c by no more than $\tilde{\mu} \cdot 2\varepsilon^\rho w_c \leq \varepsilon w_c$ by our choice of $\rho = 10$, using the definition of $\tilde{\mu} \leq \frac{2}{\varepsilon^2}$. The second rounding step does not increase the load of machines so this upper bounds on the load holds at the end of the second rounding step. Since the third rounding step and the final rounding step only move jobs to machine $\rangle$ the last bound on the load of other machines continue to hold, and we consider the impact of these rounding steps on the load of $\rangle$.

Lemma 6. *The total size of jobs scheduled to $\rangle$ in the third rounding step is at most $\varepsilon C_{\mathrm{opt}}$.*

Bounding the Total Size of Jobs Assigned to Virtual Machines. It remains to upper bound the increase of the load of machine $\rangle$ in the final rounding step. Let

$$\kappa = \frac{\varepsilon s_{\mathrm{opt}}}{(1 + 2\varepsilon) \cdot \left(\frac{1+\varepsilon}{\varepsilon^\rho} + 1\right)^{\tilde{\mu}(\lambda+1)} \cdot \left(\frac{(1+\varepsilon)^3}{\varepsilon^2}\right)},$$

and note that indeed the required property of $\kappa \leq \varepsilon^{2\rho+3} = \varepsilon^{23}$ indeed holds.

Lemma 7. *The total size of jobs on all virtual machines is at most $2\varepsilon C_{\mathrm{opt}} s_{\mathrm{opt}}$.*

The configuration assigned to machine $\rangle$ has a load bound of at least $\frac{C_{\mathrm{opt}}}{1+\varepsilon}$ and at most $C_{\mathrm{opt}} \cdot (1 + \varepsilon)$. Therefore, after the final rounding step, the load of machine $\rangle$ is at most $(1 + 3\varepsilon)C_{\mathrm{opt}} + \varepsilon C_{\mathrm{opt}} + 2\varepsilon C_{\mathrm{opt}} = (1 + 6\varepsilon) \cdot C_{\mathrm{opt}}$. Let Λ_i be the load of machine i in the output schedule and note that if i has assigned configuration c then $\Lambda_i \leq (1 + 2\varepsilon)w_c + 3\varepsilon C_{\text{opt}} \leq (1 + 6\varepsilon)w_c$. Thus the objective value of the output schedule is

$$\max_{i \in M} \Lambda_i + \sum_{i \in M} \Lambda_i^\phi \leq (1 + 6\varepsilon)C_{\text{opt}} + \sum_{c \in C} z_c \cdot (1 + 6\varepsilon)^\phi w_c \leq (1 + 6\varepsilon)^\phi obj(\text{sol})$$

and we conclude that Theorem 2 holds.

Summary of the Scheme. The algorithm initially uses the rounding to simplify the instance and uses a guessing step to guess information from the rounded instance to create the configurations and the MILP. Using the generated configurations and the MILP the algorithm finds the least cost solution from among all the MILP solutions for each value of the guessed information. This solution is then transformed into a schedule. The algorithm for solving the MILP has a complexity of $2^{O(d \log d)} \cdot poly(n)$ using Lenstra's algorithm where d is the number of variables that are required to be integer in the MILP, and

$$d = \left(\log_{(1+\varepsilon)} \frac{1}{\kappa} + 2 \right)^2 \cdot \left(\frac{1 + \varepsilon}{\varepsilon^\rho} + 1 \right)^{\tilde{\mu}(\lambda+1)} = f'\left(\frac{1}{\varepsilon} \right),$$

since κ, $\tilde{\mu}$, and λ are functions of ε. Alll other steps of the algorithm runs in polynomial time and the number of possibilities of the guessed information is also upper bounded by a polynomial of the input encoding length. Thus the complexity of the scheme is $f\left(\frac{1}{\varepsilon} \right) \cdot poly(n)$, where $f\left(\frac{1}{\varepsilon} \right)$ is a doubly exponential function in $\frac{1}{\varepsilon}$.

Theorem 1 guarantees feasibility of the solution of the algorithm and using Theorem 2 the schedule can be generated for the least cost solution with the proved approximation ratio. Thus, we have established our result stated as follows.

Theorem 3. *Problem* ums *admits an EPTAS.*

References

1. Alon, N., Azar, Y., Woeginger, G.J., Yadid, T.: Approximation schemes for scheduling on parallel machines. J. Sched. **1**(1), 55–66 (1998)
2. Belkaid, F., Maliki, F., Boudahri, F., Sari, Z.: A branch and bound algorithm to minimize makespan on identical parallel machines with consumable resources. In: Advances in Mechanical and Electronic Engineering, pp. 217–221. Springer (2012). https://doi.org/10.1007/978-3-642-31507-7_36
3. Bérczi, K., Király, T., Omlor, S.: Scheduling with non-renewable resources: minimizing the sum of completion times. In: Baïou, M., Gendron, B., Günlük, O., Mahjoub, A.R. (eds.) ISCO 2020. LNCS, vol. 12176, pp. 167–178. Springer, Cham (2020). https://doi.org/10.1007/978-3-030-53262-8_14
4. Carlier, J., Kan, A.R.: Scheduling subject to non-renewable resource constraints. Oper. Res. Lett. **1**(2), 52–55 (1982)
5. Cesati, M., Trevisan, L.: On the efficiency of polynomial time approximation schemes. Inf. Process. Lett. **64**(4), 165–171 (1997)

6. Downey, R., Fellows, M.: Parameterized Complexity. Springer, New York (1999). https://doi.org/10.1007/978-1-4612-0515-9_15
7. Epstein, L., Levin, A.: An efficient polynomial time approximation scheme for load balancing on uniformly related machines. Math. Programm. **147**(1–2), 1–23 (2014)
8. Epstein, L., Levin, A.: Minimum total weighted completion time: faster approximation schemes. CoRR, abs/1404.1059 (2014)
9. Flum, J., Grohe, M.: Parameterized Complexity Theory. Springer-Verlag. https://doi.org/10.1007/3-540-29953-X
10. Grigoriev, A., Holthuijsen, M., van de Klundert, J.: Basic scheduling problems with raw material constraints. Naval Res. Logist. **52**(6), 527–535 (2005)
11. Györgyi, P., Kis, T.: Approximation schemes for parallel machine scheduling with non-renewable resources. Eur. J. Oper. Res. **258**(1), 113–123 (2017)
12. Györgyi, P., Kis, T.: Minimizing total weighted completion time on a single machine subject to non-renewable resource constraints. J. Sched. **22**(6), 623–634 (2019). https://doi.org/10.1007/s10951-019-00601-1
13. Györgyi, P.: A PTAS for a resource scheduling problem with arbitrary number of parallel machines. Oper. Res. Lett. **45**(6), 604–609 (2017)
14. Györgyi, P., Kis, T.: Approximation schemes for single machine scheduling with non-renewable resource constraints. J. Schedul. **17**(2), 135–144 (2013). https://doi.org/10.1007/s10951-013-0346-9
15. Györgyi, P., Kis, T.: Approximability of scheduling problems with resource consuming jobs. Ann. Oper. Res. **235**(1), 319–336 (2015). https://doi.org/10.1007/s10479-015-1993-3
16. Györgyi, P., Kis, T.: Reductions between scheduling problems with non-renewable resources and knapsack problems. Theoret. Comput. Sci. **565**, 63–76 (2015)
17. Herr, O., Goel, A.: Minimising total tardiness for a single machine scheduling problem with family setups and resource constraints. Eur. J. Oper. Res. **248**(1), 123–135 (2016)
18. Hochbaum, D.S. (ed.): Approximation algorithms for NP-hard problems. PWS Pub. Co, Boston (1997)
19. Hochbaum, D.S., Shmoys, D.B.: Using dual approximation algorithms for scheduling problems theoretical and practical results. J. ACM **34**(1), 144–162 (1987)
20. Hochbaum, D.S., Shmoys, D.B.: A polynomial approximation scheme for scheduling on uniform processors: using the dual approximation approach. SIAM J. Comput. **17**(3), 539–551 (1988)
21. Jansen, K.: An EPTAS for scheduling jobs on uniform processors: using an MILP relaxation with a constant number of integral variables. SIAM J. Discrete Math. **24**(2), 457–485 (2010)
22. Jansen, K., Klein, K.-M., Verschae, J.: Closing the gap for makespan scheduling via sparsification techniques. Math. Oper. Res. **45**(4), 1371–1392 (2020)
23. Kannan, R.: Improved algorithms for integer programming and related lattice problems. In: Proceedings of the Fifteenth Annual ACM Symposium on Theory of Computing, pp. 193–206 (1983)
24. Kis, T.: Approximability of total weighted completion time with resource consuming jobs. Oper. Res. Lett. **43**(6), 595–598 (2015)
25. Kones, I., Levin, A.: A unified framework for designing EPTAS for load balancing on parallel machines. Algorithmica **81**(7), 3025–3046 (2019)
26. Lenstra, J.K., Shmoys, D.B., Tardos, É.: Approximation algorithms for scheduling unrelated parallel machines. Math. Programm. **46**(1), 259–271 (1990)
27. Lenstra, H.W., Jr.: Integer programming with a fixed number of variables. Math. Oper. Res. **8**(4), 538–548 (1983)

28. Li, J., Duan, P., Sang, H., Wang, S., Liu, Z., Duan, P.: An efficient optimization algorithm for resource-constrained steelmaking scheduling problems. IEEE Access **6**, 33883–33894 (2018)
29. Slowiński, R.: Preemptive scheduling of independent jobs on parallel machines subject to financial constraints. Eur. J. Oper. Res. **15**(3), 366–373 (1984)

Several Methods of Analysis for Cardinality Constrained Bin Packing

Leah Epstein[✉]

Department of Mathematics, University of Haifa, Haifa, Israel
lea@math.haifa.ac.il

Abstract. We consider a known variant of bin packing called *cardinality constrained bin packing*, also called *bin packing with cardinality constraints* (BPCC). In this problem, there is a parameter $k \geq 2$, and items of rational sizes in $[0, 1]$ are to be packed into bins, such that no bin has more than k items or total size larger than 1. The goal is to minimize the number of bins.

A recently introduced concept, called the price of clustering, deals with inputs that are presented in a way that they are split into clusters. Thus, an item has two attributes which are its size and its cluster. The goal is to measure the relation between an optimal solution that cannot combine items of different clusters into bins, and an optimal solution that can combine items of different clusters arbitrarily. Usually the number of clusters may be large, while clusters are relatively small, though not trivially small. Such problems are related to greedy bin packing algorithms, and to batched bin packing, which is similar to the price of clustering, but there is a constant number of large clusters. We analyze the price of clustering for BPCC, including the parametric case with bounded item sizes. We discuss several greedy algorithms for this problem that were not studied in the past, and comment on batched bin packing.

1 Introduction

The bin packing problem is defined as follows. Given items of rational sizes in $[0, 1]$, partition these items to subsets of total sizes not exceeding 1 so as to minimize the number of subsets. The subsets are called bins, and the process of assigning an item to a subset is called packing. In the classic or standard variant, there are no other attributes or conditions. This problem has been studied for fifty years [4,5,7,9,10,13,16,17,24,27–29,32–34].

Another well-known natural variant limits the number of items that one bin can receive. Given an integer parameter $k \geq 2$, bin packing with cardinality constraints (BPCC) is the variant of bin packing where every bin is required to have at most k items [3,6,11,12,15,18,23,25,30,31]. In this work we study BPCC with respect to new and old concepts. One well-studied concept is greedy algorithms. Such algorithms are frequently online, in the sense that they pack every item before seeing the future items. The problems were studied as offline problems, where an algorithm receives the entire input as a set, and as online

© Springer Nature Switzerland AG 2021
J. Koenemann and B. Peis (Eds.): WAOA 2021, LNCS 12982, pp. 117–129, 2021.
https://doi.org/10.1007/978-3-030-92702-8_8

problems, where an algorithm receives the input as a sequence. Here, we study not only algorithms but other concepts (defined below) as well, which measure the effect of input items being split into types already in the input.

Measures for Bin Packing. For an algorithm for a certain problem, that does not necessarily compute an optimal solution, we say that its asymptotic approximation ratio does not exceed R, if there exists a constant value $C \geq 0$ (this value has to be independent of the input), such that for any input J for the problem, the cost of this algorithm for input J is bounded from above by the following value: R times the optimal cost for J plus the constant C. The asymptotic approximation ratio is the infimum value R for which this inequality holds for every input J. If the constant C is equal to zero, the approximation ratio will be called strict or absolute. One specific optimal (offline) algorithm for the problem and its cost are usually denoted by OPT, and $OPT(J)$ is also used for the cost for a fixed input J. Another definition of the asymptotic approximation ratio is the supreme limit of the ratio between the cost of the algorithm and the cost of OPT, as a function of the second cost, where we take the maximum or supremum over the inputs with the same optimal cost. For online algorithms, we still use the term *approximation ratio*, and the definition is unchanged. The term *competitive ratio* has the same meaning in the literature. The asymptotic measures are considered to be more meaningful for bin packing problems, and thus we are mostly interested in those asymptotic measures. In the cases where we do not specify whether the measure is asymptotic or absolute, the result holds for both these measures. We will use such an analysis throughout the article.

Parametric Variants for Bin Packing Problems. It is often assumed that the size of an item may be very close to 1 or even equal to 1. There are applications where this may happen, but in other applications, bin capacities are much larger than item sizes. The parametric case is defined by an upper bound on item sizes. Specifically, there is a parameter β, where $0 < \beta \leq 1$, such that items sizes are rational numbers in $[0, \beta]$. For some versions of bin packing and some algorithms for it, the study of the parametric case is different from the general case, and it is of interest. There are models where the interesting cases are only those where $\frac{1}{\beta}$ is an integer, but there may be additional relevant values of β. The resulting asymptotic approximation ratio may or may not be close to 1 for small values of β [1,27,28].

Bin Packing with Item Types: Batched Bin Packing. An item type is an additional attribute of an item. This attribute corresponds to a fixed partition of the input into subsets called batches or clusters. The number of batches or clusters will be denoted by ℓ, where $\ell \geq 1$. Batched bin packing is a model that was initially defined as a semi-online input arrival scenario for bin packing problems. In this scenario, an algorithm receives all items of one type together. There is work on the version where the algorithm can pack items of different types into one bin (so the difficulty is the lack of knowledge on future batches) [8,14,19,26], and here we consider the variant where items of different batches are to be packed into separate bins [14,19]. Studies focus on the difference between solutions where items of different batches are packed separately (batched solutions), and globally optimal solutions that can pack items of different types into the same bin. We use the asymptotic approximation ratio to compare a batched optimal solution with a globally optimal solution. This is a comparison between optimal solutions, but sometimes it is easier to analyze greedy solutions instead. The problem was studied for several variants of bin packing [19,21], and it turns out that for $\ell = 2$, the tight bound on the asymptotic approximation ratio is 1.5 [19], and for larger values of ℓ the ratio increases and grows to approximately 1.69103 (see [32] for an early article where this value appeared in another context of bounded space online bin packing).

Bin Packing with Item Types: The Price of Clustering. This problem is similar to batched bin packing, only here ℓ may be large while sets of items of one type may be small. For batched bin packing, ℓ is often seen as a constant, and thus every batch typically contains a large number of items. A solution where different item types have separate bins is called a clustered solution, and it is compared to a globally optimal solution in this case as well. Thus, a clustered optimal solution is compared to a globally optimal solution using the approximation ratio once again, only here both the absolute approximation ratio and the asymptotic approximation ratio are studied. This approximation ratio is called *the price of clustering* (PoC) [2,20]. Azar et al. [2] who introduced this problem, write that (in order for the measure to be meaningful) clusters may be small, but they cannot be arbitrarily small. The assumption that an optimal solution for every cluster has at least two bins leads to tight bounds of 2 on the PoC [2]. This bound is tight even if items can be very small. For the cases where the optimal solution for every cluster requires at least q bins with $q \geq 3$, the approximation ratio is strictly smaller than 2 [2,20] and decreases to approximately 1.69103 as q grows [2,19]. The PoC was also studied for other variants of bin packing [21]. Here, in the study of BPCC, we use the smallest lower bound on q, i.e., we let $q = 2$. Our assumption for every cluster is that the optimal cost for it is at least 2. Due to the cardinality constraints, the total size of items can still be not larger than 1 for some clusters, and for such a cluster it will hold that the number of items is at least $q + 1$.

Notation for Bin Packing Problems with Types. Given an input, recall that ℓ is the number of clusters or batches. We let OPT_i be the number of bins in an optimal solution for the ith cluster or batch, and we denote the input (the set of items of type i) by I_i. We let I be the set of all items, that is, $I = \bigcup_{1 \leq i \leq \ell} I_i$, where $n = |I|$. We use the notation above for optimal solutions, and let OPT be a globally optimal solution for I, as well as its cost. We let A_i be the number of bins in the output of a fixed algorithm (which can be any algorithm, an optimal algorithm or another algorithm) for cluster i.

Greedy Algorithms and Related Work. The study of greedy algorithms for bin packing started together with the first studies of classic bin packing and [27,28].

Next Fit (NF) is an efficient algorithm that has (at most) one active bin where items are packed, and once an item cannot be added to the bin, it is closed and replaced with a new active bin. Any Fit (AF) algorithms pack an item into a non-empty bin when possible, and otherwise a new bin is used. A specific AF algorithm can be defined based on the choice of bin when there are several options. First Fit (FF) chooses the first (minimum index) bin, while Worst Fit (WF) chooses the least packed bin in terms of total size [27,28]. It is known that the asymptotic and absolute approximation ratio of FF is 1.7 [16,28]. For WF (and AF in general) and NF, this ratio is equal to 2 [27]. These algorithms (for classic bin packing) and online algorithms were sometimes analyzed for the parametric case as well [10,27,28,34]. The asymptotic approximation ratio of FF with $\beta \in (\frac{1}{t+1}, \frac{1}{t}]$ for an integer $t \geq 2$ is $\frac{t+1}{t}$ [28], and the case where the parameter $\beta \in (\frac{1}{t+1}, \frac{1}{t})$ is not different from the case $\beta = \frac{1}{t}$ (also for $t = 1$). For NF and WF, the asymptotic competitive ratio is 2 for $\beta > \frac{1}{2}$. while for $\beta \leq \frac{1}{2}$, the ratio is exactly $\frac{1}{1-\beta}$ [27].

Greedy algorithms can be applied to BPCC. The algorithm First Fit (FF) for this variant is defined as follows. Every item in the list is packed into the bin of the minimum index such that the bin has sufficient space for it and less than k items. Similarly, we can adapt other greedy algorithms, where the concept of the possibility of packing an item into a bin is tested with respect to the total size (which cannot exceed 1 together with the new item) and the number of items (which cannot exceed k together with the new item, that is it has to be at most $k - 1$ prior to the packing). BPCC was analyzed for greedy algorithms already in the article where the problem was introduced [31]. In that work, FF and its sorted version FFD were studied. For FF, the asymptotic approximation ratio tends to 2.7 as k grows. The exact asymptotic approximation ratio for every value of k was found much later, and it consists of several cases according to the value of k [15], where, in particular, for $k \geq 10$ the ratio is $2.7 - \frac{3}{k}$. Note that greedy algorithms do not typically have the best possible asymptotic approximation ratios for online algorithms for bin packing problems [3,5,27,28,31].

Structure of the Paper and Results. In this work we study two kinds of problems. The first one is BPCC with types, where we study the price of clustering and batched bin packing. The second one is greedy algorithms. In most cases, the analysis includes the parametric case. Specifically, in Sect. 2 we find the exact price of clustering for any k and β (and $q = 2$), we analyze batched bin packing with two and three batches, and we explain why large numbers of batches are uninteresting. In Sect. 3 we analyze NF and WF for the general case, and using sample cases for β we exhibit the difficulty of analysis for the parametric case. We also provide a simple analysis of the parametric case for FF, for which the general case was studied in the past [15,31].

Omitted proofs can be found in the full version of this paper [22].

2   The Price of Clustering and Batched Bin Packing

We start with a complete study of the PoC. Recall that we assume that for every cluster the optimal solution has at least two bins. For the analysis of upper bounds, we consider the action of greedy algorithms on the clusters, since the output of such an algorithm for each cluster has a simpler structure compared to an optimal solution for it, and the number of bins obviously cannot be smaller. We will use weight functions in the analysis. Such a function can relate two solutions as follows. We let the weight of a bin be the sum of weights of its items. The weight of an item is defined in the analysis, it may be different from its size, and the total weight of items for a bin may exceed 1. Weights may be defined differently for different cases. If in one solution the total weight of every bin is at most ρ and in another solution the total weight for every bin is at least 1 (where it is sufficient to prove this on average), then the ratio between the costs of the two solutions (for one input, for which the total weight is fixed) is at most ρ. If there is an additive term, that is, the requirement holds possibly excluding a constant number of bins, this results in an upper bound on the asymptotic approximation ratio.

Recall that in the parametric case we assume that item sizes are in $(0, \beta]$ for some $0 < \beta \leq 1$. We let $t = \lfloor \frac{1}{\beta} \rfloor$, so all items have sizes of at most $\beta \leq \frac{1}{t}$, but $\beta > \frac{1}{t+1}$. We are interested in all integers $t \geq 1$. As discussed earlier, for classic bin packing, even if all items are very small, the PoC for the case where optimal solutions for clusters have at least two bins is unchanged. Now, we will see that for BPCC the bounds become slightly smaller as β decreases. Specifically, we will find a function of both t and k such that the PoC is equal to this function. The bounds are between approximately 2 and approximately 4, for large values of k. As mentioned above, for certain bin packing problems and their analysis as a function of β (or t), it is sometimes the case that the ratio depends not only on t but it depends on the exact value of β. [21], but here were prove that the dependence is just on t, and the ratio is equal for all values β with the same value of t. In the next theorem we provide a complete analysis of the PoC for the general case and the parametric case.

Theorem 1. *The PoC for BPCC is equal to $\frac{4k-2}{k+1}$. For the parametric case, the PoC is equal to $4 - \frac{2t+4}{k+1} = \frac{4k-2t}{k+1}$ if $t \leq k$, and to $\frac{2k}{k+1}$ if $t \geq k$.*

Proof. The the proof of the theorem consists of four lemmas. The first two lemma provide a proof for the general case, and the next two extend the result for the parametric case. The first lemma is a proof of the upper bound for the general case.

Lemma 1. *For any input I for BPCC with the parameter $k \geq 2$, it holds that $\sum_{j=1}^{\ell} OPT_j \leq (4 - \frac{6}{k+1}) \cdot OPT = \frac{4k-2}{k+1} \cdot OPT$.*

The second lemma is a proof of the lower bound for the general case.

Lemma 2. *The PoC of BPCC with the parameter $k \geq 2$ is at least $4 - \frac{6}{k+1} = \frac{4k-2}{k+1}$.*

For any β such that $t = 1$, we already found a tight result of $\frac{4k-2}{k+1}$ on the PoC. The upper bound of Lemma 1 obviously holds, and the lower bound construction of Lemma 2 holds since we can use $N > \frac{1}{\beta-1/2}$, which means that items of sizes just above $\frac{1}{2}$ are chosen such that they are still smaller than β. Thus, we consider the case $t \geq 2$ in what follows, both for the upper bound and the lower bound.

Lemma 3. *For any input I for BPCC with the parameter $k \geq 2$, and any $t \geq 2$, it holds that $\sum_{j=1}^{\ell} OPT_j \leq (4 - \frac{2t+4}{k+1}) \cdot OPT = \frac{4k-2t}{k+1} \cdot OPT$ for $k \geq t$ and $\sum_{j=1}^{\ell} OPT_j \leq \frac{2k}{k+1} \cdot OPT$ if $k \leq t$.*

Lemma 4. *The PoC of BPCC with the parameter $k \geq 2$ and any $t \geq 2$ is at least $4 - \frac{2t+4}{k+1} = \frac{4k-2t}{k+1}$ if $t \leq k$, and at least $\frac{2k}{k+1}$ if $t \geq k$.*

Combining the last two lemma concludes the proof of the theorem.   □

Next, we consider batched bin packing with a small number of batches. We will analyze the case with two and three batches, as a function of k. For a large number of batches the problem becomes similar to batch bin packing for classic bin packing [14,19], though the bound is larger by an additive 1. Specifically, it is known that for BPCC, for large values of k, the ratios may grow by an additive factor of 1, since small items are sometimes treated almost independently [18]. For FF and other greedy algorithms, in the bad examples, these items can be presented first, and in the analysis, bins with k items can be considered separately [15,31]. This property of an additive 1 in the ratio is not true for online algorithms in general [3,11], where the tight asymptotic approximation ratio is 2, (so it is smaller than 1 plus the best possible asymptotic approximation ratio [7,10,34]). The situation for batched bin packing is similar to the case with an additive 1. As in the price of clustering, a bad situation is where all very small items are separated from other parts of the input into a separate batch. This adds 1 to the asymptotic approximation ratio for a large number of batches, which becomes approximately 2.69, since for classic bin packing this value is approximately 1.69 [19,32].

Theorem 2. *The asymptotic approximation ratio for batched bin packing and BPCC is $2 - \frac{1}{k}$ for two batches and $2.5 - \frac{2}{k}$ for three batches.*

3   Greedy Algorithms

We now analyze WF and NF. The interesting feature is that unlike standard bin packing, the performance of these two algorithms is different here.

Theorem 3. *The asymptotic approximation ratio of WF for BPCC is $3 - \frac{3}{k}$, and for NF it is $3 - \frac{2}{k}$, for $k \geq 2$.*

Proof. We start with the upper bounds. For NF, let $w(x) = \frac{2(k-1)}{k} \cdot x + \frac{1}{k}$. A bin of an optimal solution has at most k items with total size at most 1, and the total weight is at most $\frac{2(k-1)}{k} + k \cdot \frac{1}{k} = \frac{3k-2}{k}$. For NF, let X be the number of bins with k items, and let Y be the number of bins whose number of items is at least 1 and at most $k - 1$. The total number of bins is $X + Y$, and we show that the total weight is at least $X + Y - 1$. For every bin with at most $k - 1$ items that is not the last bin, the sum of its load and the load of the following bin is above 1. The sum of all these values (total loads of pairs of consecutive bins) is above $Y - 1$ for all such bins, but it is possible that some bins were considered both as a bin with at most $k - 1$ items and as a bin following such a bin (if two consecutive bins have at most $k - 1$ items, and the second one is not the last bin), and therefore, since the total size for every bin was considered at most twice, the total size of items is above $\frac{Y-1}{2}$. The number of items is at least $k \cdot X + Y$, and thus the total weight is at least $\frac{k-1}{k} \cdot (Y - 1) + \frac{kX+Y}{k} > X + Y - 1$.

For WF, let $w(x) = \frac{2(k-2)}{k} \cdot x + \frac{1}{k}$ if $x \leq \frac{1}{2}$ and otherwise $w(x) = \frac{2(k-2)}{k} \cdot x + \frac{2}{k}$. That is, an item of size above $\frac{1}{2}$ has an extra addition of $\frac{1}{k}$ to its weight. In this case the weight of at most k items of total size at most 1 does not exceed $\frac{2(k-2)}{k} + \frac{k+1}{k} = \frac{3k-3}{k}$, since at most one item has size above $\frac{1}{2}$. For NF, we use similar notation, and let X, Y, and Z denote the numbers of bins with k items, bins with at least two items and at most $k-1$ items, and bins with a single item, respectively. The number of items is at least $k \cdot X + 2 \cdot Y + Z$. Due to the action of WF, the number of items of sizes above $\frac{1}{2}$ is at least $Z - 1$, since no two items packed alone into bins by WF can be packed together, while every two items of sizes at most $\frac{1}{2}$ could have been packed together in terms of their sizes (without additional items). The argument regarding the total size is as before, and given the modified notation, the total size of items is above $\frac{Y+Z-1}{2}$. Thus, the total weight is at least

$$\frac{Z-1}{k} + \frac{kX + 2Y + Z}{k} + \frac{k-2}{k} \cdot (Y + Z - 1) > X + Y + Z - 1 .$$

For the lower bound of NF, let $N > 0$ be a large integer, which is divisible by k, and let $\varepsilon > 0$ be a small value such that $\varepsilon < \frac{1}{10N}$. There are $N - 1$ large items of sizes in $(\frac{1}{2}, \frac{1}{2} + N\varepsilon]$ (this interval is contained in $(0.5, 0.6)$), $N - 2$ medium items of sizes in $(\frac{1}{2} - N\varepsilon, \frac{1}{2})$ (this interval is contained in $(0.4, 0.5)$), and there are also $N(k-2)$ small items, each of size $\frac{\varepsilon}{k}$. The large items have sizes of $\frac{1}{2} + q \cdot \varepsilon$ for $2 \leq q \leq N$. The medium items have sizes of $\frac{1}{2} - q \cdot \varepsilon$ for $1 \leq q \leq N - 2$.

A possible solution has N bins, where every bin has $k - 2$ small items, and one or two other items, where the total size of other items is at most $1 - \varepsilon$, which are defined below. Out of these bins, there are $N - 4$ bins, where the ith bin has one item of size $\frac{1}{2} + (i + 1) \cdot \varepsilon$ and one item of size $\frac{1}{2} - (i + 2)\varepsilon$. The total size for these two items is $1 - \varepsilon$, and this leaves five unpacked items that are large or medium, with sizes $\frac{1}{2} + (N - 2)\varepsilon$, $\frac{1}{2} + (N - 1)\varepsilon$, $\frac{1}{2} + N\varepsilon$, $\frac{1}{2} - 2\varepsilon$, and $\frac{1}{2} - \varepsilon$. The first three items are packed into separate bins, and the two last items are packed together.

NF receives the items in the following order. First, all small items arrive, and they are packed into $\frac{N(k-2)}{k}$ bins that receive k items each and will not receive other items. Large and medium items arrive such that the input alternates between them, starting and ending this part of the input with large items. The large items are sorted by decreasing size and the medium items are sorted by increasing size. We get that for this part of the input (after the small items) for $i = 1, 2, \ldots, N - 2$, items of indices $2i - 1$ and $2i$ have sizes of $\frac{1}{2} + (N + 1 - i)\varepsilon$ and $\frac{1}{2} - (N - 1 - i)\varepsilon$, respectively, and the item of index $2N - 3$ has size $\frac{1}{2} + 2\varepsilon$ (which corresponds to the case $i = N - 1$, but there is no item of index $2N - 2$). The first item is packed into a new bin since the previous bin has k items. We show that every pair of items have a total size above 1, and thus every medium or large item is packed into its own bin. If this holds, the number of bins is $\frac{N(k-2)}{k} + 2N - 3 = N(3 - \frac{3}{N} - \frac{2}{k})$, which implies the lower bound by letting N grow without bound. An item in an even index in the second part of the input has size $\frac{1}{2} - (N - 1 - i)\varepsilon$, where the item before it has size $\frac{1}{2} + (N + 1 - i)\varepsilon$ and the item after it has size $\frac{1}{2} + (N - i)\varepsilon$. The total sizes are $1 + 2\varepsilon$ and $1 + \varepsilon$, respectively. Note that this construction is valid for $k = 2$, but in that case there are no small items.

The lower bound for $k = 2$ and WF is proved separately. Assume that $k \geq 3$. For a lower bound of WF, let $N > 0$ be a large integer, which is divisible by k, let $\varepsilon > 0$ be a small value such that $\varepsilon < \frac{\varepsilon}{10N}$, and let $\delta = \frac{\varepsilon}{2^{N+4}}$. The input consists of $N - 1$ huge items, each of size $\frac{1}{2} + \delta$, $N(k - 3)$ small items of size $\frac{\delta}{k}$, N large items, where the ith item (for $i = 1, 2, \ldots, N - 4$) has size $\frac{1}{2} - \frac{\varepsilon}{2^i}$, and N medium items, where the ith medium item has size $\frac{5}{3} \cdot \frac{\varepsilon}{2^i}$. A possible solution has N bins with at most three items that are not small packed into every bin, where their total size is at most $1 - \delta$, and the bin also has $k - 3$ small items. For $i = 1, 2, \ldots, N - 1$, there is a bin with one huge item, one large item of size $\frac{1}{2} - \frac{\varepsilon}{2^i}$ and one medium item of size $\frac{5}{3} \cdot \frac{\varepsilon}{2^{i+1}}$, where the total size of these three items is

$$\left(\frac{1}{2} + \delta\right) + \left(\frac{1}{2} - \frac{\varepsilon}{2^i}\right) + \left(\frac{5}{3} \cdot \frac{\varepsilon}{2^{i+1}}\right) = 1 + \delta - \frac{\varepsilon}{6 \cdot 2^i} \leq 1 + \delta - \frac{\varepsilon}{6 \cdot 2^N} < 1 - \delta,$$

since $\frac{\varepsilon}{6 \cdot 2^N} > \frac{\varepsilon}{2^{N+3}} = 2\delta$. The two remaining items that are not small have sizes of $\frac{1}{2} - \frac{\varepsilon}{2^N}$ and $\frac{5\varepsilon}{6}$, and they are packed together (with $k - 3$ small items).

For WF, the small items are presented first, and they are packed into $\frac{N(k-3)}{k}$ bins that cannot be reused. Then, pairs of a large item of size $\frac{1}{2} - \frac{\varepsilon}{2^i}$ and one medium item of size $\frac{5}{3} \cdot \frac{\varepsilon}{2^i}$, are presented for $i = 1, 2, \ldots, N$. The total size of such a pair of items is $\frac{1}{2} - \frac{\varepsilon}{2^i}$ and one medium item of size $\frac{5}{3} \cdot \frac{\varepsilon}{2^i}$ is $\frac{1}{2} + \frac{\varepsilon}{3 \cdot 2^{i-1}}$. The first pair is packed into a bin since all previous bins already have k items each. Every time that an item of size $\frac{1}{2} - \frac{\varepsilon}{2^{i+1}}$ is presented, it cannot be packed into previous bins, since the minimum load of bins with less than k items is $\frac{1}{2} + \frac{\varepsilon}{3 \cdot 2^{i-1}}$, and $\left(\frac{1}{2} - \frac{\varepsilon}{2^{i+1}}\right) + \left(\frac{1}{2} + \frac{\varepsilon}{3 \cdot 2^{i-1}}\right) = \left(\frac{1}{2} - \frac{\varepsilon}{4 \cdot 2^{i-1}}\right) + \left(\frac{1}{2} + \frac{\varepsilon}{3 \cdot 2^{i-1}}\right) > 1$. When the medium item of size $\frac{5}{3} \cdot \frac{\varepsilon}{2^{i+1}}$ arrives, the last bin has load below $\frac{1}{2}$ while other bins that can receive it have loads above $\frac{1}{2}$, and it is combined into the last bin, creating a load of $\frac{1}{2} + \frac{\varepsilon}{3 \cdot 2^i}$. Finally, all huge items are presented, and each one is packed into a new bin because previous bins either have k items or have loads above $\frac{1}{2}$. Thus, WF has $\frac{N(k-3)}{k} + N + (N - 1)$ bins, and the lower bound on the approximation ratio is $\frac{3k-3}{k}$ by letting N grow without bound.

The lower bound for the case $k = 2$ for WF follows from very simple inputs with $2N$ items of size 0.4 followed by $2N$ items of size 0.6. WF creates pairs of items of size 0.4 and the other items are packed into separate bins, while an optimal solution has N bins with one item of each size. ☐

Next, we perform an analysis for several closely related special cases of β, for NF and WF. The goal is to show the difficulty of solving the general case for different values of β. We start with the case $\beta = 0.4$.

Theorem 4. *The asymptotic approximation ratio for WF and NF for BPCC, $k \geq 4$, and $\beta = 0.4$ is $\frac{8}{3} - \frac{10}{3k}$.*

Proof. For the upper bound, we define the following weight function $w(x) = \frac{1}{k} + \frac{5(k-2)}{3k} \cdot x$. For any bin of an optimal solution, since there are at most k items whose total size is at most 1, we get a total weight of at most $k \cdot \frac{1}{k} + \frac{5(k-2)}{3k} = \frac{8}{3} - \frac{10}{3k}$.

Consider the output of WF or NF. Since no item has size above β, every bin that is not the last bin and has at most $k - 1$ items has total size above $1 - \beta = 0.6$. Since $\beta < 0.5$, every such bin has at least two items. The weight of every bin with k items is at least $k \cdot \frac{1}{k} = 1$. For other bins, the weight of every bin except for possibly the last one is at least $2 \cdot \frac{1}{k} + 0.6 \cdot \frac{5(k-2)}{3k} = 1$.

To prove a lower bound, we define a set of bins which act as an offline solution, and additionally define an ordering of the items for NF and WF. Let N, M be large positive integers, where N is divisible by k, and let $\varepsilon > 0$ be a small value such that $3^{2M} \cdot \varepsilon < \frac{1}{100}$. Every bin has $k - 4$ or $k - 3$ items of size zero, depending on the number of other items. The larger items defined next.

The offline packing is as follows. There are $3^0 \cdot N = N$ bins containing three items of size $0.2 + 3^{2M} \cdot \varepsilon$, whose total size is below 0.63. For $i = 1, 2, \ldots, M - 1$, there are $3^i \cdot N$ bins containing three items of size $0.2 + 3^{2(M-i)} \cdot \varepsilon < 0.21$ and one item of size $0.4 - 3^{2M-2i+1} \cdot \varepsilon > 0.39$, where the total size of these items is $3(0.2 + 3^{2(M-i)} \cdot \varepsilon) + (0.4 - 3^{2M-2i+1} \cdot \varepsilon) = 1$. Finally, there are $3^M \cdot N$ bins with one item of size $0.2 + \varepsilon < 0.21$, one item of size $0.4 - \frac{\varepsilon}{3}$, and one item of size $0.4 - 3\varepsilon$, whose total size is below 1. The total number of bins in this solution is $\Delta = N \cdot \sum_{i=0}^{M} 3^i = N \cdot 3^M \cdot \sum_{i=0}^{M} 3^{-i} < 1.5 \cdot N \cdot 3^M$, using the sum of an infinite geometric series. The bins that have $k - 3$ items of size zero are the first N bins and the 3^M bins with two items of sizes close to 0.4, and all other bins have $k - 4$ items of size zero. Thus, the number of items of size zero is $(k - 3) \cdot \Delta + M \cdot (3^0 + 3^M)$.

We define the order in which NF and WF receive the items. First, all items of size zero arrive, every bin will have k such items, and the number of such bins is $\frac{\Delta(k-4)+N\cdot(3^M+1)}{k}$, which is an integer since N is divisible by k and Δ is divisible by N. Every item of slightly smaller than $\beta = 0.4$ is followed by an item of size slightly smaller than 0.2. Specifically, the items of sizes approximately β are presented ordered by non-decreasing size, and every item of size $0.4 - 3^{2j-1} \cdot \varepsilon$ (the number of such items is $N \cdot 3^{M-j+1}$ for $1 \leq j \leq M$ and $N \cdot 3^M$ for $j = 0$) is followed by an item of size $0.2 + 3^{2j} \cdot \varepsilon$, for $j = M, M - 1, \ldots, 1, 0$ (the number of such items is also $N \cdot 3^{M-j+1}$ for $1 \leq j \leq M$ and $N \cdot 3^M$ for $j = 0$). We claim that both algorithms pack one item of size approximately β and one item of size approximately 0.2 (two items that arrive consecutively) into every bin. The number of additional bins is therefore

$$N \cdot (3^M + \sum_{j=1}^{M} 3^{M-j+1}) = N \cdot 3^M \cdot (1 + \sum_{j=1}^{M} 3^{1-j}) = N \cdot 3^M \cdot (1 - \frac{1}{3^M} + \sum_{j=0}^{M} 3^{-j})$$

$$= \Delta + N \cdot 3^M (1 - \frac{1}{3^M}) \ .$$

The ratio between the number of bins for WF and NF and Δ is

$$\frac{k-4}{k} + \frac{N(3^M + 1)}{k\Delta} + 1 + \frac{N \cdot 3^M (1 - \frac{1}{3^M})}{\Delta} \ .$$

By using $\Delta < 1.5 \cdot N \cdot 3^M$, we get a ratio of at least $2 - \frac{4}{k} + \frac{2(1+\frac{1}{3^M})}{3k} + \frac{2(1-\frac{1}{3^M})}{3}$. Letting M grow to infinity, we get a lower bound of $2 - \frac{4}{k} + \frac{2}{3k} + \frac{2}{3} = \frac{8}{3} - \frac{10}{3k}$. It is left to justify the calculation in the sense that the packing of NF and WF is as described. The bins with zero size items cannot receive any other items. We total sizes of pairs of items with index j for which we claim that they are packed together are $0.6 + 2 \cdot 3^{2j-1} \cdot \varepsilon$. All further items of sizes close to 0.4 that arrive later have sizes of at least $0.4 - 3^{2j-1} \cdot \varepsilon$, and are too large to be packed into earlier bins. For items of sizes close to 0.2, NF will pack every such item into the active bin and not into an earlier bin. For WF, all earlier bins will have loads above 0.6 while the currently last bin has a smaller load, and the item of size close to 0.2 will be packed there.   □

To illustrate the difficulty of analysis for parametric cases further, we focus on WF. We will prove that the same bound can be proved for WF for the case where $\beta \in (0.4, \frac{5}{12} \approx 0.41666]$ and $4 \leq k \leq 5$, but for $k \geq 6$, we now show that the bound is higher for example for $\beta = 0.41$. Consider an input with N bins (for a large integer N divisible by k), where there are $(13k - 45)N$ items of size zero, $11N$ items of size 0.41, $11N$ items of size 0.4099, $11N$ items of size 0.1801, $10N$ items of size 0.1804, and $2N$ items of size 0.0902. An optimal solution has $13N$ bins, where the first $11N$ bins have (each) one item of each size out of 0.41, 0.4099, and 0.1801, and $k - 3$ items of size zero, and the other $2N$ bins have (each) five items of size 0.1804, one item of size 0.0902, and $k - 6$ items of size zero. The input starts with items of size zero, and both algorithms have $\frac{(13k-45)N}{k}$ bins with k items of size zero. Then items of size 0.4099 alternate with items of size 0.1804 or pairs of items of sizes 0.0902, and then items of size 0.41 alternate with items of size 0.1801. In this process, similar to the construction for $\beta = 0.4$, an additional set of $22N$ bins is created. Thus, the ratio is at least $\frac{35-45/k}{13} = \frac{35}{13} - \frac{45}{13k}$. This value is indeed strictly larger than $\frac{8}{3} - \frac{10}{3k}$ for $k \geq 6$ (for example, in the case $k = 6$ we get above 2.115 rather than approximately 2.111.

Theorem 5. *The asymptotic approximation ratio for WF for BPCC and $\beta \in (0.4, \frac{5}{12} \approx 0.41666]$ is $\frac{11}{6}$ for $k = 4$ and 2 for $k = 5$.*

Proof. Let $w(x) = \begin{cases} \frac{1}{k} + \frac{2(k-2)}{3k} & \text{for} \quad x \in (\frac{1}{3}, \beta] \\ \frac{1}{k} + \frac{k-2}{3k} & \text{for} \quad x \in (\frac{1}{6}, \frac{1}{3}] \\ \frac{1}{k} & \text{for} \quad x \in (0, \frac{1}{6}] \end{cases}$ be a weight function,

which we will use in the analysis.

Any bin of an optimal solution has at most k items, where there are at most five items whose sizes are above $\frac{1}{3}$. Moreover, every item of size above $\frac{1}{3}$ can be counted as two items of sizes above $\frac{1}{6}$, and therefore the total weight is at most $k \cdot \frac{1}{k} + 5 \cdot \frac{k-2}{3k} = \frac{8}{3} - \frac{10}{3k}$, which is equal to the claimed bounds for $k = 4, 5$.

It is left to show that almost every bin of WF has weight of at least 1. This holds for bins with k items. All other bins that are not the last bin have total sizes above $1 - \beta \geq \frac{7}{12}$, and thus every bin has at least two items, since $\beta \leq \frac{5}{12}$. We claim that except for at most two such bins, every bin with at most $k - 1$ items either has a total size of items above $\frac{2}{3}$, or the bin has at least one item of size above $\frac{1}{3}$ (or both). Note that if a bin has total size above $\frac{2}{3}$ and it has exactly two items, at least one of them has size above $\frac{1}{3}$, and if it has at least four items, at least one of them has size above $\frac{1}{6}$. For that, consider the first bin with total size at most $\frac{2}{3}$ and at most $k - 1$ items. If this is the last bin or there is no such bin, we are done. Otherwise, for every bin of WF opened later, the first item has size above $\frac{1}{3}$.

Consider a bin of WF that has at least two and at most $k - 1$ items, and it is not the last one, and it is not a bin with total size at most $\frac{2}{3}$ and no item of size above $\frac{1}{3}$. Assume that the bin has exactly two items. If the bin has total size above $\frac{2}{3}$, then one of the items has size above $\frac{1}{3}$. Otherwise, the bin also has such an item. The size of the second item is above $(1 - \beta) - \beta \geq \frac{1}{6}$. Thus, the total weight is at least $2 \cdot \frac{1}{k} + \frac{3(k-2)}{3k} = 1$. If the bin has four items, since $4 \leq k-1$, this situation occurs only for $k = 5$, and we show that at least one item has size above $\frac{1}{6}$. Indeed, it either has an item with size above $\frac{1}{3}$, or otherwise the total size of $k-1 = 4$ items above $\frac{2}{3}$, in which case the largest one has size above $\frac{1}{6}$. For $k = 5$ and a bin with four items, the total weight is $(k-1) \cdot \frac{1}{k} + \frac{k-2}{3k} = \frac{3k-3+k-2}{3k} = 1$. In the case of a bin with three items, we claim that out of the three items, there is at least one of size above $\frac{1}{3}$ or else there are two items of sizes above $\frac{1}{6}$. Indeed, if the total size is above $\frac{2}{3}$, the largest item has size no larger than $\frac{1}{3}$ and the two other items have sizes at most $\frac{1}{6}$, we reach a contradiction. Thus, for three items, the total weight is at least $3 \cdot \frac{1}{k} + \frac{2(k-2)}{3k} = \frac{9+2k-4}{3k} \geq 1$ for $k = 4, 5$. □

For FF, we are able to find tight bounds for any β. Recall that FF was already fully analyzed for BPCC with all values of k [15,31]. The lower bounds hold for any $\beta \in (\frac{1}{2}, 1]$, and therefore we consider the case $\beta \leq \frac{1}{2}$. Assume that $\beta \in (\frac{1}{t+1}, \frac{1}{t}]$ for an integer $t \geq 2$. The case where $k \leq t$ is trivial since $k \cdot \beta \leq \frac{k}{t} \leq 1$, which means that FF packs k items into every bin and the packing is optimal.

Theorem 6. *The asymptotic approximation ratio of FF for BPCC and $\beta \in (\frac{1}{t+1}, \frac{1}{t}]$ for an integer $t \geq 2$ is $1 + \frac{(k-t)(t+1)}{kt}$ for $k \geq t$.*

References

1. Azar, Y., Cohen, I.R., Fiat, A., Roytman, A.: Packing small vectors. In: Proceedings of the 27th Annual ACM-SIAM Symposium on Discrete Algorithms, (SODA2016), pp. 1511–1525 (2016)
2. Azar, Y., Emek, Y., van Stee, R., Vainstein, D.: The price of clustering in bin-packing with applications to bin-packing with delays. In: The 31st ACM on Symposium on Parallelism in Algorithms and Architectures (SPAA2019), pp. 1–10 (2019)
3. Babel, L., Chen, B., Kellerer, H., Kotov, V.: Algorithms for on-line bin-packing problems with cardinality constraints. Discrete Appl. Math. **143**(1–3), 238–251 (2004)
4. Baker, B.S., Coffman, E.G., Jr.: A tight asymptotic bound for next-fit-decreasing bin-packing. SIAM J. Algebraic Discrete Methods **2**(2), 147–152 (1981)
5. Balogh, J., Békési, J., Dósa, G., Epstein, L., Levin, A.: A new and improved algorithm for online bin packing. In: Proceedings of the 26th European Symposium on Algorithms (ESA2018), pp. 5:1–5:14 (2018)
6. Balogh, J., Békési, J., Dósa, G., Epstein, L., Levin, A.: Online bin packing with cardinality constraints resolved. J. Comput. Syst. Sci. **112**, 34–49 (2020)
7. Balogh, J., Békési, J., Dósa, G., Epstein, L., Levin, A.: A new lower bound for classic online bin packing. Algorithmica **83**(7), 2047–2062 (2021)
8. Balogh, J., Békési, J., Dósa, G., Galambos, G., Tan, Z.: Lower bound for 3-batched bin packing. Discrete Optim. **21**, 14–24 (2016)
9. Balogh, J., Békési, J., Dósa, G., Sgall, J., van Stee, R.: The optimal absolute ratio for online bin packing. J. Comput. Syst. Sci. **102**, 1–17 (2019)
10. Balogh, J., Békési, J., Galambos, G.: New lower bounds for certain classes of bin packing algorithms. Theoret. Comput. Sci. **440**, 1–13 (2012)
11. Békési, J., Dósa, G., Epstein, L.: Bounds for online bin packing with cardinality constraints. Inf. Comput. **249**, 190–204 (2016)
12. Caprara, A., Kellerer, H., Pferschy, U.: Approximation schemes for ordered vector packing problems. Naval Res. Logistics **92**, 58–69 (2003)
13. Dósa, G.: The tight absolute bound of First Fit in the parameterized case. Theor. Comput. Sci. **596**, 149–154 (2015)
14. Dosa, G.: Batched bin packing revisited. J. Scheduling **20**(2), 199–209 (2015). https://doi.org/10.1007/s10951-015-0431-3
15. Dósa, G., Epstein, L.: The tight asymptotic approximation ratio of First Fit for bin packing with cardinality constraints. J. Comput. Syst. Sci. **96**, 33–49 (2018)
16. Dósa, G., Sgall, J.: First Fit bin packing: a tight analysis. In: Proceedings of the 30th International Symposium on Theoretical Aspects of Computer Science (STACS2013), pp. 538–549 (2013)
17. Dósa, G., Li, R., Han, X., Tuza, Z.: Tight absolute bound for first fit decreasing bin-packing: $FFD(L) \leq 11/9OPT(L) + 6/9$. Theoret. Comput. Sci. **510**, 13–61 (2013)
18. Epstein, L.: Online bin packing with cardinality constraints. SIAM J. Discrete Math. **20**(4), 1015–1030 (2006)
19. Epstein, L.: More on batched bin packing. Oper. Res. Lett. **44**(2), 273–277 (2016)
20. Epstein, L.: On bin packing with clustering and bin packing with delays. Discrete Optim. **41**, 100647 (2021)
21. Epstein, L.: Open-end bin packing: new and old analysis approaches. CoRR, abs/2105.05923 (2021)

22. Epstein, L.: Several methods of analysis for cardinality constrained bin packing. CoRR, abs/2107.08725 (2021)
23. Epstein, L., Levin, A.: AFPTAS results for common variants of bin packing: a new method for handling the small items. SIAM J. Optim. **20**(6), 3121–3145 (2010)
24. Fernandez de la Vega, W., Lueker, G.S.: Bin packing can be solved within $1 + \varepsilon$ in linear time. Combinatorica **1**(4), 349–355 (1981)
25. Fujiwara, H., Kobayashi, K.: Improved lower bounds for the online bin packing problem with cardinality constraints. J. Comb. Optim. **29**(1), 67–87 (2013). https://doi.org/10.1007/s10878-013-9679-8
26. Gutin, G., Jensen, T., Yeo, A.: Batched bin packing. Discrete Optim. **2**(1), 71–82 (2005)
27. Johnson, D.S.: Fast algorithms for bin packing. J. Comput. Syst. Sci. **8**, 272–314 (1974)
28. Johnson, D.S., Demers, A., Ullman, J.D., Garey, M.R., Graham, R.L.: Worst-case performance bounds for simple one-dimensional packing algorithms. SIAM J. Comput. **3**, 256–278 (1974)
29. Karmarkar, N., Karp, R.M.: An efficient approximation scheme for the one-dimensional bin-packing problem. In: Proceedings of the 23rd Annual Symposium on Foundations of Computer Science (FOCS1982), pp. 312–320 (1982)
30. Kellerer, H., Pferschy, U.: Cardinality constrained bin-packing problems. Ann. Oper. Res. **92**, 335–348 (1999)
31. Krause, K.L., Shen, V.Y., Schwetman, H.D.: Analysis of several task-scheduling algorithms for a model of multiprogramming computer systems. J. ACM **22**(4), 522–550 (1975)
32. Lee, C.C., Lee, D.T.: A simple online bin packing algorithm. J. ACM **32**(3), 562–572 (1985)
33. Ramanan, P., Brown, D.J., Lee, C.C., Lee, D.T.: Online bin packing in linear time. J. Algorithms **10**, 305–326 (1989)
34. van Vliet, A.: An improved lower bound for online bin packing algorithms. Inf. Process. Lett. **43**(5), 277–284 (1992)

Weighted Completion Time Minimization for Capacitated Parallel Machines

Ilan Reuven Cohen[1]([⊠]), Izack Cohen[1]([⊠]), and Iyar Zaks[2]([⊠])

[1] Faculty of Engineering, Bar-Ilan University, Ramat Gan, Israel
{ilan-reuven.cohen,izack.cohen}@biu.ac.il
[2] Faculty of Industrial Engineering and Management, The Technion - IIT,
Haifa, Israel

Abstract. We consider the weighted completion time minimization problem for capacitated parallel machines, which is a fundamental problem in modern cloud computing environments. We study settings in which the processed jobs may have varying duration, resource requirements and importance (weight). Each server (machine) can process multiple concurrent jobs up to its capacity. Due to the problem's $\mathcal{NP}$-hardness, we study heuristic approaches with provable approximation guarantees. We first analyze an algorithm that prioritizes the jobs with the smallest volume-by-weight ratio. We bound its approximation ratio with a decreasing function of the ratio between the highest resource demand of any job to the server's capacity. Then, we use the algorithm for scheduling jobs with resource demands equal to or smaller than 0.5 of the server's capacity in conjunction with the classic weighted shortest processing time algorithm for jobs with resource demands higher than 0.5. We thus create a hybrid, constant approximation algorithm for two or more machines. We also develop a constant approximation algorithm for the case with a single machine. This research is the first, to the best of our knowledge, to propose a polynomial-time algorithm with a constant approximation ratio for minimizing the weighted sum of job completion times for capacitated parallel machines.

1 Introduction

In this work, we study capacitated machine scheduling problems. These problems were initially encountered in production settings where jobs are processed in batches (e.g., scheduling jobs for heat treatment ovens and wafer fabrication processes [5]). Recently, interest in these problems has increased because of the search for solutions that can be used in modeling modern cloud computing environments [7]. Differently than most scheduling models in which a resource serves a single job at any given time [2,4,9,23], in modern cloud computing environments, multiple jobs can run concurrently on the same server subject to its capacity constraints (e.g., memory, cores, bandwidth). This fact can be seen in the following three examples of well-known cloud computing platforms in which resources are simultaneously shared by multiple jobs and clients: Amazon Web

© Springer Nature Switzerland AG 2021
J. Koenemann and B. Peis (Eds.): WAOA 2021, LNCS 12982, pp. 130–143, 2021.
https://doi.org/10.1007/978-3-030-92702-8_9

Services (AWS), Microsoft Azure and Google Cloud Platform (GCP). These platforms leverage virtualization technologies, such as VMware products and Xen, to allow each physical machine to be shared by multiple jobs. Virtualization also helps in reducing the costs of maintenance, operation and provisioning [20].

Managers of cloud computing environments who wish to increase the utilization of their data centers typically resort to improving the scheduling algorithms that allocate jobs to machines. In this context, we focus on minimizing the weighted sum of job completion times, which is one of most common objective functions [7]. The weights imply that some jobs may be more important than others; thus, the scheduler has to take into consideration that the delay of one job can incur a higher "cost" than the delay of another.

The non-capacitated counterparts of the considered problem have been widely researched. One of the most known results is that the Shortest Processing Time (SPT) priority rule minimizes the (non-weighted) sum of job completion times [23]. The SPT was extended to the Weighted Shortest Processing Time (WSPT), which was used in the weighted version of the completion time minimization problem. Since the latter problem is $\mathcal{NP}$-complete for more than two machines [8], various solution approaches focus on developing polynomial-time heuristic approximation algorithms that bound the worst case performance with respect to an optimal solution. For concreteness, a desired algorithm A for a minimization problem P would have a ρ approximation ratio such that for any instance I of P, $A(I) \leq \rho \cdot OPT(I)$, where $OPT(I)$ is the value of an optimal solution for I. At [6] they proved that scheduling according to the WSPT priority rule provides a constant approximation algorithm. [16] improved the WSPT approximation ratio (ρ) to $(1 + \sqrt{2})/2$ and proved that it is tight.

In the capacitated setting, [12] was the first to present a constant approximation algorithm for minimizing the *non-weighted* sum of completion times using the Smallest Volume First (SVF) priority rule combined with the SPT priority rule. Our work extends theirs into the more general, weighted case, by combining Weighted Smallest Volume First (WSVF) and WSPT with the objective of developing the first algorithm with a constant approximation ratio for the weighted completion time problem in the capacitated setting. Our analysis, moreover, improves their approximation ratio for the non-weighted case.

1.1 Contributions and Techniques

We develop a constant approximation ratio algorithm for minimizing the weighted sum of job completion times on capacitated machines. Here are the main contributions:

1. A polynomial-time scheduling algorithm with a $\left(1 + \frac{1}{1-\alpha}\right)$-approximation ratio, if the ratio between jobs' demands and the servers' capacities is at most α.

2. A polynomial-time scheduling algorithm with a $4 + o(1/M)$ approximation ratio for $M \geq 2$ machines.
3. A polynomial-time scheduling algorithm with a $12 + \epsilon$ approximation ratio for a single machine ($M = 1$).

To the best of our knowledge, our result is the first constant approximation algorithm for the weighted completion time minimization problem for the capacitated parallel machine problem. In addition, we improve the approximation guarantees for the non-weighted version. In [12], the authors proved a $\left(\frac{3\alpha}{1-\alpha} + 3 \right)$ approximation ratio, for the case when the ratio of jobs' demands and servers' capacities is at most α, and a $(5 + o(1/M))$ approximation ratio for $M > 1$. The new way we analyse the algorithm for $M > 1$ and the use of lemmas from [16] enable us to prove the approximation ratio for the weighted case and improve the ratio presented at [12] for the unweighted case. Our algorithm for $M > 1$ partitions the jobs into high- and low-resource demand classes where jobs within the former class require 50% or more of a machine's capacity and jobs in the latter class require less than 50%. Accordingly, the algorithm partitions the machines into two groups for processing the two job classes. The high-demand class is scheduled via the WSPT and the low-demand class via a WSVF priority rule. In the WSVF method, jobs are ordered in a non-decreasing order of the ratio between their processing time multiplied by the demand to the weight. Then, the jobs are assigned to a machine according to their priority, with the algorithm assigning the next unscheduled job to the earliest possible time t. Our proof involves analyzing the WSVF performance for instances where job resource demands are smaller than a constant α.

In the analysis of the WSVF, we bound the start time of each job, provided that prior to the start time, all the machines processed at least $1 - \alpha$ demand of higher priority jobs. Then, we bound the optimal cost, by the optimal cost of a non-capacitated converted instance. Finally, we develop an improved bound by using the characterization of [6] for the non-capacitated setting. For the single machine case, we can not use the partition base approach so we extend the algorithm from [12] to the weighted case (the full details for this case are presented in the appendix).

1.2 Prior Work

There is a vast amount of research about machine scheduling problems owing to their theoretical and practical importance (interested readers are referred to the following reviews [18, 21]. For the sake of brevity, we focus on recent capacitated machine scheduling studies.

Researchers explored several objective functions. One very popular objective is to minimize the processing makespan—that is, the completion time of the last job (e.g., [3, 13, 22] and [14]). Others suggested that in settings with a release time and deadline for each job, a reasonable objective is to maximize the total weight of jobs completed before their deadline (see [1] and [10]). Another line of research models the capacitated machine scheduling problem as an online problem in which jobs arrive over time [17]. In this line of research, typical objective functions are to minimize the response time (i.e., the time elapsed from the job's arrival until it is scheduled) and to maximize the throughput [11].

Three works that considered the weighted sum of flow-times or completion times are [7,12] and [19]. At [7] the researchers considered an online problem of weighted flow time minimization, assuming that jobs can be preempted with no penalty and delay. It is important to note, though, that preemption may incur significant switching costs (e.g., setup costs) and memory loss; preemption may be also forbidden due to system restrictions or client commitments.

[19], who also studied an online capacitated machine scheduling problem, assumed that a job can run at a slower rate when receiving a fraction of its demand or that a job can be processed in parallel on different machines. They then used Online Convex Optimization (OCO) to solve the scheduling optimization problem. These assumptions may hold in specialized computing environments but in standard environments it may be costly or technically infeasible to split a job between machines or to process it at a slower rate using a portion of the required resources.

[12] proposed a constant approximation algorithm for minimizing the sum of completion times. They, however, did not consider the more general weighted version of the problem. We close this gap by developing an approximation algorithm that solves the weighted version of the problem and improves the approximation ratio for the non-weighted version that was presented in [12].

2 Formal Problem Definition

We consider N jobs that need to be processed by M identical machines. Each job j has a processing time p_j, demand d_j and weight w_j, which are known in advance. We assume, without loss of generality (hereafter, w.l.o.g.), that $p_j \geq 1$, and $d_j \in (0,1]$ is a fraction of the required demand with respect to a machine's capacity. We denote $v_j = p_j \cdot d_j$ as job j's volume. We focus on a non-preemptive schedule, meaning that a started job is processed without interruption until its completion. The scheduler assigns each job to a machine and determines its start time, s_j; accordingly, the completion time of the job is $c_j = s_j + p_j$. We mark T as the upper bound on the maximal completion time of a job (details about how to set it appeared later in the article), For convenience We mark $\mathcal{N} = \{1, 2, ..., N\}, \mathcal{M} = \{1, 2, ..., M\}, \mathcal{T} = \{0, 1, ..., T\}$.

Let $G^i(t)$ be the set of jobs processed by machine i at time $t \in \mathcal{T}$. $j \in G_i(t)$ if job j is assigned to machine i and $t \in [s_j, c_j)$. A feasible schedule must ensure that the total demand of the jobs assigned to a machine does not exceed its capacity, at any given time. Mathematically,

$$\sum_{j \in G^i(t)} d_j \leq 1 \quad \forall i \in \mathcal{M}, t \in \mathcal{T}.$$

Our goal is to find a feasible solution that minimizes the weighted sum of completion times:

$$\min \sum_{j \in \mathcal{N}} w_j \cdot c_j.$$

As mentioned, the problem is $\mathcal{NP}$-complete. Accordingly, we are looking for a polynomial-time scheduling algorithm with a guarantee on the maximal ratio between the objective function value achieved by the algorithm and the optimal solution value. We want the developed approximation algorithm to provide a constant ratio. As discussed next, we base our algorithm on the WSVF priority rule.

3   WSVF Algorithm and Analysis

The WSVF algorithm orders the jobs in a non-decreasing order according to their volume over weight values, i.e., $(p_j \cdot d_j)/w_j$. The algorithm schedules the highest priority unassigned job (the one with the smallest value) at the earliest time t on a machine that is available to process the job until it is completed; see Fig. 1 for an example.

Algorithm 1: WSVF

Set $\mathcal{J}$ to be list of jobs, sorted in a non-decreasing order of their $(p_j \cdot d_j)/w_j$ values

for job $j \in \mathcal{J}$ **do**

Schedule job j on machine i that provides the earliest start time t, s.t.

$$d_j + \sum_{h \in G_i(t')} d_h \le 1, \text{ for all } t' \in [t, t + p_j)$$

end for

By definition, the assignment of Algorithm 1 is feasible. Next, we show that its approximation ratio depends on $\alpha = \max_{j \in \mathcal{N}} d_j$, the maximum resource demand of any job.

Theorem 1. *If any job requires at most* $\alpha < 1$, *WSVF is a* $\left(\frac{1}{1-\alpha} + 1\right)$-*approximation algorithm for the weighted completion time minimization problem.*

To prove Theorem 1, we need to establish bounds on a problem instance $\hat{I}$, which is a compressed instance of the original problem instance I, where I denotes the set of jobs $\mathcal{J}$ ordered by the WSVF priority rule. We "compress" the original jobs such that their demands become 1 (as in the non-capacitated setting). For this, we take the ordered set of jobs I, and for each job in the set, we fix the duration to be $\hat{p}_j = p_j \cdot d_j$, which is the so-called volume of job j in the original instance, and fix the demand as $\hat{d}_j = 1$. We note that the

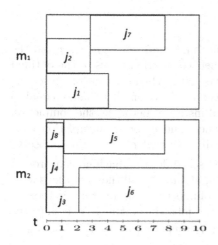

jobs	p_j	d_j	w_j	$\frac{p_j \cdot d_j}{w_j}$
1	4	0.4	8	0.2
2	3	0.4	5	0.24
3	2	0.25	1.5	0.33
4	1	0.45	1	0.45
5	7	0.4	4	0.7
6	7	0.5	4	0.875
7	5	0.45	2	1.125
8	1	0.28	0.2	1.4

Fig. 1. An illustration of the WSVF algorithm's run. In this illustration we have $M = 2$ machines and $N = 8$ jobs. Each job is defined by its parameters (p_j, d_j, w_j) as described above. The jobs are sorted by a non-decreasing order of $p_j \cdot d_j / w_j$, for example, $p_1 \cdot d_1 / w_1 = 0.2 < 0.24 = p_2 \cdot d_2 / w_2$. Then running over the jobs according to the priority order, each is scheduled on the machine with the earliest available d_j capacity for p_j time steps. Here, jobs j_1, j_2, j_3 and j_4 are scheduled at time 0 according to the available machine capacities. Job j_5 can be scheduled only after the completion of j_4 so $s_5 = c_4 = 1$, and jobs j_6 and j_7 are waiting for the completion of jobs j_3 and j_2, respectively. Job j_8, although it has a low priority, is scheduled at time 0 because machine m_2 has a resource 'window' that is large enough to process job j_8. The objective function value in the example is $\sum_{j=1}^{8} w_j \cdot c_j = 8 \cdot 4 + 5 \cdot 3 + \cdots + 2 \cdot 8 + 0.2 \cdot 1 = 135.2$.

compressed instance preserves the ordering of the jobs and the list of jobs is ordered according to the WSPT priority rule.

We bound the cost of WSVF on I by proving that the optimal solution value of the corresponding compressed instance $\hat{I}$ is smaller than or equal to the optimal solution value of I. For ease of notation, we denote by $C_M(I)$ the objective function value (cost) achieved by algorithm WSVF for instance I using M machines and by $C_M^*(I)$ the optimal cost of scheduling I on M machines.

We note the following two observations: 1) $C_N(I) = \sum_{j \in \mathcal{N}} w_j p_j$ since when using N machines and $N = |I|$, it is always optimal to assign each job to a machine, and 2) for a single machine, $C_1(\hat{I}) = C_1^*(\hat{I}) = \sum_j (w_j \cdot \sum_{h=1}^{j} \hat{p}_h)$ by Smith's rule [24].

We begin by establishing the relations between $C_M^*(I), C_M^*(\hat{I}), C_1^*(\hat{I})$ and $C_M(I)$. First, since $c_j \geq p_j$ always holds:

Observation 2. $C_N(I) \leq C_M^*(I)$.

Next, we bound the optimal cost on M machines by the optimal cost on the compressed instance on M machines.

Claim 1. $C_M^*(\hat{I}) \leq C_M^*(I)$.

Proof. Consider, w.l.o.g., one of the M machines and let $j_1, j_2, ..., j_k$ be the jobs assigned to this machine ordered by their completion times in the optimal schedule (which has a cost $C_M^*(I)$). We prove that a scheduler that processes the compressed jobs on the same machine in this order has a smaller than or equal to cost compared to the uncompressed optimal instance. Let c_j^* be the completion time of job j in the optimal original instance, and $\hat{c}_j$ be its completion time in the compressed instance schedule. We will show that $c_j^* \geq \hat{c}_j$. First, observe that $\hat{c}_j = \sum_{h=1}^{j} \hat{p}_h$, since in the compressed schedule, the machine can process a single job at each time. Next, by definition, at time c_j^*, all the first j jobs have been completed on this machine. By the volume preservation rule, it took at least $\sum_{h=1}^{j} p_h \cdot d_h$ time units to complete the jobs since at any time step, the machine can process a maximal volume of 1; therefore, $c_j^* \geq \sum_{h=1}^{j} p_h \cdot d_h = \sum_{h=1}^{j} \hat{p}_h = \hat{c}_j$. Hence, when we sum over all machines and jobs,

$$C_M^*(\hat{I}) \leq \sum_{j \in \mathcal{N}} w_j \cdot \hat{c}_j \leq \sum_{j \in \mathcal{N}} w_j \cdot c_j^* = C_M^*(I).$$

The Left-Hand-Side (LHS) inequality holds since the optimal cost C_M^* is equal to or smaller than the cost of any schedule and the Right-Hand-Side (RHS) inequality follows from extending the volume preservation argument to M machines.

Next, we use Theorem 1 from [6] to construct a lower bound for $C_M^*(\hat{I})$ using a single machine optimal cost $C_1^*(\hat{I})$.

Claim 2. $C_M^*(\hat{I}) \geq \frac{1}{M} C_1^*(\hat{I})$.

Proof. [6] proved a tight bound on the optimal cost of the non-capacitated case with M machines with respect to the optimal cost of an instance with a single machine. Specifically, they proved

$$C_M^*(\hat{I}) - \frac{1}{2} C_N(\hat{I}) \geq \frac{1}{M}(C_1^*(\hat{I}) - \frac{1}{2} C_N(\hat{I})).$$

Therefore, we have

$$C_M^*(\hat{I}) \geq \frac{1}{M} C_1^*(\hat{I}) + \frac{1}{2} C_N(\hat{I})(1 - \frac{1}{M}) \geq \frac{1}{M} C_1^*(\hat{I})$$

since $\frac{1}{2} C_N(\hat{I})(1 - \frac{1}{M}) \geq 0$.

We now compare $C_M(I)$ and $C_1(\hat{I})$ by bounding the start time s_j for each job $j \in \mathcal{N}$. We prove that s_j can be bounded by a factor that depends on α and M times the total volume of jobs preceding j in I (which is ordered by WSVF). In other words, $\sum_{h < j} v_h$, which is the start time of job j in $C_1(\hat{I})$.

Lemma 3. $\forall j \in \mathcal{N} : s_j \leq \frac{1}{(1-\alpha)M} \Sigma_{h=1}^{j-1} v_h$

Proof. To prove Lemma 3, we prove that prior to s_j, each machine processes a total demand of at least $1 - \alpha$ on jobs with a higher priority than j. This property also appeared in [12] and holds for any priority-based algorithm. For a machine $i \in \mathcal{M}$ at time $t \in \mathcal{T}$ and job $j \in \mathcal{N}$, we define $D_j^i(t) = \sum_{h \in G^i(t), h \leq j} d_h$, which is the total demand of jobs among the first j jobs that are processed on machine i at time t. By proving the following property, we prove that for all $t < s_j$, we have $D_{j-1}^i(t) \geq 1 - \alpha$.

Claim 4. *For all $i \in \mathcal{M}, j \in \mathcal{N}$, $min\{D_j^i(t), 1 - \alpha\}$ is non-increasing with t.*

Proof. Choose a machine i (we omit the superscript i when it is clear from the context), and consider w.l.o.g. only jobs assigned to this machine (with the same priority order as in I), other jobs not affected the total demand D. We prove the claim using an induction on j. First, consider the case of $j = 1$,

$$min\{D_1(t), 1 - \alpha\} = \begin{cases} min\{d_1, 1 - \alpha\} & \text{if } t \leq p_1, \\ 0 & \text{if } t > p_1. \end{cases}$$

It is a non-increasing function with t so the claim holds for $j = 1$.

Now we use the induction assumption that the claim is true for $j - 1$ and prove the claim for $j > 1$. First, we argue that according to the WSVF, $D_j(t) = D_{j-1}(t) > 1 - \alpha$, for all $0 \leq t < s_j$. Otherwise, if $0 \leq t < s_j$ exists such that $D_{j-1}(t) \leq 1 - \alpha$ (that is, $D_{j-1}(t) + \alpha \leq 1$), then for $t \leq t' < s_j$, we have $D_{j-1}(t') \leq 1 - \alpha$ by our induction assumption, and since $d_j \leq \alpha$ by our assumption on the input, the WSVF would assign job j before time s_j.

Now, we can look at the expression $min\{D_j(t), 1 - \alpha\}$ value over time t:

$$min\{D_j(t), 1 - \alpha\} = \begin{cases} 1 - \alpha & \text{if } t < s_j, \\ min\{D_{j-1}(t) + d_j, 1 - \alpha\} & \text{if } s_j \leq t \leq s_j + p_j, \\ min\{D_{j-1}(t), 1 - \alpha\} & \text{if } t > s_j + p_j. \end{cases}$$

Using the above and the induction assumption, we can conclude that $min\{D_j(t), 1 - \alpha\}$ is non-increasing with t.

Corollary 1. *For all $i \in \mathcal{M}, j \in \mathcal{N}$ and $0 \leq t < s_j$, we have $D_{j-1}^i(t) > 1 - \alpha$.*

Proof. We follow the proof of Claim 4; if at any $t < s_j$, $D_{j-1}(t) \leq 1 - \alpha$, then by the invariant of Claim 4, the WSVF would process job j earlier than s_j by its definition.

Next, we use Corollary 1 to set an upper bound on the starting time of every job j and to conclude the proof of Lemma 3. By corollary 1, we have that each machine used at least $(1 - \alpha)$ of its capacity until time s_j, to process the first $j - 1$ jobs. Thus, the following inequality holds for M machines

$$\Sigma_{h=1}^{j-1} v_h \geq M(1 - \alpha)s_j,$$

where the LHS is the sum of the first $j - 1$ job volumes that were processed on M machines and the RHS follows from the lower bound on the used volume extended to M machines. Reorganizing the above formula leads to an upper bound on the starting time of job j:

$$s_j \leq \frac{1}{(1-\alpha)M} \Sigma_{h=1}^{j-1} v_h.$$

By using the above results, we can prove the main lemma for bounding $C_M(I)$:

Lemma 5. *Given any instance I, such that for all $j \in \mathcal{N}$, $d_j \leq \alpha$ for $0 < \alpha < 1$ we have:*

$$C_M(I) \leq C_N(I) + \frac{C_1(\hat{I})}{(1-\alpha)M}$$

Proof. For every job $j \in \mathcal{N}$

$$c_j = p_j + s_j \leq p_j + \frac{1}{(1-\alpha)M} \cdot \sum_{h=1}^{j-1} v_h = p_j + \frac{1}{(1-\alpha)M} \cdot \sum_{h=1}^{j-1} \hat{p}_h, \qquad (1)$$

where the LHS equality exists by definition, the inequality follows from Lemma 3, and the RHS equality follows from the definition $v_h = d_h \cdot p_h = \hat{p}_h$.

Summing over all jobs to find the total cost, we have:

$$C_M(I) = \sum_{j=1}^{N} w_j \cdot c_j$$

$$\leq \sum_{j=1}^{N} w_j \cdot p_j + \sum_{j=1}^{N} w_j \cdot \left(\frac{1}{(1-\alpha)M} \cdot \sum_{h=1}^{j-1} \hat{p}_h \right)$$

$$= \sum_{j=1}^{N} w_j \cdot p_j + \frac{1}{(1-\alpha)M} \cdot \sum_{j=1}^{N} (w_j \cdot \sum_{h=1}^{j-1} \hat{p}_h)$$

$$\leq C_N(I) + \frac{C_1(\hat{I})}{(1-\alpha)M},$$

where the first inequality follows from Eq. 1 and the second inequality holds since $C_N(I) = \sum_{j=1}^{n} w_j \cdot p_j$, and $C_1(\hat{I}) = \sum_{j=1}^{n} (w_j \cdot \sum_{h=1}^{j} \hat{p}_h) \geq \sum_{j=1}^{n} (w_j \cdot \sum_{h=1}^{j-1} \hat{p}_h)$.

Finally, we prove Theorem 1, which follows immediately from Lemma 5 and Claims 1 and 2.

Proof. (Proof of Theorem 1).

$$C_M(I) \leq C_N(I) + \frac{C_1(\hat{I})}{(1-\alpha)M} \leq C_N(I) + \frac{C_M^*(\hat{I})}{1-\alpha} \leq C_M^*(I) \left(1 + \frac{1}{1-\alpha}\right).$$

From the above we can deduce that for high load systems ($C_N(I) << C_M^*(I)$) with low demands jobs ($\alpha << 1$) WSVF getting closer to the optimal solution.

4   Constant Approximation Algorithm for All Instances

In this section, we extend the previous results to remove the dependency on α, the maximum demand of any job, which leads to a constant approximation algorithm. We rely on the observations that, given a set of jobs with demands that are larger than $1/2$, only a single job can run on a machine at any time, and that the WSVF has a constant approximation ratio on jobs with demands equal to or smaller than $1/2$. Thus, by splitting the jobs based on their resource demands into two sets of machines, we achieve a constant approximation algorithm.

To this end, we introduce the Hybrid-WSVF:

Algorithm 2: HYBRID-WSVF

Split the jobs in I into two sets $I^l = \{j : d_j \leq \frac{1}{2}\}$ and $I^h = \{j : d_j > \frac{1}{2}\}$.

Schedule I^l on $M_1 = \lceil \frac{2(M-2)}{3} \rceil + 1$ machines using WSVF.

Schedule I^h on $M_2 = M - M_1$ using WSPT.

Theorem 3. *HYBRID-WSVF is a* $4 + o(\frac{1}{M})$*-approximation for the weighted completion time minimization problem where M is the number of machines.*

Proof. First, we schedule the low demand jobs (i.e., $d_j \leq 0.5$) on M_1 machines. Thus, from Lemma 5, we have

$$C_{M_1}(I^l) \leq C_N(I^l) + \frac{1}{(1-\alpha)M_1} \cdot C_1(\hat{I}^l).$$

For the low demand jobs, α values are within the interval $(0, 0.5]$ with a specific value per I^l. Thus,

$$C_N(I^l) + \frac{1}{(1-\alpha)M_1} \cdot C_1(\hat{I}^l) \leq C_N(I^l) + \frac{2}{M_1} \cdot C_1(\hat{I}^l)$$

$$= C_N(I^l) + \frac{2M}{M_1} \cdot \frac{C_1(\hat{I}^l)}{M} \leq C_M^*(I^l) + 2 \cdot \frac{M}{M_1} C_M^*(I^l),$$

where the last RHS inequality follows from Observation 2, Claim 1 and Claim 2.

Next, we deal with the high demand jobs ($d_j > 0.5$), which are scheduled on M_2 machines. Since any two high demand jobs cannot run simultaneously on the same machine, the scheduling problem is equivalent to the weighted non-capacitated setting. Therefore, we can use a result from [6] who proved the following bound for the WSPT with M machines:

$$C_M(\hat{I}) \leq C_N(\hat{I}) + \frac{1}{M} C_1(\hat{I}).$$

Following their result, we state that:

$$C_{M_2}(I^h) \leq C_N(I^h) + \frac{1}{M_2} C_1(I^h) = C_N(I^h) + \frac{M}{M_2} \frac{C_1(I^h)}{M} \leq C_M^*(I^h) + \frac{M}{M_2} C_M^*(I^h),$$

where the RHS inequality follows from Observation 2 and Claim 2.

By setting $M_1 = \lceil \frac{2(M-2)}{3} \rceil + 1$ and $M_2 = M - M_1 = \lfloor \frac{M-2}{3} \rfloor + 1$, and by observing $C_M^*(I) \geq C_M^*(I^h) + C_M^*(I^l)$, we conclude the proof for Theorem 3.

5   Conclusions and Future Work

Algorithms for scheduling jobs on capacitated machines can significantly affect the performance of cloud computing environments. While such environments handle jobs of varying importance (e.g., cost), there were no constant approximation algorithms known, to the best of our knowledge, for the related problem of minimizing the weighted sum of job completion times. This paper closes this gap. The suggested algorithm also improves the best-known approximation ratio for the non-weighted problem (for $M \geq 2$).

The results presented in this paper may be enhanced by going in several research directions, of which we mention three. The first significant theoretical extension may consider capacitated problems with multiple capacitated resources. In real-world cloud computing environments, CPU, memory, storage and bandwidth may be scarce resources. To realize such an extension, one would need to consider a multidimensional demand for each job. A second extension would be to use our methods to find performance guarantees in stochastic environments in which, for example, processing duration can be characterized using probabilistic knowledge. The third research direction would be to develop a model that accommodates jobs with release dates and deadlines. Such research may also be considered to be a natural extension of the current model.

Appendix: The Single Machine Case

In this section, we introduce a different algorithm for the special case of $M = 1$, and conclude that for the weighted completion time minimization for capacitated machine, there exists a constant competitive algorithm for any number of machines. Note that Algorithm 2 is well defined for $M \geq 2$ machines; for $M = 1$ machines, the partition approach cannot be applied. Instead, we extend the ideas of the non-weighted case for $M = 1$ in [12] to the general weighted case.

The algorithm uses two main tools, a knapsack algorithm that determines a subset of highest weight jobs with a total volume restriction, and a 2D-strip packing algorithm for scheduling this set, where each job j is represented by a rectangle r_j defined with width p_j and height d_j.

The algorithm will work in iterations until it assigns all jobs, where in iteration ℓ the algorithm will execute the following steps:

1. Solve a knapsack packing problem: Compute J_ℓ, a maximum weighted set of jobs with a total volume less than 2^ℓ using a $1 + \epsilon$ resource augmentation knapsack algorithm.
2. Pack the jobs in J_ℓ into 3 strips of width $(1 + \epsilon)2^\ell$ and height 1
3. Concatenate the strips from the earlier step and schedule the job on the machine.

Jobs can be schedule more than once when using the above algorithm, of course it is unneeded and practically the job will be scheduled only according to the first strip it appear in. Note that after $\ell^{\max} = \log(\sum_{j \in \mathcal{N}} v_j)$ iterations, all the jobs are packed, and the algorithm can discard jobs that were scheduled in previous iterations. We will show that the first two steps of finding the set of jobs and packing it into strips can be done in polynomial time. Finding a maximum weighted set of jobs ($\arg\max_J\{\sum_{j\in J} w_j : \sum_{j\in J} v_j \leq B\}$) is equivalent to the knapsack problem: Given a set of items, each with a size and a value, determine which item to include in a collection so that the total size is less than or equal to a given limit and the total value is as large as possible. We apply the resource augmentation solution of [25] for this problem, which proves that given a bound on the total size L, there exists a polynomial time in which a set of total size $(1 + \epsilon) \cdot L$ with a total profit of at least the optimal profit of a set with total size L can be computed. Second, we utilize the algorithm from [15], which packs rectangles (without rotations) with total volume of L and maximal height of 1 into 3 strips of width L and height 1.

Algorithm 3: PackAndSchedule(I)

---
for $\ell = 0, 1, 2, \ldots, \ell^{\max}$ **do**

   Compute $J_\ell \subseteq I$, a maximal weighted set with total volume of at most $2^\ell \cdot (1 + \epsilon)$.

   Pack J_ℓ into strip S of height 1 and width $3 \cdot (1 + \epsilon) \cdot 2^\ell$.

   Schedule S in the interval $[3(1 + \epsilon) \sum\limits_{h=0}^{\ell-1} 2^h, 3(1 + \epsilon) \sum\limits_{h=0}^{\ell} 2^h)]$

end for

Theorem 4. *PackAndSchedule is a $12(1 + \epsilon)$ approximation polynomial-time algorithm for total weighted completion time minimization in the single machine case.*

Proof. The algorithm provides a feasible scheduling allocation since each of the 2D packing solution strips having a maximal height (demand) of 1 is assigned to a disjoint time interval.

Let $W_\ell = \sum_{j\in J_\ell} w_j$ be the sum of weights that are processed in iteration ℓ. Note that jobs in J_ℓ are scheduled until time $3(1+\epsilon)\cdot(2^{\ell+1}-1)$. In addition, when jobs in J_ℓ are processed, the total weight of jobs that are not yet completed is at most $W - W_{\ell-1}$. Therefore, we can bound the cost of the algorithm 3 by summing over the iterations and multiplying the unprocessed weight by the completion time:

$$C_1(I) \leq 3(1 + \epsilon) \sum_{\ell\geq 0}(2^{\ell+1} - 1)(W - W_{\ell-1})$$

The optimal scheduler can only complete jobs with a total weight of at most W_ℓ up to time 2^ℓ, because by volume preservation, it is not possible to pack more than 2^ℓ volume until this point, and by the correctness of the knapsack algorithm, this is the maximum profit (weight) that can be packed by W_ℓ if the total volume is at most 2^ℓ; therefore, there are jobs with a total weight of at least $W - W_\ell$ that are unprocessed at time 2^ℓ. Hence, we can give a lower bound for the cost of the optimal scheduler:

$$C_1^*(I) \geq W + \sum_{\ell \geq 0} 2^\ell (W - W_\ell)$$

By combining the two above inequalities, we have that PackAndSchedule is a $12(1 + \epsilon)$ approximation algorithm as required.

References

1. Albagli-Kim, S., Shachnai, H., Tamir, T.: Scheduling jobs with dwindling resource requirements in clouds. In: IEEE INFOCOM 2014-IEEE Conference on Computer Communications, pp. 601–609. IEEE (2014)
2. Balouka, N., Cohen, I.: A robust optimization approach for the multi-mode resource-constrained project scheduling problem. Eur. J. Oper. Res. **291**, 457–470 (2019)
3. Bougeret, M., Dutot, P.F., Jansen, K., Robenek, C., Trystram, D.: Approximation algorithms for multiple strip packing and scheduling parallel jobs in platforms. Disc. Math. Algorithms Appl **3**(04), 553–586 (2011)
4. Cohen, I., Postek, K., Shtern, S.: An adaptive robust optimization model for parallel machine scheduling. arXiv preprint arXiv:2102.08677 (2021)
5. Damodaran, P., Ghrayeb, O., Guttikonda, M.C.: Grasp to minimize makespan for a capacitated batch-processing machine. Int. J. Adv. Manuf. Technol. **68**(1–4), 407–414 (2013)
6. Eastman, W.L., Even, S., Isaacs, I.M.: Bounds for the optimal scheduling of n jobs on m processors. Manag. Sci. **11**(2), 268–279 (1964)
7. Fox, K., Korupolu, M.: Weighted flowtime on capacitated machines. In: Proceedings of the Twenty-Fourth Annual ACM-SIAM Symposium on Discrete Algorithms, pp. 129–143. SIAM (2013)
8. Garey, M.R., Johnson, D.S.: Computers and intractability. In: A Guide to the Theory of NP-Completeness (1979)
9. Graves, S.C.: A review of production scheduling. Oper. Res. **29**(4), 646–675 (1981)
10. Guo, L., Shen, H.: Efficient approximation algorithms for the bounded flexible scheduling problem in clouds. IEEE Trans. Parallel Distrib. Syst. **28**(12), 3511–3520 (2017)
11. Hota, A., Mohapatra, S., Mohanty, S.: Survey of different load balancing approach-based algorithms in cloud computing: a comprehensive review. In: Behera, H.S., Nayak, J., Naik, B., Abraham, A. (eds.) Computational Intelligence in Data Mining. AISC, vol. 711, pp. 99–110. Springer, Singapore (2019). https://doi.org/10.1007/978-981-10-8055-5_10
12. Im, S., Naghshnejad, M., Singhal, M.: Scheduling jobs with non-uniform demands on multiple servers without interruption. In: IEEE INFOCOM 2016-The 35th Annual IEEE International Conference on Computer Communications, pp. 1–9. IEEE (2016)
13. Jansen, K., Rau, M.: Linear time algorithms for multiple cluster scheduling and multiple strip packing. In: Yahyapour, R. (ed.) Euro-Par 2019. LNCS, vol. 11725, pp. 103–116. Springer, Cham (2019). https://doi.org/10.1007/978-3-030-29400-7_8
14. Jansen, K., Trystram, D.: Scheduling parallel jobs on heterogeneous platforms. Electron. Notes Disc. Math. **55**, 9–12 (2016)
15. Jansen, K., Zhang, G.: Maximizing the total profit of rectangles packed into a rectangle. Algorithmica **47**(3), 323–342 (2007)

16. Kawaguchi, T., Kyan, S.: Worst case bound of an LRF schedule for the mean weighted flow-time problem. SIAM J. Comput. **15**(4), 1119–1129 (1986)
17. Kumar, M., Sharma, S.C., Goel, A., Singh, S.P.: A comprehensive survey for scheduling techniques in cloud computing. J. Netw. Comput. Appl **143**, 1–33 (2019)
18. Liu, S.: A review for submodular optimization on machine scheduling problems. In: Du, D.-Z., Wang, J. (eds.) Complexity and Approximation. LNCS, vol. 12000, pp. 252–267. Springer, Cham (2020). https://doi.org/10.1007/978-3-030-41672-0_16
19. Liu, Y., Xu, H., Lau, W.C.: Online job scheduling with resource packing on a cluster of heterogeneous servers. In: IEEE INFOCOM 2019-IEEE Conference on Computer Communications, pp. 1441–1449. IEEE (2019)
20. Malhotra, L., Agarwal, D., Jaiswal, A., et al.: Virtualization in cloud computing. J. Inf. Tech. Softw. Eng. **4**(2), 1–3 (2014)
21. Meiswinkel, S.: Mechanism design and machine scheduling: literature review. In: On Combinatorial Optimization and Mechanism Design Problems Arising at Container Ports. PL, pp. 15–30. Springer, Wiesbaden (2018). https://doi.org/10.1007/978-3-658-22362-5_2
22. Muter, İ: Exact algorithms to minimize makespan on single and parallel batch processing machines. Eur. J. Oper. Res. **285**(2), 470–483 (2020)
23. Pinedo, M.: Scheduling, vol. 5. Springer, Heidelberg (2012). https://doi.org/10.1007/978-3-319-26580-3
24. Smith, W.E.: Various optimizers for single-stage production. Naval Res. Logist. Q. **3**(1–2), 59–66 (1956)
25. Williamson, D.P., Shmoys, D.B.: The Design of Approximation Algorithms. Cambridge University Press, Cambridge (2011)

Server Cloud Scheduling

Marten Maack⬤, Friedhelm Meyer auf der Heide, and Simon Pukrop[(✉)]⬤

Heinz Nixdorf Institute & Department of Computer Science, Paderborn University,
Paderborn, Germany
{martenm,fmadh,simonjp}@mail.uni-paderborn.de

Abstract. Consider a set of jobs connected to a directed acyclic task graph with a fixed source and sink. The edges of this graph model precedence constraints and the jobs have to be scheduled with respect to those. We introduce the Server Cloud Scheduling problem, in which the jobs have to be processed either on a single local machine or on one of many cloud machines. Both the source and the sink have to be scheduled on the local machine. For each job, processing times both on the server and in the cloud are given. Furthermore, for each edge in the task graph, a communication delay is included in the input and has to be taken into account if one of the two jobs is scheduled on the server, the other in the cloud. The server can process jobs sequentially, whereas the cloud can serve as many as needed in parallel, but induces costs. We consider both makespan and cost minimization. The main results are an FPTAS with respect for the makespan objective for a fairly general case and strong hardness for the case with unit processing times and delays.

Keywords: Scheduling · Cloud · Precedence constraints · Communication delays · Approximation · NP-hardness

1 Introduction

Scheduling with precedence constraints with the goal of makespan minimization is widely considered a fundamental problem. It has already been studied in the 1960s by Graham [6] and receives a lot of research attention up to this day (see e.g. [5,7,9]). One problem variant that has received particular attention recently, is the variant with communication delays (e.g. [3,4,7]). Another, more contemporary topic concerns scheduling using external resources like, for instance, machines from the cloud and several models in this context have been considered of late (e.g. [1,10,12]). In this paper, we introduce and study a model closely connected to both settings, where jobs with precedence constraints may either be processed on a single server machine or on one of many cloud machines. Here, communication delays may occur only if the computational setting is changed,

This work was partially supported by the German Research Foundation (DFG) within the Collaborative Research Centre "On-The-Fly Computing" under the project number 160364472—SFB 901/3.

J. Koenemann and B. Peis (Eds.): WAOA 2021, LNCS 12982, pp. 144–164, 2021.
https://doi.org/10.1007/978-3-030-92702-8_10

the server and cloud machines may behave heterogeneously, i.e., jobs may have different processing times on the server and in the cloud, and scheduling in the cloud incurs costs proportional to the computational load performed in this context. Both makespan and cost minimization is considered. We believe that the present model provides a useful link between scheduling with precedence constraints and communication delays on the one hand and cloud scheduling on the other.

Problem. We consider a scheduling problem SCS in which a task graph $G = (\mathcal{J}, E)$ has to be scheduled on a combination of a local machine (server) and a limitless number of remote machines (cloud). The task graph is a directed, acyclic graph with exactly one source $\mathcal{S} \in \mathcal{J}$ and exactly one sink $\mathcal{T} \in \mathcal{J}$. Each job $j \in \mathcal{J}$ has a processing time on the server p_s and on the cloud p_c. We consider $p_s(\mathcal{S}) = p_s(\mathcal{T}) = 0$ and $p_c(\mathcal{S}) = p_c(\mathcal{T}) = \infty$. For every other job the values of p_s and p_c can be arbitrary in $\mathbb{N}$, meaning that the server and the cloud are unrelated machines in our default model. An edge $e = (i, j)$ denotes precedence, job i has to fully processed before job j can start. Furthermore an edge $e = (i, j)$ has a communication delay of $c(i, j) \in \mathbb{N}$, which means, that after job i finished, j has to wait an additional $c(i, j)$ time steps before it can start, if i and j are not both scheduled on the same type of machine (server or cloud).

A schedule π is given as the tuple $(\mathcal{J}^s, \mathcal{J}^c, C)$. $\mathcal{J}^s$ and $\mathcal{J}^c$ are a proper partition of $\mathcal{J}$: $\mathcal{J}^s \cap \mathcal{J}^c = \emptyset$ and $\mathcal{J}^s \cup \mathcal{J}^c = \mathcal{J}$. $\mathcal{J}^s$ denotes jobs that are processed on the server in π, $\mathcal{J}^c$ respectively for the cloud. $C : \mathcal{J} \mapsto \mathbb{N}$ maps jobs to their completion time.

We define a bit of notations before we formally define the validity of a schedule. Let $p^{\pi}(j)$ be equal to $p_s(j)$ iff $j \in \mathcal{J}^s$, and $p_c(j)$ iff $j \in \mathcal{J}^s$. $p^{\pi}(j)$ denotes the actual processing time of job j in π. Let $E^* := \{(i, j) \in E | (i \in \mathcal{J}^s \wedge j \in \mathcal{J}^c) \vee (i \in \mathcal{J}^c \wedge j \in \mathcal{J}^s)\}$ be the set of edges between jobs on different machines. Intuitively, for all the edges in E^* we have to take the communication delays into consideration, for all edges in $E \setminus E^*$ we only care about the precedence.

We call a schedule π valid if and only if the following conditions are met:

a) There is always at most one job processing on the server:
$\forall_{i \in \mathcal{J}^s} \forall_{j \in \mathcal{J}^s \setminus \{i\}} : (C(i) \leq C(j) - p^{\pi}(j)) \vee (C(i) - p^{\pi}(i) \geq C(j))$

b) Tasks are not started before the previous tasks has been finished/ the required communication is done:
$\forall_{(i,j) \in E \setminus E^*} : (C(i) \leq C(j) - p^{\pi}(j))$
$\forall_{(i,j) \in E^*} : (C(i) + c(i, j) \leq C(j) - p^{\pi}(j))$

The makespan of a schedule is given by the completion time of the sink $\mathcal{T}$: $C(\mathcal{T})$. The cost of a schedule is given by the time it spends processing tasks on the cloud: $\sum_{i \in \mathcal{J}^c} p^{\pi}(i)$. Note here, that by requiring $p_s(\mathcal{S}) = p_s(\mathcal{T}) = 0$ and $p_c(\mathcal{S}) = p_c(\mathcal{T}) = \infty$, we assume every job to start and end on the server. This is done only for convenience as it defines a clear start and end state for each schedule.

Naturally two different optimization problems arise from the definition. First, given a deadline d, find a schedule with lowest cost and $C(\mathcal{T}) \leq d$. Second, given a cost budget b, find a schedule with smallest makespan and $\sum_{i \in \mathcal{J}^{\circ}} p^{\pi}(i) \leq b$. In both instances the d, respectively the b, is strict. The natural decision variant is: given both d and b find a schedule that adheres to both, if one exists. Note here, that the decision variant is naturally easier (or at least as easy) as the optimization problems.

Remark 1. Instances of SCS might contain schedules with a makespan (and therefore cost) of 0. We can check for those in polynomial time: First, remove all edges with communication delay 0, we get a set of connected components K. Iff $\forall_{k \in K} (\forall_{j \in k} \ p_s(j) = 0) \vee (\forall_{j \in k} \ p_c(j) = 0)$, then there is a schedule with makespan of 0. For the rest of the paper we will assume that our algorithms check that beforehand and are only interested in schedules with makespan > 0.

Results. We start by establishing (weak) NP-hardness already for the case without communication delays and very simple task graphs. More precisely, for the case in which the task graph forms one chain starting with the source and ending with the sink and the case in which the graph is fully parallel, i.e., each job $j \in \mathcal{J} \setminus \{\mathcal{S}, \mathcal{T}\}$ is only preceded by the source and succeeded by the sink. On the other hand, we establish FPTAS results for both the chain and fully parallel case with arbitrary communication delays and with respect to both objective functions. Furthermore, we present a 2-approximation for the case without delays and identical server and cloud machines ($p_c = p_s$) but arbitrary task graph and the makespan objective and show that the respective algorithm can also be used to solve the problem optimally with respect to both objectives in the case of unit processing times. These results are all relatively simple and are discussed in Sect. 2. In Sect. 3 we aim to generalize the FPTAS result for chain graphs regarding the makespan as much as possible. We are able to show that an FPTAS can be achieved as long as the *maximum cardinality source and sink dividing cut* ψ is constant. Intuitively, this parameter upper bounds the number of edges that have to be considered together in a dynamic program and in many relevant problem variants it can be bounded or replaced by the longest anti-chain length. We provide a formal definition in Sect. 3. Next, we turn our attention to strong NP-hardness results in Sect. 4. We are able to show, that a classical reduction due to Lenstra and Rinnooy Kan [8] can be adapted to prove NP-hardness already for the variant of SCS without communication delays and processing times equal to one or two. Now, in the case of unit processing times we know that problem can be solved in polynomial time, and hence we are interested in the case with unit processing times and communication delays. We design an intricate reduction to show that this very basic case is NP-hard as well. Note that in this setting the server and cloud machines are implicitly identical. Furthermore, we are able to show that a slight variation of this reduction implies that no constant approximation with respect to the cost objective can be achieved regarding the general problem. Lastly, in Sect. 5, we consider approximation algorithms for the case with unit processing times and delays. We show that a relatively simple approach yields a $\frac{1+\varepsilon}{2\varepsilon}$-approximation for $\varepsilon \in (0, 1]$ regarding the cost objective if we allow a makespan of $(1 + \varepsilon)d$. Table 1 provides an overview of the results.

Table 1. An overview of the results of this paper.

Algorithmic Results	
Fully parallel or chain task graph	FPTAS w.r.t. cost and makespan
Task graph with constant ψ	FPTAS w.r.t. makespan
$c = 0$, $p_c = p_s$ (no delays, identical machines)	2-approximation w.r.t. makespan
$c = 0$, $p_c = p_s = 1$	polynomial w.r.t. makespan and cost
$c = p_c = p_s = 1$ (unit delays, unit sizes)	$\frac{1+\varepsilon}{2\varepsilon}$-approximation w.r.t. cost with makespan at most $(1+\varepsilon)d$
Hardness Results	
Fully parallel or chain task graph, $c = 0$	(weakly) NP-hard
$\forall j \in \mathcal{J} : c(j) = 0, p_c(j), p_s(j) \in \{1, 2\}$	(strongly) NP-hard
$c = p_c = p_s = 1$ (unit delays, unit sizes)	(strongly) NP-hard
General problem	no constant approximation w.r.t. cost

Related Work. Probably the closest related model to the one considered in this paper was studied by Aba et al. [1]. In this paper the input is very similar, however, in both computational settings an unbounded number of machines may be used and the goal is makespan minimization. The authors show NP-hardness on the one hand, and identify cases that can be solved in polynomial time on the other. In the conclusion of this paper a model very similar to the one studied in this work is mentioned as an interesting research direction. For a detailed discussion of related models, we refer to the preprint version of the above work [2].

The present model is closely related to the classical problem of makespan minimization on parallel machines with precedence constraints, where a set of jobs with processing times, a precedence relation on the jobs (or a task graph), and a set of m machines are given. The goal is to assign the jobs to starting times and machines such that the precedence constraints are met and the last job finishes as soon as possible. In the 1960's, Graham [6] introduced the list scheduling heuristic for this problem and proved it to be a $(2 - \frac{1}{m})$-approximation. Interestingly, to date, this is essentially the best result for the general problem. On the other hand, Lenstra and Rinnooy Kan [8] showed that no better than $\frac{4}{3}$-approximation can be achieved for the problem with unit processing times, unless P = NP. In more recent days, there has been a series of exciting new results for this problem starting with a paper by Svensson [13] who showed that no better than 2-approximation can be hoped for assuming a variant of the unique games conjecture. Furthermore, Levey and Rothvoss [9] presented an approximation scheme with nearly quasi-polynomial running time for the variant with unit processing times and a constant number of machines, and Garg [5] improved the running time to quasi-polynomial shortly thereafter. These results utilized so called LP-hierarchies to strengthen linear programming relaxations of the problems. This basic approach has been further explored in a series of subsequent works, e.g. [3,4,7], which in particular also investigate the problem variant where a communication delay is incurred for pairs of precedence-constrained jobs running on different machines. The latter problem variant is closely related to our setting as well.

Lastly, there is at least a conceptual relationship to problems where jobs are to be executed in the cloud. For example, a problem was considered by Saha [12] in which cloud machines have to be rented in fixed time blocks in order to schedule a set of jobs with release dates and deadlines minimizing the costs which are proportional to the rented time blocks. Another example is a work by Mäcker et al. [10] in machines of different types can be rented from the cloud and machine dependent setup times have to be payed before they can be used. Jobs arrive in an online fashion and the goal is again cost minimization. Both papers reference further work in this context.

2   Preliminary Results

In this section we briefly discuss some results that can be considered low hanging fruits and give a first overview concerning the complexity and approximability of our problem. We first consider NP-hardness and are able to show:

Theorem 1. *The SCS problem is weakly NP-hard already for fully parallel or chain task graphs and without communication delays.*

In fully parallel task graphs each job that is not the source or sink has exactly two edges (connecting it to the source and sink). The results are achieved using fairly straight-forward reductions from the subset sum or knapsack problem, respectively.

In the following, we present complementing FPTAS results for the variants of *SCS* with fully parallel and chain task graphs. Furthermore, in both of the above reductions we did have no communication delays and in one of them the jobs had the same processing time on the server and the cloud. Hence, we take a closer look at this case as well and present a simple 2-approximation even for arbitrary task graphs and with respect to the makespan objective.

Fully Parallel Case. We show that the variant of *SCS* with fully parallel task graph can be dealt with using straight-forward applications of well-known results and techniques. In particular, we can design two simple dynamic programs for the search version of the problem that consider for each job the two possibilities of scheduling them on the cloud or on the server and compute for each possible budget or deadline the lowest makespan or cost, respectively, that can be achieved with the jobs considered so far. These dynamic programs can then be combined with suitable rounding procedures that reduce the number of considered states and search procedures for approximate values for the optimal cost or makespan, respectively, yielding:

Theorem 2. *There is an FPTAS for SCS with fully parallel task graph with respect to both the cost and the makespan objective.*

Proof. We start by designing the dynamic programs for the search version of the problem with budget b and deadline d. Without loss of generality, we assume $\mathcal{J} = [n] \cup \{\mathcal{S}, \mathcal{T}\}$ and set $c(j) = c(\mathcal{S}, j) + c(j, \mathcal{T})$.

For each deadline $d' \in \{0, 1, \dots, d\}$ and $j \in \{0, 1, \dots, n\}$, we want to compute the smallest cost $C[j, d']$ of all the schedules of the jobs $1, \dots, j$ adhering to the deadline d' on the server. We may do so by setting $C[0, d'] = 0$ for each d' and considering the two possibilities of scheduling job j on the cloud or server. In particular, let $C_1[j, d'] = C[j - 1, d'] + p_c(j)$ if $p_c(j) + c(j) \leq d$ and $C_1[j, d'] = \infty$ otherwise, and, furthermore, $C_2[j, d'] = C[j - 1, d' - p_s(j)]$ if $p_s(j) \leq d'$ and $C_2[j, d'] = \infty$ otherwise. Then, we may set $C[j, d'] = \min\{C_1(j, d'), C_2(j, d')\}$. Now, if $C[n, d] > b$, we know that there is no feasible solution for the search version, and otherwise we can use backtracking starting from $C[n, d]$ to find one. The time and space complexity is polynomial in d and n.

In the second dynamic program, we compute the smallest makespan $M[j, b']$ of all the schedules of the jobs $1, \dots, j$ adhering to the budget b', for each budget $b' \in \{0, 1, \dots, b\}$ and $j \in \{0, 1, \dots, n\}$. Again, we set $M[0, b'] = 0$ for each b' and consider the two possibilities of scheduling job j on the cloud or server. To that end, let $M_1[j, b'] = M[j - 1, b' - p_c(j)]$ if $p_c(j) + c(j) \leq d$ and $M_1[j, b'] = \infty$ otherwise, and, furthermore, $M_2[j, b'] = M[j - 1, b'] + p_s(j)$. Then, we may set $M[j, b'] = \min\{M_1(j, b'), M_2(j, b')\}$. Again, if $M[n, b] > d$, we know that there is no feasible solution for the search version, and otherwise we can use backtracking starting from $M[n, b]$ to find one. The time and space complexity is polynomial in b and n.

For both programs, we can use rounding and scaling approaches to trade the complexity dependence in d or b with a dependence in $poly(n, \frac{1}{\varepsilon})$ incurring a loss of a factor $(1 + O(\varepsilon))$ in the makespan or cost, respectively, if a solution is found. This can then be combined with a suitable search procedure for approximate values of the optimal makespan or cost. For details, we refer to Sect. 3, where such techniques are used and described in more detail. In addition to the techniques mentioned there, the possibility of a cost zero solution has to be considered which can easily be done in this case. □

Chain Graph Case. We present FPTAS results for the variant of *SCS* with chain task graph. The basic approach is very similar to the fully parallel case.

Theorem 3. *There is an FPTAS for SCS with chain task graph with respect to both the cost and the makespan objective.*

Proof. We again start by designing dynamic programs for the search version of the problem with budget b and deadline d. Without loss of generality, we assume $\mathcal{J} = \{0, 1, \dots, n + 1\}$ with $\mathcal{S} = 0$, $\mathcal{T} = n + 1$, and $j \in [n]$ being the j-th job in the chain.

For each deadline $d' \in \{0, 1, \dots, d\}$, job $j \in \{0, 1, \dots, n + 1\}$, and location $\ell \in \{s, c\}$ (referring to the server and cloud) we want to compute the smallest cost $C[d', j, \ell]$ of all the schedules of the jobs $1, \ldots, j$ adhering to the deadline d' and with the job j being scheduled on ℓ. p_c. To that end, we set $C[d', 0, s] = 0$, $C[d', 0, c] = \infty$, and with slight abuse of notation use the convention $C[z, j, \ell] = \infty$ for $z < 0$. Further values can be computed via the following recurrence relations:

$$C[d', j, s] = \min\{C[d' - p_s(j) - c(j - 1, j), c], C[d' - p_s(j), s]\}$$
$$C[d', j, c] = \min\{C[d' - p_c(j), c] + p_c(j), C[d' - p_c(j) - c(j - 1, j), s] + p_c(j)\}$$

If $C[d, n + 1, s] > b$, we know that there is no feasible solution for the search version, and otherwise we can use backtracking starting from $C[d, n + 1, s]$ to find one. The time and space complexity is polynomial in d and n.

In the second dynamic program, we compute the smallest makespan $M[j, b', \ell]$ of all the schedules of the jobs $0, \ldots, j$ adhering to the budget b' and with job j placed on location ℓ, for each $b' \in \{0, 1, \ldots, b\}$, $j \in \{0, 1, \ldots, n+1\}$ and $\ell \in \{s, c\}$. We set $M[b', 0, s] = 0$, $M[b', 0, c] = \infty$, use the convention $M[z, j, \ell] = \infty$ for $z < 0$, and the recurrence relations:

$$M[b', j, s] = \min\{M[b', c] + p_s(j) + c(j - 1, j), M[b', s] + p_s(j)\}$$
$$M[b', j, c] = \min\{M[b' - p_c(j), c] + p_c(j), M[b' - p_c(j), s] + p_c(j) + c(j - 1, j)\}$$

If $M[b, n + 1, s] > d$, we know that there is no feasible solution for the search version, and otherwise we can use backtracking starting from $M[b, n + 1, s]$ to find one. The time and space complexity is polynomial in b and n.

Like in the fully parallel case, we can use rounding and scaling approaches to trade the complexity dependence in d or b with a dependence in $poly(n, \frac{1}{\varepsilon})$ incurring a loss of a factor $(1 + O(\varepsilon))$ in the makespan or cost, respectively, if a solution is found. This can then be combined with a suitable search procedure for approximate values of the optimal makespan or cost. For details, we refer to Sect. 3, where such techniques are used and described in more detail. In addition to the techniques mentioned there, the possibility of a cost zero solution has to be considered which can easily be done in this case as well.   □

No Delays and Identical Machines. We design a simple heuristic for the case in which the server and the cloud machines behave the same, that is, $p_c(j) = p_s(j)$ for each job j (except for the source and sink), and the communication delays all equal zero. In this case, we may define the length of a chain in the task graph as the sum of the processing times of the jobs in the chain. The first step in the algorithm is to identify a longest chain in the task graph, which can be done in polynomial time. The jobs of the longest chain are scheduled on the server and the remaining jobs on the cloud each as early as possible. Now, the makespan of the resulting schedule is the length of a longest chain, which is optimal (or better) and there are no idle times on the server. However, the schedule may not be feasible since the budget may be exceeded. Hence, we repeatedly do the following: If the budget is still exceeded, we pick a job scheduled on the cloud with maximal starting time and move it on to the server right before its first successor (which may be the sink). Some jobs on the server may be delayed by this but we can do so without causing idle times. If all the processing times are equal this procedure produces an optimal solution and otherwise there may be an additive error of up to the maximal job size. Hence, we have:

Theorem 4. *There is a 2-approximation for SCS without communication delays and identical server and cloud machines.*

It is easy to see, that the analysis is tight considering an instance with three jobs: One with size b, one with size $b + \varepsilon$, and one with size 2ε. The first jobs precedes the last one. Our algorithm will place everything on the server, while the first job is placed on the cloud in the optimal solution.

Note that we can take a similar approach to find a solution with respect to the cost objective by placing more and more jobs on the server as long as the deadline is still adhered to. However, an error of one job can result in an unbounded multiplicative error in the objective in this case. On the other hand, it is easy to see that in the case with unit processing times, there will be no error at all in both procedures yielding:

Corollary 1. *The variant of SCS without communication delays and unit processing times can be solved in polynomial time with respect to both the makespan and the cost objective.*

3   Constant Cardinality Source and Sink Dividing Cut

We introduce the concept of a *maximum cardinality source and sink dividing cut*. For $G = (\mathcal{J}, E)$, let $\mathcal{J}_S$ be a subset of jobs, such that $\mathcal{J}_S$ includes S and there are no edges (j, k) with $j \in \mathcal{J} \setminus \mathcal{J}_S$ and $k \in \mathcal{J}_S$. In other words, in a running schedule $\mathcal{J}_S$ and $\mathcal{J} \setminus \mathcal{J}_S$, could represent already processed jobs and still to be processed jobs respectively. Denote by $\mathcal{J}_S^G$ the set of all such sets $\mathcal{J}_S$. We define

$$\psi := \max_{\mathcal{J}_S \in \mathcal{J}_S^G} |\{(j, k) \in E | j \in \mathcal{J}_S \wedge k \in \mathcal{J} \setminus \mathcal{J}_S\}|,$$

the maximum number of edges between any set $\mathcal{J}_S$ and $\mathcal{J} \setminus \mathcal{J}_S$ in G. In a series-parallel task graph ψ is equal to the maximum anti-chain size of the graph.

In this chapter we discuss how to solve or approximate SCS problems with a constant size ψ, but otherwise arbitrary task graphs. We first consider the deadline confined cost minimization, in Theorem 6 we show how to adapt this to the budget confined makespan minimization. We give a dynamic program to optimally solve instances of SCS with arbitrary task graphs. At first we will not confine the algorithm to polynomial time. Consider a given problem instance with $G = (\mathcal{J}, E)$, its source S and sink $\mathcal{T}$, processing times $p_s(j)$ and $p_c(j)$ for each $j \in \mathcal{J}$, communication delays $c(i, j)$ for each $(i, j) \in E$ and a deadline d.

We define intermediate states of a (running) schedule, as the states of our dynamic program. Such a state contains two types of variables. First we have two global variables, the timestamp t and the number of time steps the server

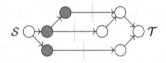

Fig. 1. State of a schedule, *open edges* are marked, loc_j and f_j kept for gray jobs

has been unused f_s. In other words, the server has not finished processing a job since $t - f_s$. The second type is defined per *open edge*. An open edge is a $e = (j, k)$ where j has already been processed, but k has not. For each such edge add the variables $e = (j, k)$ (the edge itself), $loc_j \in \{s, c\}$ denoting if j was processed on the server (s) or the cloud (c) and f_j denoting the number of time steps that have passed since j finished processing. If a job j is contained in multiple open edges, loc_j and f_j are still only included once. Write the state as $[t, f_s, e^1 = (j^1, k^1), loc_{j^1}, f_{j^1}, \ldots, e^m = (j^m, k^m), loc_{j^m}, f_{j^m}]$, where $e^1, \ldots, e^m$ denote all open edges (Fig. 1). Note here, that there are informations that we purposefully drop from a state: the completion time and location of every processed job without open edges, as those are not important for future decisions anymore. There might be multiple ways to reach a specific state, but we only care about the minimum possible cost to achieve that state, which is the *value* of the state.

We iteratively calculate the value of every reachable state with $t = 0, 1, 2, \ldots$. We start with the trivial state $[t = 0, f_s = 0, e^1, \ldots, e^m, loc_{\mathcal{S}} = s, f_{\mathcal{S}} = 0] = 0$, where $e^1, \ldots, e^m \in E$ with $e^i = (\mathcal{S}, j)$. This state forms the beginning of our *(sorted) state list*. We keep this list sorted in an ascending order of state values (costs) at all times. We exhaustively calculate every state that is reachable during a specific time step, given the set of states reachable during the previous time step. Intuitively, we try every possible way to "fill up" the still undefined time windows f_s and f_j.

Finally, we give the actual dynamic program in Algorithm 1. After the dynamic program finished, we iterate through the state list one last time and take the first state $[t = d, f_s]$. The value of that state is the minimum cost possible to schedule G in time d. One can easily adapt this procedure to also yield such a schedule, by keeping a list of all processed jobs per state containing their location and completion time.

Lemma 1. DPFGG *'s runtime is bounded in* $O(d^{2\psi+3} \cdot n^{2\psi+1})$.

Proof. At any point there are a maximum of $O(d \cdot (d \cdot n)^\psi)$ states in the state list. For every t we look at every state. Since we never insert a state in front of the state we are currently inspecting (costs can only increase), this traverses the list exactly once. For each of those states we calculate every possible successor, of which there are $O(\psi)$ and traverse the state list an additional time to correctly insert or update the state. We iterate from $t = 0$ to d and therefore get a runtime of: $O(d \cdot ((d \cdot (d \cdot n)^\psi) \cdot \psi \cdot (d \cdot (d \cdot n)^\psi))) = O(d^3 \cdot n \cdot (d \cdot n)^{2\psi}) \leq O(d^{2\psi+3} \cdot n^{2\psi+1})$. $\square$

Algorithm 1. DPFGG: Dynamic Program for General Graphs

1: initialize state list SL with start state (as defined above)
2: **for all** $state \in SL$ **do**
3:     let $\mathcal{J}^{state}$ be the set of all jobs that are endpoints in open edges from $state$
4:     **for all** $j \in \mathcal{J}^{state}$ **do**
5:         **if** $\forall (k,j) \in E : (k,j)$ also open edge in $state$ **then**
6:             *can j be processed on the server?*
7:             **if** $f_s \geq p_s(j)$ **then**
8:                 $jFits \leftarrow TRUE$
9:                 **for all** $(k,j) \in E$ **do**
10:                     **if** $loc_k = s \wedge f_k < p_s(j)$ or $loc_k = c \wedge f_k < p_s(j) + c(k,j)$ **then**
11:                         $jFits \leftarrow FALSE$
12:                 **if** $jFits = TRUE$ **then**
13:                     *calculate resulting state* $state'$*, value equal to state*
14:                     **for all** $(k,j) \in E$ **do**
15:                         remove (k,j) from $state'$
16:                         **if** j is last open successor of k **then**
17:                             remove f_k and loc_k from $state'$
18:                         add $f_j = 0$, $loc_j = s$ and all new open edges to $state'$
19:                     **if** $state' \in SL$ **then**
20:                         update value of $state'$ in SL if new value lower
21:                         then move $state'$ to correct position in SL
22:                     **else**
23:                         add $state'$ to correct position in SL (always after $state$)
24:             *can j be processed on the cloud?*
25:             *analogously to the previous case, cost value of state' increased by* $p_c(j)$
26: *check end condition*
27: **if** a state $[t = d, f_s] \in SL$ **then**
28:     return lowest value of such states
29: **if** $t < d$ **then**
30:     *move from t to* $t + 1$
31:     **for all** each $state \in SL$ **do**
32:         increase t, f_s and each f_j in $state$ by 1
33:     Back to step 2

3.1 Rounding the Dynamic Program

We use a rounding approach on DPFGG to get a program that is polynomial in $n = |\mathcal{J}|$, given that ψ is constant. We scale d, c, p_c, and p_s by a factor $\varsigma := \frac{\varepsilon \cdot d}{2n}$. Denote by $\hat{d} := \lceil \frac{d}{\varsigma} \rceil \leq \frac{2n}{\varepsilon} + 1$, $\hat{p}_s(j) := \lfloor \frac{p_s(j)}{\varsigma} \rfloor$, $\hat{p}_c(j) := \lfloor \frac{p_c(j)}{\varsigma} \rfloor$ and $\hat{c}(x) := \lfloor \frac{c(x)}{\varsigma} \rfloor$. Note here, that we round up d but everything else down. We run the dynamic program with the rounded values, but still calculate the cost of a state with the original unscaled values.

We transform the output π' to the unscaled instance, by trying to start every job j at the same (scaled back up) point in time as in the scaled schedule.

Since we rounded down, there might now be points in the schedule where a job j can not start at the time it is supposed to. This might be due to the server not being free, a parent node of j that has not been fully processed or an unfinished communication delay. We look at the first time this happens and call the mandatory delay on j Δ and increase the start time of every remaining job by Δ. Repeat this process until all jobs are scheduled. We introduce no new conflicts with this procedure, since we always move everything together as a block. Call this new schedule π.

Theorem 5. *Assuming a constant number ψ DPFGG combined with the scaling technique finds a schedule π with at most optimal cost and a makespan $\leq (1+\varepsilon) \cdot d$ in time $poly(n, \frac{1}{\varepsilon})$, for any $\varepsilon > 0$.*

Proof. We start by proving the runtime of our algorithm. We can scale the instance in polynomial time, this holds for both scaling down and scaling back up. The dynamic program now takes time in $O(\hat{d}^{2\psi+3} \cdot n^{2\psi+1})$, where $\hat{d} \leq \frac{2n}{\varepsilon} + 1$. Since ψ is constant this results in an dynamic program runtime in $poly(n, \frac{1}{\varepsilon})$. In the end we transform the schedule as described above, for that we go trough the schedule once and delay every job no more than n times. Trivially, this can be done in polynomial time as well.

Secondly we show that the makespan of π is at most $(1 + \varepsilon) \cdot d$. Every valid schedule for the unscaled problem is also valid in the scaled problem, meaning that there is no possible schedule we overlook due to the scaling. In the other direction this might not hold. First, while scaling everything down we rounded the deadline up. This means, that scaled back we might actually work with a deadline of up to $d + \varsigma$. Secondly, we had to delay the start of jobs to make sure that we only start jobs when it is actually possible. In the worst case we delay the sink $\mathcal{T}$ a total of $n - 2$ times, once for every job other than $\mathcal{S}$ and $\mathcal{T}$. Each time we delay all remaining jobs we can bound the respective $\Delta < 2 \cdot \varsigma$. This is due to the fact that each of the delaying options can not delay by more than ς (as that is the maximum timespan not regarded in the scaled problem) and only a direct predecessor job and the communication from it needing longer can coincide to a non-parallel delay. Taking both of these into account, a valid schedule for the scaled problem might use time up to

$$d + \varsigma + (n - 2) \cdot (2\varsigma) \leq d + 2n\varsigma = (1 + \varepsilon) \cdot d$$

in the unscaled instance.

Lastly, we take a look at the cost of π. While rounding, we did not change the calculation of a states value, and with every valid schedule of the unscaled instance being still valid in the scaled instance we can conclude that the cost of π is smaller or equal to an optimal solution of the original problem.   □

Theorem 6. *DPFGG combined with the scaling technique and a binary search over the deadline yields an FPTAS for the cost budget makespan problem, for graphs with a constant number ψ.*

Proof. Theorem 5 can be adapted to solve this, assuming that we know a reasonable makespan estimate of an optimal solution to use in our scaling factor. During the algorithm discard any state with costs bigger than the budget and terminate when the first state $[t, f_s]$ is reached. The t gives us the makespan.

Using a makespan estimate that is too big will lead to a rounding error that is not bounded by $\varepsilon \cdot mspan_{OPT}$, a too small estimate might not find a solution. To solve this, we start with an estimate that is purposefully large. Let $d^{max} = \sum_{j \in \mathcal{J}} p_s(j)$ be the sum over all processing times on the server. There is always a schedule with 0 costs and makespan d^{max}. We run our algorithm with the scaling factor $\varsigma^0 := \frac{\varepsilon \cdot d^{max}}{4n}$. Iteratively repeat this process with scaling factor $\varsigma^i = \frac{1}{2^i} \varsigma^0$ for increasing i starting with 1. At the same time half the original deadline estimate in each step, which leads to $\hat{d}$, and therefore the runtime, to stay the same in each iteration. End the process when the algorithm does not find a solution for the current i and deadline estimation. This infers that there is no schedule with the wanted cost budget and a makespan smaller or equal to $\frac{1}{2^i} d^{max}$ (in the unscaled instance), therefore $\frac{1}{2^i} d^{max} < mspan_{OPT}$. We look at the result of the previous run $i - 1$: The scaled result was optimal, therefore the unscaled version has a makespan of at most

$$mspan_{ALG} \leq mspan_{OPT} + 2n \cdot \varsigma^{i-1} \tag{1}$$

$$= mspan_{OPT} + 2n \cdot \frac{1}{2^{i-1}} \cdot \frac{\varepsilon \cdot d^{max}}{4n} \tag{2}$$

$$= mspan_{OPT} + \varepsilon \cdot \frac{1}{2^i} d^{max} \leq (1 + \varepsilon) mspan_{OPT}. \tag{3}$$

It should be easy to infer from Lemma 1 that each iteration of this process has polynomial runtime. Combined with the fact that we iterate at most $\log d^{max}$ times we get a runtime that is in $poly(n, \frac{1}{\varepsilon})$. □

Remark 2. The results of this chapter work, as written, for a constant ψ. Note here, that for series parallel digraphs, this is equivalent to a constant anti-chain size. The algorithms can also be adapted to work on any graph with constant anti-chain size, if the communication delays are *locally small*. Delays are locally small, if for every $(j, k) \in E$, $c(j, k)$ is smaller or equal than every $p_c(k')$, $p_s(k')$, $p_c(j')$ and $p_s(j')$, where k' is every direct successor of j and j' every direct predecessor of k [11].

4   Strong NP-Hardness

In this section, we consider more involved reductions then in Sect. 2 in order to gain a better understanding for the complexity of the problem. First, we show that a classical result due to Lenstra and Rinnooy Kan [8] can be adapted to prove that already the variant of *SCS* without communication delays and processing times equal to one or two is NP-hard. This already implies strong NP-hardness. Remember that we did show in Sect. 2 that *SCS* without communication delays and with unit processing times can be solved in polynomial time. Hence, it seems natural to consider the problem variant with unit processing times and communication delays. We prove this problem to be NP-hard as well via an intricate reduction from $3SAT$ that can be considered the main result of this section. Lastly, we show that the latter reduction can be easily modified to get a strong inapproximability result regarding the general variant of SCS and the cost objective.

No Delays and Two Sizes. We show strong hardness for the case without communication delays and $p_c(j), p_s(j) \in \{1, 2\}$ for each job j. The reduction is based on a classical result due to Lenstra and Rinnooy Kan [8].

Let $G = (V, E)$, k be a clique instance with $|E| > \binom{k}{2}$, and let $n = |V|$ and $m = |E|$. We construct an instance of the cloud server problem in which the communication delays all equal zero and both the deadline and the cost bound is $2n + 3m$. There is one vertex job $J(v)$ for each node $v \in V$ and one edge job $J(e)$ for each edge $e \in E$ and $J(\{u, v\})$ is preceded by $J(u)$ and $J(v)$. The vertex jobs have size 1 and the edge jobs size 2 both on the server and on the cloud.

Furthermore there is a dummy structure. First, there is a chain of $2n + 3m$ many jobs called the anchor chain. The i-th job of the anchor chain is denoted $A(i)$ for each $i \in \{0, \dots 2n+3m-1\}$ and has size 1 on the cloud and size 2 on the server. Next, there are gap jobs each of which has size 1 both on the server and the cloud. Let $k^* = \binom{k}{2}$ and $v \prec w$ indicate that an edge from v to w is included in the task graph. There are four types of gap jobs, namely $G(1, i)$ for $i \in \{0, \dots k-1\}$ with edges $A(2i) \prec G(1, i) \prec A(2(i + 1))$, $G(2, i)$ for $i \in \{0, \dots k^* - 1\}$ with $A(2k + 3i + 1) \prec G(2, i) \prec A(2k + 3(i + 1))$, $G(3, i)$ for $i \in \{0, \dots (n - k) - 1\}$ with $A(2k + 3k^* + 2i) \prec G(3, i) \prec A(2k + 3k^* + 2(i + 1))$, and $G(4, i)$ for $i \in \{0, \dots (m-k^*)-1\}$ with $A(2n+3k^*+3i+1) \prec G(4, i) \prec A(2n+3k^*+3(i+1))$ for $i < (m - k^*) - 1$ and $A(2n + 3m - 2) \prec G(4, (m - k^*) - 1)$. Lastly, there are the source and the sink which precedes or succeeds all of the above jobs, respectively.

Lemma 2. *There is a k-clique, if and only if there is a schedule with length and cost at most $2n + 3m$.*

Proof. First note that in a schedule with deadline $2n + 3m + 1$ the anchor chain has to be scheduled completely on the cloud. If the schedule additionally satisfies the cost bound, all the other jobs have to be scheduled on the server. Furthermore, for the gap and anchor chain jobs there is only one possible time slot due to the deadline. In particular, $A(i)$ starts at time i, $G(1, i)$ at time $2i + 1$, $G(2, i)$ at time $2k + 3i + 2$, $G(3, i)$ at time $2k + 3k^* + 2i + 1$, and $G(4, i)$ at time $2n + 3k^* + 3i + 2$. Hence, there are k length 1 slots positioned directly before the $G(1, i)$ jobs left on the server, as well as, k^* length 2 slots directly before the $G(2, i)$ jobs, $n - k$ length 1 slots directly before the $G(3, i)$ jobs, and $m - k^*$ length 2 slots directly before the $G(2, i)$ jobs (see also Fig. 2). The m edge jobs have to be scheduled in the length 2 slots, and hence the vertex jobs have to be scheduled in the length 1 slots.

$\implies$: Given a k-clique, we can position the k clique vertices in the first k length 1 slots, the corresponding k^* edges in the first length 2 slots, the remaining vertex jobs in the remaining length 1 slots, and the remaining edge jobs in the remaining length 2 slots.

$\impliedby$: Given a feasible schedule, the vertices corresponding to the first length 1 slots have to form a clique. This is the case, because there have to be k^* edge jobs in the first length 2 slots and all of their predecessors are positioned in the first length 1 slots. This is only possible if these edges are the edges of a k-clique. $\square$

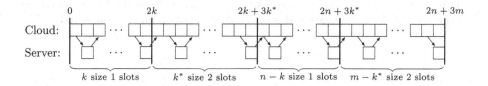

Fig. 2. The dummy structure for the reduction from the clique problem to a special case of *SCS*. Time flows from left to right, the anchor chain jobs are positioned on the cloud, and the gap jobs on the server.

Hence, we have:

Theorem 7. *The SCS problem with job sizes 1 and 2 and without communication delays is strongly NP-hard.*

In the above reduction the server and the cloud machines are unrelated relative to each other due to different sizes of the anchor chain jobs. However, it is easy to see that the reduction can be modified to a uniform setting where the cloud machines have speed 2 and the server speed 1. If we allow communication delays, even identical machines can be achieved.

Unit Size and Unit Delay. We consider a unit time variant of our model in which all $p_c = p_s = 1$ and all $c = 1$. Note here, that this also implies that the server and the cloud are identical machines (the cloud still produces costs, while the server does not). As usual for reductions we look at the decision variant of the problem: Is there a schedule with cost smaller or equal to b while adhering to the deadline d.

Theorem 8. *The SCS^1 problem is strongly NP-hard.*

We give a reduction $3SAT \leq_p SCS^1$. Let ϕ be any boolean formula in 3-CNF, denote the variables in ϕ by $\mathcal{X} = \{x_1, x_2, \ldots, x_m\}$ the clauses by $\mathcal{C} = \{C_1^\phi, C_2^\phi, \ldots, C_n^\phi\}$. Before we define the reduction formula we want to give an intuition and a few core ideas used in the reduction.

The main idea is that we ensure that nearly everything has to be processed on the cloud, there are only a few select jobs that can be handled by the server. For each variable there will be two jobs, of which one can be processed on the server, the selection will represent an assignment. For each clause there will be a job per literal in that clause, only one of which can be processed on the server, and only if the respective variable job is 'true'. Only if for each variable and for each clause one job is handled by the server the schedule will adhere to both the cost and the time limits.

A core technique of the reduction is the usage of an anchor chain (Fig. 3). An anchor chain of length l consists of two chains of the same length $l := d - 2$, where we interlock the chains by inserting (a_i, b_{i+1}) and (b_i, a_{i+1}) for two parallel edges (a_i, a_{i+1}) and (b_i, b_{i+1}). The source $\mathcal{S}$ is connected to the two start nodes of the anchor chain, the two nodes at the end of the chain are connected to $\mathcal{T}$.

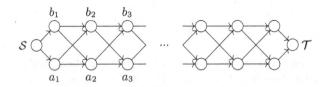

Fig. 3. Schematic representation of an anchor chain

Lemma 3. *If the task graph of a SCS^1 problem contains an anchor chain, every valid schedule has to schedule all but one of a_1, b_1 and one of a_l, b_l on the cloud. For every job a_i, b_i $1 < i < l$ the time step in which it will finish processing on the cloud in every valid schedule is $i + 1$.*

Finally we give the reduction function $f(\phi) = G, d, b$, where $G = (\mathcal{J}, E)$. Set $d = 12 + m + n$ and $b = |\mathcal{J}| - (2 + m + n)$. We define G by constructively giving which jobs and edges are created by f. Create an anchor chain of length $d - 2$, this will be used to limit parts of a schedule to certain time frames. Note that by Lemma 3 we know that every valid schedule of $G = (\mathcal{J}, E), d, \mathbf{k}$ has every node pair of the anchor chain (besides the first and last) on the cloud at a specific fixed timestamp. More specifically, the completion time of a_i and a_{i+j} differ by exactly j time units. For each variable $x_i \in \mathcal{X}$ create two jobs j_{x_i} and $j_{\bar{x}_i}$ and edges $(a_{1+i}, j_{x_i}), (a_{1+i}, j_{\bar{x}_i})$ and $(j_{x_i}, a_{5+i}), (j_{\bar{x}_i}, a_{5+i})$. For each clause C_p^ϕ create a clause job $j_{C_p^\phi}$ and edges $(a_{7+m+p}, j_{C_p^\phi})$ and $(j_{C_p^\phi}, a_{9+m+p})$. Let L_1^p, L_2^p, L_3^p be the literals in C_p^ϕ. Create jobs $j_{L_1^p}, j_{L_2^p}, j_{L_3^p}$ and edges $(j_{L_1^p}, C_p^\phi), (j_{L_2^p}, C_p^\phi), (j_{L_3^p}, C_p^\phi)$ for these literals. For every literal job $j_{L_i^p}$ connect it to the corresponding variable job j_{x_i} or $j_{\bar{x}_i}$ by a chain of length $1 + (m - i) + p$. Also create an edge from a_{3+i} to the start of the chain and an edge from the end of the chain to a_{6+m+p} (Fig. 4).

It remains to show that there is a schedule of length at most d with costs at most b in $f(\phi) = G, d, b$ if and only if there is a satisfying assignment for ϕ.

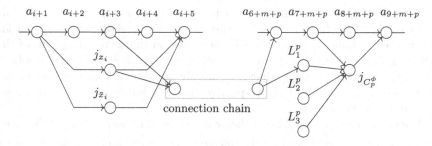

Fig. 4. Schematic representation of the variable and clause gadgets and their connection

Lemma 4. *In a deadline adhering schedule for $f(\phi) = G, d, b$ every job in the anchor chain (except on at the front and one at the end), every job in the variable and clause literal connecting chains and every clause job has to be scheduled on the cloud.*

Proof. By Lemma 3 we already know that every node in the anchor chain except one of v_1, w_1 and one of v_l, w_l has to be scheduled on the cloud. We also know, that the jobs in the anchor chain have fixed time steps in which they have to be processed. We look at some chain and its connection to the anchor chain. The start of the chain of length $1 + (m-i) + p$ is connected to a_{3+i}, the end to a_{6+m+p}. Between the end of a_{3+i} and the start of a_{6+m+p} are $6 + m + p - 1 - (3+i) = 2 + m + p - i$ time steps. So with the processing time required to schedule all $1 + (m-i) + p$ jobs of the chain, there is only one free time step, but we would need at least 2 free time steps to cover the communication cost to and from the server. (Recall here that both a_{3+i} and a_{6+m+p} have to be processed on the cloud). The same simple argument fixes each clause job to a specific time step on the server. □

Lemma 5. *In a deadline adhering schedule for $f(\phi) = G, d, b$ only one of j_{x_i} and $j_{\bar{x}_i}$ can be processed on the server for every variable $x_i \in \mathcal{X}$. The same is true for $j_{L_1^p}, j_{L_2^p}, j_{L_3^p}$ of clause C_p^ϕ.*

Proof. j_{x_i} and $j_{\bar{x}_i}$ are both fixed to the same time interval via the edges (a_{1+i}, j_{x_i}), $(a_{1+i}, j_{\bar{x}_i})$ and (j_{x_i}, a_{5+i}), $(j_{\bar{x}_i}, a_{5+i})$. Since a_{1+i} and a_{5+i} will be processed on the cloud and keeping communication delays in mind, only the middle of the three time steps in between can be used to schedule j_{x_i} or $j_{\bar{x}_i}$ on the server. Since the server is only a single machine only on of them can be processed on the server. Note here that the other job can be scheduled a time step earlier which we will later use. The argument for $j_{L_1^p}, j_{L_2^p}, j_{L_3^p}$ works analogously to the statement above. □

Lemma 6. *There is a deadline adhering schedule for $f(\phi) = G, d, b$ with costs of $|\mathcal{J}| - (2 + m + n)$ if and only if there is a satisfying assignment for ϕ. The variable jobs processed on the cloud represent this satisfying assignment*

Proof. From Lemma 3, Lemma 4 and Lemma 5 we can infer that a schedule with costs of $|\mathcal{J}| - (2 + m + n)$ has two jobs of the anchor chain, one job for each pair of variable jobs and one job per clause on the server. Two jobs of the anchor chain can always be placed on the server, the choice of variable jobs is also free. It remains to show, that we can only schedule a literal job per clause on the server if and only if the respective clause is fulfilled by the assignment inferred by the variable jobs.

The clause job $j_{C_p^\phi}$ of C_p^ϕ has to be processed in time step $9 + m + p$ (between a_{7+m+p} and a_{9+m+p}). Therefore, $j_{L_1^p}$ has to be processed no later than $8 + m + p$ or $7 + m + p$ if it is processed on the cloud or server respectively. Let j_{x_i} be the variable job connected to $j_{L_1^p}$ via a connection chain.

If j_{x_i} is *true* (scheduled on the cloud), it can finish processing at time step $3 + i$, which does not delay the start of the connection chain (which is connected to a_{3+i}, finishing in time step $4 + i$). This means that the chain can finish in time step $4 + i + 1 + (m - i) + p = 5 + m + p$, the time step $6 + m + p$ can be used for communication, allowing $j_{L_1^p}$ to be processed by the server in $7 + m + p$.

If j_{x_i} is *false* (scheduled on the server), it finishes processing at time step $4 + i$, which, combined with the induced communication delay, delays the start of the chain by 1. Therefore, the chain only finishes in time step $6 + m + p$, and $j_{L_1^p}$ has to be processed on the cloud, since there is not enough time for the communication back and forth.

Trivially, the same argument holds true for $j_{L_2^p}$ and $j_{L_3^p}$. □

It should be easy to see that the reduction function f is computable in polynomial time. Combined with Lemma 6 this concludes the proof of our reduction $3SAT \leq_p SCS^1$. The correctness of Theorem 8 trivially follows from that.

The General Case. Adapting the previous reduction we can show an even stronger result for the general case of SCS. Basically we are able to degenerate the reduction output in a way, that a satisfying assignment results in a schedule with cost 0, while every other assignment (schedule) has costs of at least 1. It should be obvious, that this also means that there is no approximation algorithm for this problem with a fixed multiplicative performance guarantee, if $P \neq NP$.

In this version we will use processing times and communication delays of 0, ∞ and values in between. Note here that ∞ can be replaced by $d + 1$, and it is also possible to realize this construction without 0, for readability and space we will only do the version using 0. To keep the following part readable we will substitute "an edge (j, j') with cost $c(j, j') = k$" simply by "an edge $c(j, j') = k$"

We follow the same general structure (an anchor chain, variable-, clause- and connection gadgets). The *anchor chain* now looks as follows: For every time step create two jobs a_i and a_i' with $p_s(a_i) = 0$, $p_c(a_i) = \infty$, $p_s(a_i') = \infty$, $p_c(a_i') = 0$ and an edge $c(a_i, a_i') = 0$. These chain links are than connected by an edge $c(a_i', a_{i+1}) = 1$. Finally we create $c(\mathcal{S}, a_1) = 1$ and $c(a_d, \mathcal{T}) = 0$. It should be easy to see, that every schedule will process a_i and a_i' in time step i on the server and the cloud respectively. This gives us anchors to the server and to the cloud for every time step, without inducing congestion or costs. Since the anchor jobs themselves have processing time of 0, the "usable" time interval between some a_i and a_{i+1} is one full time step.

For each variable $x_i \in X$ create two jobs j_{x_i}, $j_{\bar{x}_i}$ with $p_s(j_{x_i}) = p_s(j_{\bar{x}_i}) = 1$ and $p_c(j_{x_i}) = p_c(j_{\bar{x}_i}) = 0$. Create edges $c(a_i, j_{x_i}) = 1$, $c(a_i, j_{\bar{x}_i}) = 1$ and $c(j_{x_i}, a_{i+1}) = 0$, $c(j_{\bar{x}_i}, a_{i+1}) = 0$. In short, only one of them can be processed on the server, the other on the cloud. Both will finish in time step $i + 1$, the one processed on the server is "true", therefore processing both on the cloud is possible, but not helpful.

For each clause C_p^ϕ create a clause job $j_{C_p^\phi}$ with $p_s(j_{C_p^\phi}) = \infty$, $p_c(j_{C_p^\phi}) = 0$ and edges $c(a'_{5+m+3p}, j_{C_p^\phi}) = \infty$ and $c(j_{C_p^\phi}, a'_{6+m+3p}) = \infty$. This means, that $j_{C_p^\phi}$ has to finish processing by time step $6 + m + 3p$. Let L_1^p, L_2^p, L_3^p be the literals in C_p^ϕ. Create jobs $j_{L_1^p}, j_{L_2^p}, j_{L_3^p}$ each with $p_c = p_s = 1$ and edges $c(j_{L_1^p}, C_p^\phi) = 0$, $c(j_{L_2^p}, C_p^\phi) = 0$, $c(j_{L_3^p}, C_p^\phi) = 0$ for these literals. Create edges $c(a_3 + m + 3p, j_{L_1^p}) = 0$, $c(a_3 + m + 3p, j_{L_2^p}) = 0$ and $c(a_3 + m + 3p, j_{L_3^p}) = 0$, so that, in theory, all three of the literal jobs can be processed on the server, finishing in time steps $4 + m + 3p$, $5 + m + 3p$ and $6 + m + 3p$ respectively. Lastly, for every literal job $j_{L_1^p}$ connect it to the corresponding variable job j_{x_i} (or $j_{\bar{x}_i}$) by a an edge with communication delay of $m - i + 3p + 3$. Since j_{x_i} (or $j_{\bar{x}_i}$) finish processing in time step $i+1$, this means that $j_{L_1^p}$ can start no earlier than $m + 3p + 4$ (and therefore finish processing in $5 + m + 3p$), if j_{x_i} (or $j_{\bar{x}_i}$) were processed on the cloud. Recall here, that a variable job being scheduled on the server denotes that it is "true". So only a literal job that evaluates to true, can be scheduled so that it finishes processing in time step $4 + m + 3p$ on the cloud.

It follows directly, that a schedule for this construction will have costs of 0 if and only if the assignment derived from the placement of the variable jobs fulfills every clause.

Theorem 9. *There is no approximation algorithm for SCS that has a fixed performance guarantee, assuming that $P \neq NP$.*

5   Unit Size and Unit Delay

As the last step of this paper we explore simple algorithms on unit size instances with arbitrary task graphs. Recall that we proved these to be strongly NP-hard. We use resource augmentation and ask: given a SCS^1 problem instance with deadline d, find a schedule in poly. time that has a makespan of at most $(1+\varepsilon) \cdot d$ that *approximates* the optimal cost in regards to the actual deadline d. If there is a chain of length d or $d - 1$, that chain has to be scheduled on the server, since there is no time for the communication delay. For instances with a chain of size d that is trivially optimal, for those with $d - 1$ we can check in polynomial time if any other job also fits on the server, again, finding an optimal solution. From now we assume that there is no chain of length more than $d - 2$.

First, construct a schedule which places every job on the cloud, as fast as possible. The resulting schedule from time step (ts) 1 to $(1+\varepsilon) \cdot d$ looks as follows:

one ts of communication, at most $d-2$ ts of processing on the server, another ts for communication followed by at least εd empty ts. Now pull (one of) the last job(s) that is processed on the cloud to the last empty ts and process it on the server instead. Repeat this process until the last job can not be moved to the server anymore. Do the whole procedure again, but this time starting with the cloud schedule in the end of the schedule, and each time pulling the first job to the beginning. Keep the result with lower costs. Note that one can always fill the ts being used solely for communicating from the server to the cloud with processing one job on the server, that otherwise would be one of the first jobs being processed on the cloud (the same holds for the other direction).

Theorem 10. *The described algorithm yields a schedule with approximation factor of $\frac{1+\varepsilon}{2\varepsilon}$ while having a makespan of at most $(1+\varepsilon)\cdot d$.*

Proof. Case $n \le (1+\varepsilon)d$: The algorithm places all jobs on the server, the cost is 0 and therefore optimal.

Case $(1+\varepsilon)d < n < (1+2\varepsilon)d$: Assume that the preliminary cloud-only schedule needs $d-2$ ts on the cloud, if that is not the case, we *stretch* the schedule to that length. There are n jobs distributed onto $d-2$ ts. Therefore, either from the front or from the end, there is an interval of length $\frac{d}{2}-1$ with at least $\frac{d}{2}-1$ and at most $\frac{n}{2} < \frac{(1+2\varepsilon)d}{2} = \frac{d}{2} + \varepsilon d$ many jobs. It should be easy to see, that the algorithm will schedule those at most $\frac{d}{2} + \varepsilon d - 1$ jobs to the $\frac{d}{2}-1$ plus the free εd many time slots. If the interval included less than $\frac{d}{2} + \varepsilon d - 1$ jobs, it will simply continue until the $\frac{d}{2} - 1 + \varepsilon d$ ts are filled with jobs being processed on the server. With the one job we can process on the server during the communication ts we process $\frac{d}{2} + \varepsilon d$ jobs on the server and have costs of $n - (\frac{d}{2} + \varepsilon d)$. An optimal solution has costs of at least $n-d$. For $\varepsilon \ge 0.5$ it holds that: $cost_{ALG} = n - (\frac{d}{2} + \varepsilon d) \le n - d \le cost_{OPT}$, otherwise:

$$\frac{cost_{ALG}}{cost_{OPT}} \le \frac{n - (\frac{d}{2} + \varepsilon d)}{n - d} \le \frac{(1+\varepsilon)d - (\frac{d}{2} + \varepsilon d)}{(1+\varepsilon)d - d} \le \frac{0.5d}{\varepsilon d} = \frac{1}{2\varepsilon}$$

Case $(1+2\varepsilon)d \le n$: In this case we simply observe that our algorithm places at least εd many jobs on the server. For $\varepsilon \ge 1$ it holds that: $cost_{ALG} = n - \varepsilon d \le n - d \le cost_{OPT}$, otherwise:

$$\frac{cost_{ALG}}{cost_{OPT}} \le \frac{n - \varepsilon d}{n - d} \le \frac{(1+2\varepsilon)d - \varepsilon d}{(1+2\varepsilon)d - d} = \frac{d + \varepsilon d}{2\varepsilon d} = \frac{1+\varepsilon}{2\varepsilon}$$

$\square$

6   Conclusion

We give a small overview over the future research directions that emerge from our work. **SCS**: Sect. 4 gives a strong inapproximability result for the general case with regards to the cost function. For two easy cases (chain and fully parallel graphs) we could establish FPTAS results, for graphs with a constant ψ we have an algorithm that finds optimal solutions with a $(1 + \varepsilon)$ deadline augmentation. Here one could explore if there are FPTAS results for different assumptions, are there approximation algorithms without resource augmentation for constant ψ instances, and lastly are there approximation algorithms with resource augmentation for the general case. For the makespan function we already have a FPTAS for graphs with a constant ψ. It remains to explore approximation algorithms or inapproximability results for the general case of this problem. **SCS**[1]: We show strong NP-hardness even for this simplified problem. Since this is a special case of the general problem all constructive results still hold, additionally we were able to give a first simple algorithm for cost optimization in general graphs. Here it would be interesting to look into more involved approximation algorithms that give better performance guarantees, maybe without resource augmentation.

Finally one could think about expanding the model. Natural first steps might be multiple (identical or different) server machines and/or different cloud environments. But keep in mind here, that additional simplifying assumptions might be necessary to keep constructive results possible.

References

1. Ait Aba, M., Munier Kordon, A., Pallez (Aupy), G.: Scheduling on two unbounded resources with communication costs. In: Yahyapour, R. (ed.) Euro-Par 2019. LNCS, vol. 11725, pp. 117–128. Springer, Cham (2019). https://doi.org/10.1007/978-3-030-29400-7_9

2. Ait Aba, M., Aupy, G., Munier-Kordon, A.: Scheduling on Two Unbounded Resources with Communication Costs. Research Report RR-9264, Inria, March 2019. https://hal.inria.fr/hal-02076473

3. Davies, S., Kulkarni, J., Rothvoss, T., Tarnawski, J., Zhang, Y.: Scheduling with communication delays via LP hierarchies and clustering. In: 61st IEEE Annual Symposium on Foundations of Computer Science, FOCS 2020, Durham, NC, USA, 16–19 November 2020, pp. 822–833. IEEE (2020). https://doi.org/10.1109/FOCS46700.2020.00081

4. Davies, S., Kulkarni, J., Rothvoss, T., Tarnawski, J., Zhang, Y.: Scheduling with communication delays via LP hierarchies and clustering II: weighted completion times on related machines. In: Marx, D. (ed.) Proceedings of the 2021 ACM-SIAM Symposium on Discrete Algorithms, SODA 2021, Virtual Conference, 10–13 January 2021, pp. 2958–2977. SIAM (2021). https://doi.org/10.1137/1.9781611976465.176

5. Garg, S.: Quasi-ptas for scheduling with precedences using LP hierarchies. In: Chatzigiannakis, I., Kaklamanis, C., Marx, D., Sannella, D. (eds.) 45th International Colloquium on Automata, Languages, and Programming, ICALP 2018, July 9–13, 2018, Prague, Czech Republic. LIPIcs, vol. 107, pp. 59:1–59:13. Schloss Dagstuhl - Leibniz-Zentrum für Informatik (2018). https://doi.org/10.4230/LIPIcs.ICALP.2018.59

6. Graham, R.L.: Bounds for certain multiprocessing anomalies. Bell Syst. Tech. J. **45**(9), 1563–1581 (1966). https://doi.org/10.1002/j.1538-7305.1966.tb01709.x

7. Kulkarni, J., Li, S., Tarnawski, J., Ye, M.: Hierarchy-based algorithms for minimizing makespan under precedence and communication constraints. In: Chawla, S. (ed.) Proceedings of the 2020 ACM-SIAM Symposium on Discrete Algorithms, SODA 2020, Salt Lake City, UT, USA, 5–8 January 2020, pp. 2770–2789. SIAM (2020). https://doi.org/10.1137/1.9781611975994.169

8. Lenstra, J.K., Kan, A.H.G.R.: Complexity of scheduling under precedence constraints. Oper. Res. **26**(1), 22–35 (1978). https://doi.org/10.1287/opre.26.1.22

9. Levey, E., Rothvoss, T.: A (1+epsilon)-approximation for makespan scheduling with precedence constraints using LP hierarchies. In: Wichs, D., Mansour, Y. (eds.) Proceedings of the 48th Annual ACM SIGACT Symposium on Theory of Computing, STOC 2016, Cambridge, MA, USA, 18–21 June 2016, pp. 168–177. ACM (2016). https://doi.org/10.1145/2897518.2897532

10. Mäcker, A., Malatyali, M., auf der Heide, F.M., Riechers, S.: Cost-efficient scheduling on machines from the cloud. J. Comb. Optim. **36**(4), 1168–1194 (2018). https://doi.org/10.1007/s10878-017-0198-x

11. Möhring, R.H., Schäffter, M.W., Schulz, A.S.: Scheduling jobs with communication delays: Using infeasible solutions for approximation. In: Diaz, J., Serna, M. (eds.) ESA 1996. LNCS, vol. 1136, pp. 76–90. Springer, Heidelberg (1996). https://doi.org/10.1007/3-540-61680-2_48

12. Saha, B.: Renting a cloud. In: Seth, A., Vishnoi, N.K. (eds.) IARCS Annual Conference on Foundations of Software Technology and Theoretical Computer Science, FSTTCS 2013, December 12–14, 2013, Guwahati, India. LIPIcs, vol. 24, pp. 437–448. Schloss Dagstuhl - Leibniz-Zentrum für Informatik (2013). https://doi.org/10.4230/LIPIcs.FSTTCS.2013.437

13. Svensson, O.: Hardness of precedence constrained scheduling on identical machines. SIAM J. Comput. **40**(5), 1258–1274 (2011) https://doi.org/10.1137/100810502

FIFO and Randomized Competitive Packet Routing Games

Bjoern Tauer[1,2] and Laura Vargas Koch[1(✉)]

[1] School of Business and Economics, RWTH Aachen University, Aachen, Germany
{bjoern.tauer,laura.vargas}@oms.rwth-aachen.de
[2] Department of Computer Science, RWTH Aachen University, Aachen, Germany
tauer@algo.rwth-aachen.de

Abstract. In competitive packet routing games, packets route over time through a network from a start to a destination node. The packets selfishly minimize their arrival time, which depends on one hand on the transit times of the edges and on the other hand on the suffered waiting times. These occur whenever several packets try to enter an edge simultaneously. In those situations, scheduling policies determine which packet is allowed to enter this edge first and which has to wait.

We analyze the impact of two different scheduling policies on the inefficiency of equilibria, namely of a FIFO policy and a randomized policy. In a FIFO policy conflicts between interacting packets are resolved by their arrival time. In a randomized policy, an ordering of the packets is drawn uniformly at random and in any conflict the packet with the lower position is prioritized. While the existence of pure Nash equilibria in single-commodity games with the FIFO policy is already known, we provide a constructive proof for the existence of pure Nash equilibria in games with a randomized policy. In contrast, in multi-commodity games, there are examples without a pure Nash equilibrium for both policies. We provide several bounds on the price of anarchy and stability in single-commodity games and analyze the total price of anarchy in multi-commodity games. Surprisingly, the price of anarchy in single-commodity games is linear in the number of packets for both examined priority policies. For randomized priorities, we furthermore show a connection of parallel networks to singleton congestion games with affine linear cost functions. From this, we derive a tight constant bound on the price of stability and the price of anarchy.

Keywords: Packet routing games · Coordination policy ·
Randomization · First-in-first-out · Price of anarchy · Price of stability

1 Introduction

Packet Routing is a well-studied problem [LMRR94, LMR94, LMR99, Rot13] with applications in various areas including logistics, road traffic and data traffic.

We want to thank the Deutsche Forschungsgemeinschaft (DFG, German Research Foundation) – UnRAVeL-Research Training Group 2236/1 for funding this research.

J. Koenemann and B. Peis (Eds.): WAOA 2021, LNCS 12982, pp. 165–187, 2021.
https://doi.org/10.1007/978-3-030-92702-8_11

To model a road traffic or data traffic situation remarkably large and complex networks are needed. Moreover, the lack of a central authority makes it impossible to control the streams through the network. Thus, a competitive view on packet routing is of interest. In competitive packet routing, every packet tries to minimize its arrival time independently and selfishly. Since packets travel through the same network simultaneously they influence each other's arrival times. Accordingly, clear rules are needed for the movements of packets in the network and their interaction. To analyze the impact of certain regulations, a common approach is to investigate their influence on stable states in the network, e.g., equilibrium states. A network designer can reasonably choose these rules to influence the arising (equilibrium) states. In road traffic, this corresponds to a city planner choosing, e.g., a speed limit or a give-of-way road to reduce congestion. In this paper, we concentrate on the latter. We are interested in whether classes of priority rules exist that guarantee good equilibrium states. In the past already several different rules have been studied. The FIFO rule, where packets are ordered by their arrival time seems to naturally follow from traffic and is the most widely studied rule [CCCW17, Ism17, SST18, WHK14]. Moreover, it is the base of the traffic simulation software MATSim [ANH16, HNA16]. We show that for this rule the price of anarchy is linear in the number of players. This result appears to be unforeseen since the price of anarchy is as high as in rules where the power of players is not as equally distributed as in the case of the FIFO rule. Besides FIFO, player priorities [HPS+18], edge priorities [SSV18], the highest density first rule [HMRT11] as well as fair cost-sharing policies [HMRT09] are investigated. For none of these variants, a constant bound on the price of anarchy could be shown, if considered at all. To be precise, the price of anarchy of all deterministic model variants turns out to be at least linear in the number of players or in the network dilation. Motivated by this observation, the natural question arises if the bounds can be improved incorporating randomization. In store-and-forward packet routing optimization problems, randomness turned out to be a very useful tool [LMRR94, Rot13]. Although the approximation algorithms can be derandomized this approach simplified the analysis and improved the understanding. Therefore, we examine a randomized priority policy and investigate whether we can also achieve better equilibrium states. Furthermore, in related models, the ordering of the players (as a priority rule, as a tie-breaker or as the network entrance time) leads to high social costs and also to an unbounded price of anarchy. See e.g.[HPS+18] and [SST18], where Scarsini et al.observe that 'the selfish behavior of the first players has consequences that cannot be undone and are paid by all future generations'. To circumvent this effect random priorities seem to be the right choice. Unfortunately, it turns out that for randomized priorities the price of anarchy is again linear in the number of players. Nevertheless, we give a constructive proof for the existence of pure Nash equilibria in single-commodity networks. Moreover, for parallel networks, we point out a connection to singleton congestion games with affine costs and show a tight bound on the price of anarchy and stability.

1.1 Related Work

Vickrey [Vic69] was the first one to propose a competitive model to predict traffic. Such a model is the deterministic queuing model by Koch and Skutella [KS11]. It is a continuous competitive flow over time model, which was analyzed extensively [CCO19,CCL11,CCL15,CCO17,KS11,SS18]. A discrete equivalent to this model corresponds to competitive packet routing games with FIFO. Variants of this have been examined from different perspectives [CCCW17,Ism17,SST18]. Cao et al. [CCCW17] concentrate on the existence of subgame perfect equilibria, while Ismaili [Ism17] focuses on computational complexity results for questions as computing a best response or a Nash equilibrium. Scarsini et al. [SST18] determine the dependency of the price of anarchy on seasonal inflows. Werth, Holzhauser and Krumke [WHK14] analyze variants of players' cost and the social cost functions. For a game with weighted players and where player's costs are bottleneck costs, they present a tight bound on the price of anarchy which is linear in the number of players.

While in the models presented so far, conflicts between players are solved with respect to their arrival time, there are also models where this is handled differently. Harks et al.introduced a competitive packet routing model with a priority ordering of all players [HPS+18]. A similar model where priority is assigned to edges, not to players is presented in [SSV18]. While in these models both time and packet size are discrete, there are also variants with discrete time steps but a continuous flow [KNS11] and discrete packets but a continuous time [HMRT11,KM15].

1.2 Our Contribution

For FIFO competitive packet routing games our main result is a lower bound example on the price of anarchy (PoA) for single-commodity games, i.e., games where all players share the same origin and destination. We achieve a lower bound linear in the number of players by nesting several Braess-like graphs. After shortly recapping the existence results of pure Nash equilibria in FIFO competitive packet routing games, we present further inefficiency bounds for this model. We show with a simple example where even in single-commodity games, the optimal solution might not be a pure Nash equilibrium, more precisely, the price of stability (PoS) is at least $\frac{19}{17}$. Furthermore, the PoA is upper bounded by $\frac{1}{2}(n+1)$ for single-commodity games with n players. Together with our lower bound example, it follows that the PoA bound is tight. In multi-commodity instances, existence of pure Nash equilibria cannot be guaranteed. Here, we apply the LP-technique introduced by Kulkarni et al. [KM15] to show an upper bound of $6D^2 - D$ on the price of total anarchy. The network dependent constant D used in this bound denotes the dilation of the network including 0-cost edges and is properly defined in Sect. 3.2. Our main example also serves as lower bound example here with PoA = $\sqrt{D}$.

For games with randomized priorities, our main result is the existence proof of pure Nash equilibria for every single-commodity instance. We show that there is always a Nash equilibrium where players distribute to a set of disjoint paths. Using this, we present a greedy algorithm for both choosing these paths and distributing the players to the chosen paths. This gives a nice and simple algorithm to compute a pure Nash equilibrium. For multi-commodity instances, we provide an example where no pure Nash equilibrium exists. Furthermore, we investigate the inefficiency of randomized games. The efficiency analysis of parallel networks leads to a tight bound of $\frac{4}{3}$ both for the price of stability as well as the price of anarchy. This result relies on the fact that parallel network games correspond to linear congestion games with affine cost functions. Thus, we can use a bound provided by [Fot10]. For general single-commodity games, we show an upper bound of $\frac{1}{2}(n+1)$ on the price of stability. To do so, we use the structure of the Nash equilibria constructed by the algorithm. Moreover, we provide a Braess graph as an example with a price of anarchy of $\frac{1}{2}(n+1)$. Utilizing the LP-technique by Kulkarni et al. [KM15] also for randomized multi-commodity games, we show an upper bound of $6D^2 - D$ on the price of total anarchy. Furthermore, the Braess graph provides a lower bound of $\frac{1}{4}(D+1)$.

2   Model

In a competitive packet routing game, we consider a directed graph $G = (V, E)$ with a finite set of vertices V and a finite set of edges E. Each edge $e \in E$ is equipped with a free flow transit time $\tau_e \in \mathbb{N}_0$. We have a set of players $N = \{1, \ldots, n\}$, representing packets, and each of them has an origin vertex $o_i \in V$ and a destination vertex $d_i \in V$. We assume that the minimal distance between the origin and the destination is at least 1.

Furthermore, we are given a priority policy, which organizes the interplay of the packets. In every time step, where two or more players aim to enter the same edge, the policy determines which players have to wait and which player can enter the edge. In this work, we consider a FIFO policy and a randomized policy, which we introduce below.

Every player $i \in N$ has a strategy space $\mathcal{P}_i$ which is the set of all simple paths connecting o_i and d_i, and from which she can choose a strategy $P_i \in \mathcal{P}_i$. We limit ourselves to the set of simple paths since it is never advantageous to choose a non-simple path for the considered policies. In case that all players share the same origin and the same destination vertex, we call the game *single-commodity*. Otherwise, it is a *multi-commodity* game.

A state of the game is a tuple of strategies such that every player has chosen a strategy from her strategy space $P = (P_1, \ldots, P_n)$ with $P_i \in \mathcal{P}_i$. The space of all states is $\mathcal{P}$, i.e., $\mathcal{P} = \mathcal{P}_1 \times \ldots \times \mathcal{P}_n$.

The cost of a player i in a state P is her arrival time $C_i(P)$ at the destination vertex. The arrival time consists of two parts. The first is the free flow transit time along P_i given by $\tau(P_i) = \sum_{e \in P_i} \tau_e$, which is independent of P_{-i}. The second part is the waiting time which occurs along the path and which highly depends on the strategies chosen by the other players and on the priority policy. Let $w_{i,e}(P)$ be the (expected) waiting time of player i on edge e in state P. With this definition $C_i(P) = \sum_{e \in P_i} (w_{i,e}(P) + \tau_e)$.

While every player i aims to minimize $C_i(P)$ in a game with FIFO as priority policy, in a game with randomized priorities players minimize their expected arrival time $\mathbb{E}[C_i(P)]$. The social cost of state P is $\sum_{i \in N} C_i(P)$ or $\sum_{i \in N} \mathbb{E}[C_i(P)]$, respectively.

To analyze the efficiency of pure Nash equilibria we consider the famous concepts of the *price of stability* (PoS) [ADK+04] and the *price of anarchy* (PoA) [KP99]. In games where pure Nash equilibria do not always exist, we bound the ratio of the maximal social cost of a coarse correlated equilibrium and the socially optimal cost. This is known as the *price of total anarchy* [BHLR08].

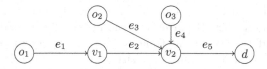

Fig. 1. Network with free flow transit time 1 for all edges to illustrate the FIFO policy and the randomized policy.

2.1   Coordination Policies

The edge model we consider is motivated by the deterministic queuing model [KS11, Vic69]. Every edge has a bottleneck capacity at the beginning. Packets aiming to enter this edge line up and at every time step one of those packets is determined by a coordination policy to enter the edge while the others need to wait for at least one time step. After traversing the bottleneck, packets do not influence the transit time of the other packets anymore, instead all need a fixed free flow transit time.

The coordination policy is the rule, which uniquely determines the next packet to be processed. In this work, we examine a FIFO policy and a randomized policy.

FIFO Policy. In a *first-in-first-out policy* players can enter an edge in order of their arrival times, i.e., a player trying to enter an edge is always preferred over all later arriving players. Only if more than one player arrives at the same point in time, a global priority list $\pi = (1, \ldots, n)$ is applied as a tie-breaking mechanism. If two players i and j with $i <_\pi j$ are in conflict with each other and arrived at the same time, then player i is preferred over player j.

Example 1. Consider the network depicted in Fig. 1 with three players, originating in o_i for $i \in \{1, 2, 3\}$ and heading towards a shared sink d. Every player has a unique o_i-d path P_i. In profile $P = (P_1, P_2, P_3)$ players 2 and 3 arrive at vertex v_2 after one unit of time. Thus, they are ordered regarding to the priority list $\pi = (1, 2, 3)$. Accordingly, player 2 enters edge e_5 at time 1, while player 3 has to wait in the queue of edge e_5. At the next point in time, also player 1 arrives at vertex v_2. Even though the priority of player 1 is higher than the one of player 3, player 3 enters edge e_5 first since she arrived at vertex v_2 earlier. Thus, the arrival times are $C_1(P) = 4$, $C_2(P) = 2$ and $C_3(P) = 3$.

Randomized Policy. In a *randomized policy*, a total ordering of the players $\pi = (1, \ldots, n)$ is drawn uniformly at random at the beginning of the game and after the choice of the strategies of the players. In case of a conflict, the players are handled as described for the tie-breaking situation in the FIFO part. This means the player with the lowest position in the ordering is always preferred.

Example 2. Consider the example depicted in Fig. 1 with three players taking their unique o_i-d path for $i \in \{1, 2, 3\}$. The arrival time of a player depends on the drawn priority list. If player 1 is the first or the second in the priority list, she arrives at time 3, else at time 4. This yields an expected arrival time $\mathbb{E}[C_1(P)] = \frac{10}{3}$. For players 2 and 3 the situation is identical. If player 2 is first, she arrives at time 2. If she is second she arrives at time 2 or time 3 each with probability $\frac{1}{2}$, since it depends on the position of players 1 and 3 in the priority list. If she is third, she arrives at time 4. All together this gives an expected arrival time of $\mathbb{E}[C_i(P)] = \left(\frac{1}{3} + \frac{1}{6}\right) \cdot 2 + \frac{1}{6} \cdot 3 + \frac{1}{3} \cdot 4 = \frac{17}{6}$ for $i \in \{2, 3\}$.

2.2   Making the Game Well-Defined

In a competitive packet routing game, players choose paths and afterwards they are embedded into the network. This embedding process repeats a procedure of two steps for every integral time step. First, for every edge e all packets that can potentially enter this edge are determined. These are the packets that fully traversed the preceding edge e' of their path, i.e. spent $\tau_{e'}$ time units on e' or have the edge e as the first edge of their path. In the second step, according to the coordination mechanism, the packet that is actually allowed to enter the edge e at the current time step is determined. Note that a packet traversing some 0-cost edges might enters multiple edges at the same time step. But only one packet can enter an edge at every point in time. All remaining packets have to wait and are considered again in the next integral time step of the process until they are allowed to enter edge e.

Depending on the state, i.e., the paths of all players, a player evaluates her strategy and also the alternatives. Thus, for a well-defined game, we need to convert the paths into a sequence of movements in the network. This allows then to define a mapping $C : \mathcal{P} \to \mathbb{R}_{\geq 0}^n$ that, given a state, assigns (expected) arrival times.

The expected arrival time in a game with randomized priorities can be computed by enumerating all $n!$ permutations of player priority lists. For a player priority list $\pi = (1, \ldots, n)$, we can handle one player after another. We begin by embedding player 1 into the network and thereby calculating the cost of P_1 and deleting the edges along P_1 at the respective times. We proceed with P_2 until we embedded all paths. Since player i is not influenced by any player $j >_\pi i$, this is possible. For FIFO, the existence of an embedding is less obvious. An Algorithm is presented in [Ism17, Algorithm 1, Theorem 2]. The idea of the proof is to embed the players time step by time step. For each time step the graph of 0-cost edges is considered and on this graph on every component the players are forwarded in order of the tie-breaking rule.

3   FIFO Policy

In related models [HPS+18, SSV18], the price of anarchy in single-commodity instances is linear in the number of players. Especially in the worst-case examples for global player priority lists [HPS+18], the total ordering of the players is responsible for the occurring delay. This allows the player with the highest remaining priority to choose a shortest path intersecting with any other possible path. An approach to address this disadvantageous effect is the FIFO policy. We first shortly recap existence results for pure Nash equilibria and afterward provide bounds on the prices of stability and anarchy. The main result of this section is the tight bound of $\frac{1}{2}(n+1)$ on the price of anarchy in single-commodity games.

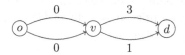

Fig. 2. Single-commodity game, the transit time τ_e is depicted for all $e \in E$.

3.1   Existence of Pure Nash Equilibria

In single-commodity games, pure Nash equilibria are guaranteed to exist and can be computed with a greedy type approach.

Theorem 1 ([WHK14])**.** *Every single-source FIFO competitive packet routing game with arbitrary player priorities as tie-breaking rule has a pure Nash equilibrium.*

In contrast, in multi-commodity games, there are instances without pure Nash equilibrium.

Theorem 2 ([Ism17, WHK14])**.** *In a multi-commodity FIFO competitive packet routing game with player priorities as tie-breaking rule, there may not exist a pure Nash equilibrium. This holds true even for planar, acyclic graphs.*

3.2   Inefficiency of Pure Nash Equilibria in Single-Commodity Games

Within this section we first show that there exist instances where the price of stability is at least $\frac{19}{17}$. Second, we prove that the price of anarchy in single-commodity games with FIFO policy is still linear in the number of players.

Theorem 3. *There is a single-commodity FIFO competitive packet routing game instance, with a price of stability of $\frac{19}{17}$.*

Proof. Consider the network depicted in Fig. 2 with 6 players. In an optimal profile, the players coordinate their strategies in such a way that the upper v-d edge is used twice (at time 0 and time 1). This results in the optimal arrival pattern $(1, 2, 3, 3, 4, 4)$, and thus in $C(OPT) = 17$. On the contrary, in any pure Nash equilibrium the first two players use the lower v-d edge. Thus, the upper v-d edge will be used firstly at time 1. This results in an arrival pattern $(1, 2, 3, 4, 4, 5)$ with cost $C(NE) = 19$. In total, the price of stability is at least PoS $\geq \frac{19}{17}$. It can be shown that this is maximal for the given network by computing the PoS in dependency of n and taking the derivative.   □

The problem to derive an upper bound on the price of stability below the following price of anarchy bound remains open. Next, we present a sophisticated lower bound example with a price of anarchy of $\frac{1}{2}(n + 1)$. Analogous to [SSV18] it follows that this bound is tight.

Theorem 4. *For single-commodity FIFO competitive packet routing games, the price of anarchy is upper bounded by $\frac{1}{2}(n + 1)$. This bound is tight.*

Proof. We present an instance with PoA $= \frac{1}{2}(n + 1)$, by cleverly connecting several Braess-like graphs. There exists a pure Nash equilibrium where players 1 to i block the network for all points in time $\theta \in \{0, \ldots, i - 1\}$ for all $i \in N$. Thus, the players arrive one after another at the destination. To construct the example, we first describe the network, then the strategies of all players and finally we show that the described state is indeed a pure Nash equilibrium.

Our network, the extended Braess graph EBG, is a generalization of the Braess graph. It results from a consecutive alignment of sub-graphs $BG^*(2)$, $BG^*(3), \ldots, BG^*(n)$. A sub-graph $BG^*(i)$ corresponds to a single path of $2 \cdot i$ vertices, e.g., see $BG^*(3)$ in Fig. 3. We visualize all vertices with odd index as one column and all vertices with even index as a second column. For $i = 2, \ldots, n-1$, every sub-graph $BG^*(i)$ is extended to the *height* of $BG^*(n)$ by adding another sub-graph $CG(n - i)$ above $BG^*(i)$, see Fig. 4. A graph $CG(i)$ consists of two columns of i vertices each, connected by an edge, see Fig. 3. Every vertex of the right column of a combined stack $(CG(n - i - 1), BG^*(i + 1))$ is connected to the vertex of the same height of the left column of stack $(CG(n - i), BG^*(i))$ by an edge. We denote the vertices of EBG by v_ℓ^k, where $\ell \in \{1, \ldots, 2n - 2\}$ denotes the columns of the stacks counting from left to right and $k \in \{1, \ldots, n\}$ the layer from top to bottom.

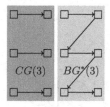

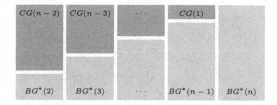

Fig. 3. Sub-graphs $CG(n)$ and $BG^*(n)$ for $n = 3$.

Fig. 4. Consecutive alignment of sub-graph stacks $(CG(i), BG^*(i))$ for $i = 2, \ldots, n$.

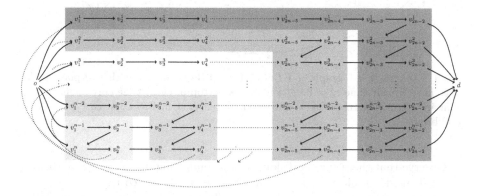

Fig. 5. Single-commodity instance EBG with n players and n disjoint o-d paths. All edges have cost 0, except for the edges (v_{2n-2}^i, d) which have cost 1 for all $i \in [n]$.

The single source o is connected by a path to each vertex of the first column, i.e., to the left-hand side of $BG^*(2)$ and $CG(n-2)$. More precisely, there exists a path $p_k = (o, \ldots, v_1^k)$ for all $k \in [n]$. Note that these paths will be intersected by so-called *backloop* paths, which are introduced in more detail after the basic structure of the graph is explained. Sink d is connected to the last column, i.e., the right-hand side of $BG^*(n)$. More precisely, we have edges (v_{2n-2}^k, d) for all $k \in [n]$. Finally, the vertex v_{2i-2}^n, which is the end of the zig-zag path of sub-graph $BG^*(i)$, is connected with the first vertex v_1^{n-i} of layer $n - i$ for all $i = 2, \ldots, n - 1$ via the backloop paths. To increase readability we depict the paths p_k as single edges and the backloop paths as dotted edges in Fig. 5. All edges have transit time 0, except the ones that are adjacent to d which have transit time 1. The construction of EBG is depicted in Fig. 5. Now, it is time to revisit the backloop paths. They are constructed such that first, the final graph is planar; second, all backloop paths are edge and vertex disjoint; and third, a backloop path with destination v_1^ℓ, $\ell \in \{1, \ldots, n - 2\}$ intersects all o-v_1^k paths, where $k < \ell$, and has exactly one edge which each of those paths in common. See Fig. 6 for an exemplary construction. Note that the vertices denoted with b are introduced to guarantee the above-mentioned properties.

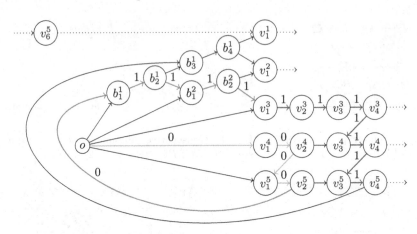

Fig. 6. Exemplary sub-graph of $EBG(5)$ for a FIFO instance with five players. All edges have transit time $\tau_e = 0$. The numbers on top of every edge correspond to the point in time when this edge is traversed by player 4 in the worst-case pure Nash equilibrium.

Observe that there exist n disjoint paths connecting o and d, each path with cost 1. Thus, for n players, the optimal state in EBG has cost n.

In contrast, in a worst pure Nash equilibrium, the players arrive one after another. The strategy of player 1 is to go along the straight upper path and then through $BG^*(n)$, i.e., she uses the blue path in Fig. 5. Player 2 goes straight along the second highest path, then through $BG^*(n-1)$, followed by the backloop path, then the highest path and afterward through $BG^*(n)$. This corresponds to the red path and afterward the blue path in Fig. 5.

In general player i takes the i-th path from the top, traverses $BG^*(n-i+1)$, and then copies the strategy of player $i-1$, where the initial path $(o, \ldots, v_1^{i-1})$ of the copied strategy is replaced with the *backloop* path $(v_{2n-2i}^n, \ldots, v_1^{i-1})$. This is depicted in Fig. 6 representatively.

In this strategy profile, no player can be overtaken by any player with lower priority. Observe that the constructed graph has a planar embedding. So, whenever all the edges that are used by a player i at time θ are deleted the graph and especially o and d are disconnected. This implies that all players $1, \ldots, i$ block the graph at time $\theta \in \{0, \ldots, i-1\}$ and thus player i cannot be overtaken by a subsequent player. To see this more concretely, consider Fig. 6. Here, player 4 starts with the orange path. The highlighted red edges correspond to the backloop path of player 4. Afterward, she follows the strategy of player 3, which is highlighted in green. Observe that by deleting all edges used at time 1 by player 4 the graph is separated into two parts, i.e., one containing o and the other containing d. Furthermore, no player with priority below player 4 can reach any of the start points of these edges before time 1, since either the edges are already used by players with higher priority or are blocked by player 4. Thus, by playing the defined strategy player 4 is prevented from being overtaken by any subsequent player. Since the same argumentation holds true for every predecessor of player 4, player 4 has also no improving move. Since every player uses her predecessor's path one time step later, this argument holds for all time steps. Hence, no player has an incentive to deviate at all.

Hence the described state is a pure Nash equilibrium. As a consequence, we obtain the claimed bound $\text{PoA} = \frac{1}{n} \sum_{i=1}^{n} i = \frac{1}{2}(n+1)$.

The proof of the upper bound relies on similar ideas as a proof presented in [SSV18]. Consider a single-commodity game instance with the FIFO policy and global tie-breaking priority list π, say $\pi = (1, \ldots, n)$. The distance of source o and sink d is denoted by ℓ, thus no player can arrive before time ℓ. Accordingly, the cost of any optimal strategy P^* is bounded from below by $n \cdot \ell$. Let $P = (P_1, \ldots, P_n)$ be the path vector of any pure Nash equilibrium strategy. It suffices to show that the arrival time of player i in the order of the global tie-breaking priority list π does not exceed $\ell - 1 + i$ time units. We prove this statement by contradiction. Assume there exists a player $i \in \{1, \ldots, n\}$ who is arriving late, i.e., after $\ell - 1 + i$. If there are several of those, let j be the first one arriving late in the order of the tie-breaking priority list. It follows that this player has an improving move. She can start at the source o and follow a shortest path of length ℓ. If she has to wait at the entry of any edge $e = (u, v)$, there are two cases. If all players who delay player j in front of e have a higher priority, she follows the last player with higher priority. If she has to wait for a player h with a lower priority, there exists a faster o-u path to reach u, especially the one player h chooses. Since player j has a higher priority than player h, she can displace h from this sub-path without affecting any player with a higher priority. This replacement of sub-paths can be repeated until player j arrives at sink d. In total player 1 cannot be delayed at all and reaches d at time ℓ. A player j arrive at most one time unit after the last player of $\{1, \ldots, j-1\}$. Since the constructed move is improving for j it is a contradiction to the assumption that P was a pure Nash equilibrium. This yields the following equation which is maximized for $\ell = 1$,

$$\text{PoA} \leq \frac{C(P)}{C(P^*)} \leq \frac{1}{n \cdot \ell} \sum_{i=1}^{n} (\ell - 1 + i) = 1 + \frac{n-1}{2\ell}.$$

$\square$

In multi-commodity instances, we cannot guarantee the existence of pure Nash equilibria. Thus, the analysis of the inefficiency is much more involved. We present an upper bound relying on an LP-technique introduced by Kulkarni et al.[KM15]. The proof is presented in the next section, see Theorem 9. The analysis transfers directly by choosing the deterministic equivalents. For example, a variable of the LP is set one when an edge is used in the deterministic variant, while it is set to the probability that an edge is used in the randomized variant. Note that in contrast to the bound presented for the single-commodity instance this bound does not depend on the number of players but only on the dilation, which is a property of the network. The dilation D

of a graph is the number of edges on a longest simple o_i-d_i path for $i \in N$, where every edge of length $\tau_e > 1$ is replaced by τ_e edges of length 1, i.e., $D := \max_{i \in N} \max_{P \in \mathcal{P}_{o_i,d_i}} \sum_{e \in P} \max\{\tau_e, 1\}$.

Theorem 5. *The price of total anarchy in multi-commodity FIFO games is bounded by $6D^2 - D$, where D is the dilation.*

Since the price of total anarchy generalizes the price of anarchy this implies that $6D^2 - D$ is an upper bound on the price of anarchy in every instance which admits a pure Nash equilibrium.

Corollary 1. *In a game with FIFO policy, where a pure Nash equilibrium exists, the price of anarchy is upper bounded by $6D^2 - D$, where D is the dilation.*

By bounding $D \leq 3n^2$, the example presented in the proof of Theorem 4 provides an instance with $PoA = \sqrt{D}$.

Remark 1. There is a FIFO competitive packet routing game with a price of anarchy of $\Omega(\sqrt{D})$.

4   Randomized Priority Policies

In this section, we analyze the existence of pure Nash equilibria in randomized competitive packet routing games. As in FIFO games it turns out that pure Nash equilibria in single-commodity instances exist and can be computed, while pure Nash equilibria in multi-commodity instances are not guaranteed to exist. Furthermore, we analyze the efficiency. For parallel networks we show a tight bound of $\frac{4}{3}$ on the price of anarchy and the price of stability. As soon as we allow general single-commodity networks, there are instances where the price of anarchy is $\frac{1}{2}(n+1)$. Moreover, this is an upper bound on the price of stability in symmetric games. For multi-commodity instances, as for FIFO games, we bound the price of total anarchy by $6D^2 - D$ which implies a bound on the price of anarchy in every instance where a pure Nash equilibrium exists. The lower bound example gives a price of anarchy linear in D.

4.1   Existence and Computability of Pure Nash Equilibiria

In randomized competitive packet routing games, interferences of players get very fast very complicated since we get conditional probabilities for the expected arrival times. So far this is also the reason why we do not completely understand multi-commodity games. In single-commodity instances, we can restrict ourselves to disjoint paths which simplifies analysis a lot.

In the following, we present an algorithm that computes a pure Nash equilibrium in a single-commodity randomized competitive packet routing game and an example of a multi-commodity instance where no pure Nash equilibrium exists.

Algorithm 1: COMPUTES A PURE NASH EQUILIBRIUM IN A GAME WITH RANDOMIZED PRIORITIES.

Input: $G = (V, E)$ with $o, d \in V$ and a set of players $N = \{1, \ldots, n\}$
Output: A path for every player $i \in N$
/* Construct a set of possible paths. */
1 **while** *There is a feasible path* **do**
2   Compute shortest path p in G
3   Add tuple of the transit time and the path $(at(p) := \tau(p), p)$ to the set of potentially used paths $\mathcal{P}_{pot}$.
4   Delete p from G.
5 **end**
/* Assign paths to the players. */
6 **for** $i \in N$ **do**
7   Choose path P_i from $\mathcal{P}_{pot}$ with $at(P_i)$ minimal.
8   Increase $at(P_i)$ by $\frac{1}{2}$.
9 **end**

Single-commodity Games

Theorem 6. *In single-commodity competitive packet routing games with the randomized policy, we can compute a pure Nash equilibrium in time polynomial in n and $|E|$.*

Proof. First, we describe how to compute a pure Nash equilibrium and then we show why the described procedure achieves the desired goal. A pure Nash equilibrium is computed in two phases.

In phase 1, the set of possible paths is built up. A shortest path p with respect to $\tau(p) = \sum_{e \in p} \tau_e$ is computed, added to the set $\mathcal{P}_{pot}$ and afterward all edges contained in p are deleted from the network. For each path, we introduce a variable $at(p)$, where we store the expected transit time along the path. In this step, we initialize it with the transit time of path p. This means we gather disjoint paths of increasing length in $\mathcal{P}_{pot}$.

In phase 2, players are distributed to the paths. To do this a path with minimal value $at(p)$ is selected from $\mathcal{P}_{pot}$. The player is assigned to this path and $at(p)$, the expected arrival time, is increased by $\frac{1}{2}$. This increase relies on the fact that in every realization of a uniform distributed priority list every other player has a 50% chance to be positioned behind the added player and thus to encounter an additional waiting time of 1, i.e., $\frac{1}{2}$ on expectation. A pseudo-code of the algorithm is given in Algorithm 1.

To see that this algorithm works correctly first note that, under the assumption that all players select disjoint paths, the computed state is a pure Nash equilibrium. This follows immediately when $at(p)$ describes indeed the expected arrival time. Given that k players use a path p, the expected arrival time of each player is $\tau(p) + \frac{1}{2}(k - 1)$ and thus coincides with $at(p)$ at the moment when the last player selecting path p made her choice.

It remains to show that it is enough to consider only disjoint paths. Assume for contradiction the algorithm computes a state P that is not a pure Nash equilibrium. This means, there is a player i that can switch to a path P_i' and improve her expected arrival time. From the construction of the algorithm, we know that P_i' is no path in $\mathcal{P}_{pot}$ and thus intersects at least one path of this set. The described algorithm adds the paths in order of increasing length to $\mathcal{P}_{pot}$, let p be the path added first of all the paths that intersect P_i'. Note again that p is a shortest path when it was chosen to be added to $\mathcal{P}_{pot}$ and that all edges of P_i' have been available at this moment. Let furthermore e be the first edge that is shared by p and P_i'. Since p is a shortest path player i can reach e not earlier than the first player using p. Thus, player i will be inserted at her position or after her position given by the random draw compared to all other players using e and thus p. After using edge e, player i cannot improve her costs by deviating again, since p is at most as long as P_i' concerning the free flow transit time and she will never be delayed using p. Thus, if P_i' was an improving move switching to p is an improving move as well, which contradicts the construction of P. Thus, the algorithm works correctly and it suffices to consider only disjoint paths. □

Multi-Commodity Games. In multi-commodity instances, the situation is more complex. Especially, there are examples where no pure Nash equilibrium exists.

Theorem 7. *There is a multi-commodity competitive packet routing game with randomized policy without pure Nash equilibrium.*

To prove the theorem, we use an idea presented in [Ism17], where the FIFO model is analyzed. We transfer the idea from the deterministic model to the randomized setting.

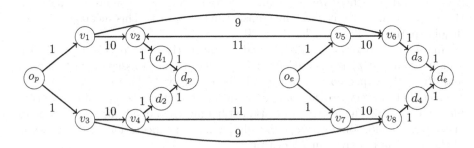

Fig. 7. Network of a game without pure Nash equilibrium.

Proof. The game presented in Fig. 7 is a Pursuer-Evader-Game. The idea is that the pursuer profits from playing the same strategy as the evader and the evader profits from playing a different strategy.

To construct such a game, we consider the network depicted in Fig. 7. We introduce six players: a pursuer going from o_p to d_p, an evader going from o_e to d_e and players to transfer the information from one side to the other, player 1 goes from o_e to d_1, player 2 from o_e to d_2 and player 3 and 4 from o_p to d_3 respectively d_4.

The pursuer and the evader choose between strategy *up* and *down*, while all other players have a unique path from their source to their sink. To understand the basic idea observe that the pursuer likes to be on the same path as the evader since with positive probability the evader will delay player 1 respectively 2 such that the pursuer will never meet this player. On the other hand, the evader wants to be on the different option since in this case player 3 respectively 4 passes before the evader wants to use the respective edge. While, when the pursuer is on the same path, there is some probability that the player is delayed and meets the evader. More concrete, if the evader and the pursuer both play the strategy *up*, the expected arrival time of the pursuer, as well as the evader, is 13.625. By switching to the strategy *down*, the evader reduces her expected costs to 13.5, while the expected arrival time of the pursuer increases to 13.75. Since all strategies are highly symmetric this proves that there is always an improving move and thus no pure Nash equilibrium.   □

4.2   Efficiency Analysis

For the randomized priority policy, we distinguish again between single-commodity and multi-commodity networks. Moreover, we investigate the special case of single-commodity games, where origin o and destination d are only connected by parallel edges. For these *parallel network randomized competitive games* we establish a constant and tight inefficiency bound. For more general single-commodity networks, we show a lower bound on the price of anarchy that is linear in the number of players. Moreover, we show that this bound is an upper bound on the price of stability. Furthermore, for multi-commodity networks, we bound the price of total anarchy by using the LP-technique introduced in [KM15] which we present in detail in this section.

Parallel Network Games. Randomized competitive games on parallel networks can be seen as singleton congestion games with affine linear cost functions. By using this fact we can show that $\frac{4}{3}$ is not only a tight bound on the price of stability but also on the price of anarchy.

Theorem 8. *The price of anarchy in randomized competitive packet routing games on parallel networks is upper bounded by $\frac{4}{3}$. This bound is tight for the price of anarchy as well as for the price of stability.*

Proof. To prove the upper bound we use a correspondance of a parallel network game to a singleton congestion game with affine linear costs. Given a parallel network game, define a congestion game with players N, resources E and cost functions $c_e(x) = \frac{1}{2}x + (\tau_e - \frac{1}{2})$. To see that the player's cost functions coincide in these games, let P be a state with $P_i = \{e\}$ and $n_e(P)$ be the number of players using edge e in state P. The players' cost in P in both games are

$$C_i(P) = c_e(n_e(P)) = \frac{1}{2}n_e(P) + \tau_e - \frac{1}{2} = \tau_e + \frac{1}{2}(n_e(P) - 1) = \mathbb{E}(C_i(P)).$$

Also the social cost functions coincide since the players are identical and the expected value is linear. In detail,

$$\begin{aligned}
C(P) &= \sum_{i \in N} c_{P_i}(n_{P_i}(P)) = \sum_{i \in N} \sum_{e \in E} \left(\left(\tau_e + \frac{1}{2}(n_e(P) - 1) \right) \cdot \mathbb{1}_{P_i = e} \right) \\
&= \sum_{i \in N} \sum_{e \in E} \left(\left(\tau_e + \frac{1}{n_e(P)} \left(\sum_{j=1}^{n_e(P)} (j - 1) \right) \right) \cdot \mathbb{1}_{P_i = e} \right) \\
&= \sum_{e \in E} \left(n_e(P)\tau_e + \left(\sum_{j=1}^{n_e(P)} (j - 1) \right) \right) = \sum_{e \in E} \sum_{j=1}^{n_e(P)} (\tau_e + j - 1) = \mathbb{E}(C(P)).
\end{aligned}$$

Due to this correspondence the price of anarchy in singleton congestion games with affine linear cost functions directly translates to our model.

As shown in [Fot10, Theorem 10] the price of anarchy in singleton games with affine cost functions can be bounded by $\frac{4}{3}$. See also Sect. 2.5 in [Rou09] for further insights. This directly translates to our model.

To show that this bound is tight we use the fact that players prefer shorter paths, since the contribution of a player to the expected cost is only $\frac{1}{2}$. Consider the network depicted in Fig. 8. It has n parallel edges, one of cost 1 and $n - 1$ of cost $k > 1$. In a game with $n = 2(k - 1)$ players, in a pure Nash equilibrium (NE), all players go along the edge with transit time 1. Since the expected cost for every player is $1 + \frac{1}{2} \cdot (2(k - 1) - 1) = k - \frac{1}{2}$ which is less than going along an edge of cost k. This yields an expected social cost of $2(k - 1)(k - \frac{1}{2})$. In an optimal solution (OPT), the first k players use the edge with cost 1 and afterward all players use different edges of cost k. This results in the following social cost $C(OPT) = \sum_{i=1}^{k} i + (k - 2)k = \frac{3}{2}k(k - 1)$. For the price of stability this means

$$\text{PoS} = \frac{\mathbb{E}[C(NE)]}{C(OPT)} = 2 \cdot \frac{2(k - 1)(k - \frac{1}{2})}{3k(k - 1)} = \frac{4}{3k}(k - \frac{1}{2}) = \frac{4}{3} - \frac{2}{3k},$$

which converges to $\frac{4}{3}$ for $k \to \infty$.   □

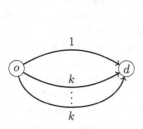 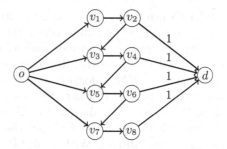

Fig. 8. Network of a game with $PoS = \frac{4}{3} - \frac{2}{3k}$

Fig. 9. The 4-Braess graph $BG(4)$, where every edge has cost $\tau_e = 0$ if not depicted otherwise.

Single-commodity Games. When introducing randomized priorities we expected that players distribute better along the available paths, but this turned out to be wrong. In the following, we show a lower bound on the price of anarchy of $\frac{1}{2}(n + 1)$ reusing the Braess-like graph applied in [HPS+18]. While, for the price of stability, we present an upper bound of $\frac{1}{2}(n + 1)$.

Proposition 1. *There is a competitive packet routing game with randomized priorities, that has a price of anarchy of at least $\frac{1}{2}(n + 1)$.*

Proof. Consider $BG(n)$ (see Fig. 9) where all players use the zigzag path (in the example it is $(o, v_1, v_2, v_3, \ldots, v_8, d)$). The expected cost of every player on every (alternative) path is $\frac{1}{2}(n + 1)$. Thus, the state is a pure Nash equilibrium with expected social cost of $\frac{1}{2}(n(n + 1))$, while the optimal solution has cost n.   □

By using the structure of a pure Nash equilibrium presented in the proof of Theorem 6, we can show an upper bound on the price of stability.

Proposition 2. *The price of stability in competitive packet routing games with randomized priorities is upper bounded by $\frac{1}{2}(n + 1)$.*

Proof. We bound the cost of the pure Nash equilibrium constructed in the algorithm of Theorem 6. The algorithm starts by assigning a player to a shortest path of the network. By always choosing a path of minimal arrival time, the algorithm only includes further paths in this assignment procedure if they improve the estimated arrival time. Thus, we can upper bound the expected arrival time of every player in the computed pure Nash equilibrium by the arrival time in the state where all players travel along the shortest path. Let ℓ be the length of the shortest path. Then, it holds for every player i in a pure Nash equilibrium P constructed by Algorithm 1 that $\mathbb{E}[C_i(P)] \leq \ell + \frac{1}{2}(n - 1)$, while the cost of the optimal solution is at least $n\ell$. For the price of stability, this leads to the following computation where we use $\ell \geq 1$ in the last inequality:
$PoS \leq \frac{1}{n\ell}(n \cdot (\ell + \frac{n-1}{2})) = 1 + \frac{n-1}{2\ell} \leq \frac{n+1}{2}$.   □

Multi-Commodity Games. In multi-commodity networks, we cannot guarantee the existence of pure Nash equilibria, but we can show that if a pure Nash equilibrium exists, the price of anarchy is bounded. Furthermore, we can bound the price of total anarchy. To do so, we use an idea originally introduced by Kulkarni et al. [KM15] for a related model. The optimal solution of an instance is described by an LP, where the cost function is a multiple of the costs of an optimal solution. The dual of the LP corresponds to a coarse correlated equilibrium, where again the cost function is a multiple of the costs of a coarse correlated equilibrium. With weak LP duality, we can estimate the cost of the coarse correlated equilibrium by a multiple of the cost of the optimal solution. This yields a bound on the price of total anarchy.

Theorem 9. *In a game with randomized priorities, the price of total anarchy is upper bounded by $6D^2 - D$, where D is the dilation.*

Proof. We start with the following preprocessing to ensure binary travel times. For a given instance, we substitute each edge $e \in E$ of transit time $\tau_e > 1$ by τ_e-many edges of length 1. This results in an instance with transit times $\tau_e \in \{0, 1\}$ for all edges $e \in E$. Note that splitting up an edge to shorter edges does not influence the set of coarse correlated equilibria.

The road-map of the proof is as follows. We define a primal-dual LP pair with primal feasible solution x, objective value $C_{\text{primal}}(x)$ and dual feasible solution (α, β, ν) with objective function value $C_{\text{dual}}(\alpha, \beta, \nu)$. We introduce some coefficients $c_1 = 2$ and $c_2 = 3D^2 - \frac{D}{2}$ such that

$$\mathbb{E}_{\bar{P} \sim \sigma}[C(\bar{P})] \leq c_1 C_{\text{dual}}(\alpha, \beta, \nu) \leq c_1 C_{\text{primal}}(x) \leq c_1 c_2 \mathbb{E}[C(OPT)],$$

where σ is a coarse correlated equilibrium, with respect to which states of the game $\bar{P}$ are drawn, x and (α, β, ν) are special primal and dual solutions defined as below and OPT denotes a social optimal solution.

We start by defining the primal linear program. Let $\mathcal{P}_i$ be the strategy set of player i in the following.

$$\min \sum_{i \in N} \sum_{P'_j \in \mathcal{P}_i} \sum_{e \in P'_j} \sum_{\theta \in \mathbb{N}} x_{eij\theta} \cdot (\theta + \tau_e)$$

$$\text{s.t.} \quad \sum_{P'_j \in \mathcal{P}_i} x_{ij} \geq 1, \qquad \forall i \in N,$$

$$\sum_{\theta \in \mathbb{N}} x_{eij\theta} \geq x_{ij}, \qquad \forall e \in E, i \in N, j : P'_j \in \mathcal{P}_i,$$

$$\sum_{i \in N} \sum_{P'_j \in \mathcal{P}_i} x_{eij\theta} \leq \frac{1}{2D}, \qquad \forall e \in E, \theta \in \mathbb{N},$$

$$x_{ij}, x_{eij\theta} \geq 0, \qquad \forall e \in E, \theta \in \mathbb{N}, i \in N, j : P'_j \in \mathcal{P}_i.$$

We claim that, given any socially optimal solution to a competitive packet routing instance, we can construct a feasible LP-solution x with objective value

$$C_{\text{primal}}(x) \leq (3D^2 - \tfrac{D}{2})\mathbb{E}[C(OPT)].$$

Let $OPT = P = (P_1, \ldots, P_n)$ be a socially optimal profile with expected arrival times $\mathbb{E}[C_i(P)]$ for all $i \in N$. We assign $x_{ij} = 1$ if player i chooses path $P'_j \in \mathcal{P}_i$, i.e., $P_i = P'_j$ and $x_{ij} = 0$ otherwise. This trivially fulfills the first inequality.

Furthermore, for every edge on every path we compute the probability $p_{eij\theta}$ that player i enters edge e on path P'_j at time θ. This is trivially 0, if a player does not select the respective path, i.e., $(x_{ij} = 0) \Rightarrow (p_{eij\theta} = 0) \quad \forall e, \theta$.

We assign $x_{eij(2D\theta)} = x_{eij(2D\theta+1)} = \cdots = x_{eij(2D\theta+2D-1)} = \frac{1}{2D} \cdot p_{eij\theta}$. To check that this is a feasible solution for the LP, first note that $x_{eij\theta}$ is only positive if $x_{ij} = 1$. Furthermore $\sum_{\theta \in \mathbb{N}} p_{eij\theta} = 1$, if $x_{ij} = 1$ since edge e is used in the optimal solution and thus the probabilities for all times need to sum up to one. Given this, it is easy to observe that the defined solution is indeed a feasible LP-solution.

For bounding the objective function we first observe that

$$\sum_{k=0}^{2D-1} x_{eij(2D\theta+k)} \cdot ((2D\theta + k) + \tau_e) = \sum_{k=0}^{2D-1} p_{eij\theta} \cdot \frac{1}{2D} \cdot (2D\theta + k + \tau_e)$$

$$= p_{eij\theta} \left(2D\theta + \tau_e + \frac{1}{2D} \sum_{k=0}^{2D-1} k \right) = p_{eij\theta} \left(2D\theta + \tau_e + \frac{2D-1}{2} \right).$$

Using this computation we can reformulate the objective function of the defined LP solution in the following way.

$$\sum_{i \in N} \sum_{P'_j \in \mathcal{P}_i} \sum_{e \in P'_j} \sum_{\theta \in \mathbb{N}} x_{eij\theta} \cdot (\theta + \tau_e)$$

$$= \sum_{i \in N} \sum_{P'_j \in \mathcal{P}_i} \sum_{e \in P'_j} \sum_{\theta \in \mathbb{N}} p_{eij\theta} \left(2D\theta + \tau_e + \frac{2D-1}{2} \right) = \star$$

For the next step, we use that the time a player enters an edge together with the transit time is a lower bound on the overall cost, which is also true in expectation. Applying this and the fact that $\frac{2D-1}{2}$ is constant and the probabilities sum up to one over all times, leads to the following computation.

$$\star \leq \sum_{i \in N} \sum_{P'_j \in \mathcal{P}_i} \sum_{e \in P'_j} \sum_{\theta \in \mathbb{N}} p_{eij\theta} \left(2D(\theta + \tau_e) + \frac{2D-1}{2} \right)$$

$$\leq \sum_{i \in N} \sum_{e \in P_i} \left(2D(\mathbb{E}[C_i(P)]) + \frac{2D-1}{2} \right)$$

Now, we estimate the objective value of the primal LP by a multiple of the expected cost of the optimal solution. Note that by definition of D we can bound the number of edges of every path P_i by D. The last inequality follows since the expected cost in every state is at least one by assumption.

$$\sum_{i \in N} \sum_{P'_j \in \mathcal{P}_i} \sum_{e \in P'_j} \sum_{\theta \in \mathbb{N}} x_{eij\theta} \cdot (\theta + \tau_e)$$

$$\leq \sum_{i \in N} \left(D \cdot \left(2D(\mathbb{E}[C_i(P)]) + \frac{2D-1}{2} \right) \right)$$

$$\leq \left(3D^2 - \frac{D}{2} \right) \sum_{i \in N} \mathbb{E}[C_i(P)] = \left(3D^2 - \frac{D}{2} \right) \mathbb{E}[C(OPT)].$$

In the remainder of the proof, we show that every coarse correlated equilibrium has expected social cost of at most $2 \cdot C_{\text{dual}}(\alpha, \beta, \nu)$ for a dual feasible solution (α, β, ν) constructed as described below. This yields the desired bound of $6D^2 - D$ on the price of total anarchy. Consider the dual LP

$$\max \sum_{i \in N} \alpha_i - \frac{1}{2D} \sum_{e \in E} \sum_{\theta \in \mathbb{N}} \beta_{e\theta}$$

$$\text{s.t. } \alpha_i - \sum_{e \in P'_j} \nu_{eij} \leq 0, \qquad \forall i \in N, j : P'_j \in \mathcal{P}_i,$$

$$\nu_{eij} - \beta_{e\theta} \leq \theta + \tau_e, \qquad \forall e \in E, i \in N, j : P'_j \in \mathcal{P}_i, \theta \in \mathbb{N},$$

$$\alpha_i, \beta_{e\theta}, \nu_{eij} \geq 0, \qquad \forall e \in E, i \in N, j : P'_j \in \mathcal{P}_i, \theta \in \mathbb{N}.$$

By weak LP duality, we know that any feasible dual solution (α, β, ν) has objective value at most $C_{\text{primal}}(x)$. It suffices to show that any coarse correlated equilibrium σ induces a feasible dual solution with objective value at least $\frac{1}{2} \sum_{i \in N} \mathbb{E}_{\bar{P} \sim \sigma}[C_i(\bar{P})]$.

We define a dual solution (α, β, ν) as follows. For each $i \in N$ let $\alpha_i = \mathbb{E}_{\bar{P} \sim \sigma}[C_i(\bar{P})]$, be the expected cost of player i in the coarse correlated equilibrium σ. Furthermore, for each $\theta \in \mathbb{N}$, $e \in E$ and state $\bar{P}$, let $\beta_{e\theta}^{\bar{P}} = |\{i \in N \mid e \in \bar{P}_i, \theta \leq \mathbb{E}[C_i(\bar{P})]\}|$ denote the number of players who use edge e in their selected path in $\bar{P}$ and have in expectation not yet arrived at their sink at time θ in the profile $\bar{P}$. To compute $\beta_{e\theta}$ we compute the expected value with respect to σ, i.e. $\beta_{e\theta} = $ (probability of $\bar{P}$ in σ)$\cdot\beta_{e\theta}^{\bar{P}}$. Finally, let $\nu_{eij} = \tau_e + \mathbb{E}_{\bar{P} \sim \sigma}[w_{ie}(P_j, \bar{P}_{-i})]$ denote the expected waiting time at the entry of e plus the traversing time τ_e in case player i switches to P_j instead of following the proposal from σ.

The first dual constraint is obviously satisfied by this definition of the dual variables, since $\alpha_i - \sum_{e \in P_j} \nu_{eij} \leq 0$ is equivalent to the following equation, which follows by the definition of a coarse correlated equilibrium

$$\mathbb{E}_{\bar{P} \sim \sigma}[C_i(\bar{P})] \leq \sum_{e \in P_j} (\tau_e + \mathbb{E}_{\bar{P} \sim \sigma}[w_{ie}(P_j, \bar{P}_{-i})]) = \mathbb{E}_{\bar{P} \sim \sigma}[C_i(P_j, \bar{P}_{-i})].$$

The second constraint $\nu_{eij} - \beta_{e\theta} \leq \theta + \tau_e$ can equivalently be written as the following equation, which is satisfied, for every outcome $\bar{P}$ and thus especially for the expected value with respect to σ.

$$\mathbb{E}_{\bar{P} \sim \sigma}[w_{ie}(P_j, \bar{P}_{-i})] \leq \theta + \beta_{e\theta}.$$

Given an outcome $\bar{P}$, this holds true since the expected waiting time of a player i switching to a path P'_j on some edge e on this path is bounded by the number of players using edge e in expectation after her expected arrival time. Since at most one player can use edge e per time step, the term θ upper bounds the number of players that passed edge e up to time θ and thus also the time of maximum delay until θ, while $\beta^{\bar{P}}_{e\theta}$ upper bounds the number of conflicts which might occur afterward.

It remains to show that $\mathbb{E}_{\bar{P} \sim \sigma}[C(\bar{P})] \leq 2C_{\mathrm{dual}}(\alpha, \beta, \nu)$. This holds true by the following computation, where we use in the first inequality that measuring the expected arrival time of a player is equivalent to counting the expected time a player spends in the network.

$$C_{\mathrm{dual}}(\alpha, \beta, \nu) = \sum_{i \in N} \alpha_i - \frac{1}{2D} \sum_{e \in E} \sum_{\theta \in \mathbb{Z}_{\geq 0}} \beta_{e\theta}$$

$$\geq \sum_{i \in N} \mathbb{E}_{\bar{P} \sim \sigma}\left[C_i(\bar{P})\right] - \frac{D}{2D} \sum_{i \in N} \mathbb{E}_{\bar{P} \sim \sigma}\left[C_i(\bar{P})\right] = \frac{1}{2}\mathbb{E}_{\bar{P} \sim \sigma}\left[C(\bar{P})\right].$$

Corollary 2. *In a game with randomized priorities, where a pure Nash equilibrium exists the price of anarchy is upper bounded by $6D^2 - D$, where D is the dilation.*

By considering the example presented in the proof of Proposition 1 and substituting $D = 2n + 1$, we get a lower bound on the price of anarchy.

Remark 2. There is a game with randomized priorities where the price of anarchy is linear in D.

5   Discussion

Motivated by the high price of anarchy in related models, we analyzed the efficiency of the FIFO priority mechanism and introduce a randomized priority mechanism. The randomized model turns out to be a theoretically interesting model which provides some nice results. A possible extension would be to draw a priority list not only once in the beginning but instead throw a coin for every conflict occurring between players. Note that in this extension of the model it gets very fast very complicated to compute the cost of a player since there are a lot of scenarios and conditional probabilities which need to be taken into account.

Unfortunately it turned out that also for both policies, FIFO and random, the price of anarchy is linear in the number of players. One reason could be that by the use of 0-cost edges there is a construction such that one player blocks all the following players without getting high costs herself. By only allowing strictly positive transit times it could be possible to achieve better bounds on the price of anarchy and stability.

References

[ADK+04] Anshelevich, E., Dasgupta, A., Kleinberg, J.M., Tardos, É., Wexler, T., Roughgarden, T.: The price of stability for network design with fair cost allocation. In: 45th Symposium on Foundations of Computer Science (FOCS 2004), Rome, Italy, 17–19 October 2004, Proceedings, pp. 295–304. IEEE Computer Society (2004)

[ANH16] Axhausen, K., Nagel, K., Horni, A. (eds.): Multi-Agent Transport Simulation MATSim. Ubiquity Press, London (2016)

[BHLR08] Blum, A., Hajiaghayi, M., Ligett, K., Roth, A.: Regret minimization and the price of total anarchy. In: Proceedings of the Fortieth Annual ACM Symposium on Theory of Computing, pp. 373–382 (2008)

[CCCW17] Cao, Z., Chen, B., Chen, X., Wang, C.: A network game of dynamic traffic. In: Proceedings of the 2017 ACM Conference on Economics and Computation, pp. 695–696 (2017)

[CCL11] Cominetti, R., Correa, J.R., Larré, O.: Existence and uniqueness of equilibria for flows over time. In: Aceto, L., Henzinger, M., Sgall, J. (eds.) ICALP 2011. LNCS, vol. 6756, pp. 552–563. Springer, Heidelberg (2011). https://doi.org/10.1007/978-3-642-22012-8_44

[CCL15] Cominetti, R., Correa, J., Larré, O.: Dynamic equilibria in fluid queueing networks. Oper. Res. 63(1), 21–34 (2015)

[CCO17] Cominetti, R., Correa, J., Olver, N.: Long term behavior of dynamic equilibria in fluid queuing networks. In: Integer Programming and Combinatorial Optimization, pp. 161–172 (2017)

[CCO19] Correa, J.R., Cristi, A., Oosterwijk, T.: On the price of anarchy for flows over time. In: Karlin, A., Immorlica, N., Johari, R. (eds.) Proceedings of the 2019 ACM Conference on Economics and Computation, EC 2019, Phoenix, AZ, USA, 24–28 June 2019, pp. 559–577. ACM (2019)

[Fot10] Fotakis, D.: Stackelberg strategies for atomic congestion games. Theory Comput. Syst. 47(1), 218–249 (2010)

[HMRT09] Hoefer, M., Mirrokni, V.S., Röglin, H., Teng, S.-H.: Competitive routing over time. In: Leonardi, S. (ed.) WINE 2009. LNCS, vol. 5929, pp. 18–29. Springer, Heidelberg (2009). https://doi.org/10.1007/978-3-642-10841-9_4

[HMRT11] Hoefer, M., Mirrokni, V.S., Röglin, H., Teng, S.-H.: Competitive routing over time. Theor. Comput. Sci. 412(39), 5420–5432 (2011)

[HNA16] Horni, A., Nagel, K., Axhausen, K.W.: Introducing matsim. In :The multi-agent transport simulation MATSim, pp. 3–7. Ubiquity Press, London (2016)

[HPS+18] Harks, T., Peis, B., Schmand, D., Tauer, B., Koch, L.V.: Competitive packet routing with priority lists. ACM Trans. Econ. Comput. 6(1), 4:1-4:26 (2018)

[Ism17] Ismaili, A.: Routing games over time with FIFO Policy. In: Devanur, N.R., Lu, P. (eds.) WINE 2017. LNCS, vol. 10660, pp. 266–280. Springer, Cham (2017). https://doi.org/10.1007/978-3-319-71924-5_19

[KM15] Kulkarni, J., Mirrokni, V.S.: Robust price of anarchy bounds via LP and fenchel duality. In: Indyk, P. (ed.) Proceedings of the Twenty-Sixth Annual ACM-SIAM Symposium on Discrete Algorithms, SODA 2015, San Diego, CA, USA, 4–6 January 2015, pp. 1030–1049. SIAM (2015)

[KNS11] Koch, R., Nasrabadi, E., Skutella, M.: Continuous and discrete flows over time - a general model based on measure theory. Math. Meth. OR 73(3), 301–337 (2011)

[KP99] Koutsoupias, E., Papadimitriou, C.: Worst-case equilibria. In: Meinel, C., Tison, S. (eds.) STACS 1999. LNCS, vol. 1563, pp. 404–413. Springer, Heidelberg (1999). https://doi.org/10.1007/3-540-49116-3_38

[KS11] Koch, R., Skutella, M.: Nash equilibria and the price of anarchy for flows over time. Theory Comput. Syst. **49**(1), 71–97 (2011)

[LMR94] Leighton, F.T., Maggs, B.M., Rao, S.: Packet routing and job-shop scheduling in O(congestion + dilation) steps. Combinatorica **14**(2), 167–186 (1994)

[LMR99] Leighton, F.T., Maggs, B.M., Richa, A.W.: Fast algorithms for finding o(congestion + dilation) packet routing schedules. Combinatorica **19**(3), 375–401 (1999)

[LMRR94] Leighton, F.T., Maggs, B.M., Ranade, A.G., Rao, S.: Randomized routing and sorting on fixed-connection networks. J. Algorithms **17**(1), 157–205 (1994)

[Rot13] Rothvoß, T.: A simpler proof for O(congestion+dilation) packet routing. In: Goemans, M., Correa, J. (eds.) IPCO 2013. LNCS, vol. 7801, pp. 336–348. Springer, Heidelberg (2013). https://doi.org/10.1007/978-3-642-36694-9_29

[Rou09] Roughgarden, T.: Intrinsic robustness of the price of anarchy. In: Mitzenmacher, M. (ed.) Proceedings of the 41st Annual ACM Symposium on Theory of Computing, STOC 2009, Bethesda, MD, USA, 31 May–2 June 2009, pp. 513–522. ACM (2009)

[SS18] Sering, L., Skutella, M.: Multi-source multi-sink Nash flows over time. In: 18th ATMOS Workshop, vol. 65, pp. 12:1–12:20 (2018)

[SST18] Scarsini, M., Schröder, M., Tomala, T.: Dynamic atomic congestion games with seasonal flows. Oper. Res. **66**(2), 327–339 (2018)

[SSV18] Scheffler, R., Strehler, M., Vargas Koch, L.: Equilibria in routing games with edge priorities. In: Christodoulou, G., Harks, T. (eds.) WINE 2018. LNCS, vol. 11316, pp. 408–422. Springer, Cham (2018). https://doi.org/10.1007/978-3-030-04612-5_27

[Vic69] Vickrey, W.S.: Congestion theory and transport investment. Am. Econ. Rev. **59**(2), 251–260 (1969)

[WHK14] Werth, T.L., Holzhauser, M., Krumke, S.O.: Atomic routing in a deterministic queuing model. Oper. Res. Perspect. **1**(1), 18–41 (2014)

Improved Online Algorithm for Fractional Knapsack in the Random Order Model

Jeff Giliberti[1,2]([⊠]) [iD] and Andreas Karrenbauer[2] [iD]

[1] ETH Zurich, Zürich, Switzerland
jgiliberti@student.ethz.ch
[2] MPI for Informatics, Saarland Informatics Campus, Saarbrücken, Germany
andreas.karrenbauer@mpi-inf.mpg.de

Abstract. The fractional knapsack problem is one of the classical problems in combinatorial optimization, which is well understood in the offline setting. However, the corresponding online setting has been handled only briefly in the theoretical computer science literature so far, although it appears in several applications. Even the previously best known guarantee for the competitive ratio was worse than the best known for the integral problem in the popular random order model. We show that there is an algorithm for the online fractional knapsack problem that admits a competitive ratio of 4.39. Our result significantly improves over the previously best known competitive ratio of 9.37 and surpasses the current best 6.65-competitive algorithm for the integral case. Moreover, our algorithm is deterministic in contrast to the randomized algorithms achieving the results mentioned above.

Keywords: Online algorithms · Random order model · Fractional knapsack

1 Introduction

The knapsack problem is well-studied and has a long history in the literature, both the offline and the online versions, where in the latter the items are revealed one after the other and an irrevocable decision whether to pick the current item has to be made immediately. In the online setting, one typically considers the *random order model*, in which the adversary controls the instance, i.e., the items with their weights and profits, but a permutation, chosen uniformly at random, determines the arrival order of the items. In this model, the performance measure of an algorithm $\mathcal{A}$ is its *competitive ratio*. We say that $\mathcal{A}$ is r-competitive with respect to the profit $\mathrm{OPT}(\mathcal{I})$ of an optimal offline algorithm for input $\mathcal{I}$, if $\mathbb{E}[\mathcal{A}(\mathcal{I})] \geq (1/r - o(1)) \cdot \mathrm{OPT}(\mathcal{I})$ holds for all inputs $\mathcal{I}$. Here, the expectation is taken over the random permutation as well as over random choices of the algorithm. The $o(1)$-term vanishes asymptotically with respect to the number of

The work was carried out while Jeff Giliberti was a virtual intern at MPI for Informatics

J. Koenemann and B. Peis (Eds.): WAOA 2021, LNCS 12982, pp. 188–205, 2021.
https://doi.org/10.1007/978-3-030-92702-8_12

items in the input instance. In a series of papers [1,5,24], the competitive ratio for online knapsack has been improved to 6.65, achieved by a randomized algorithm in [1]. The online version of the fractional knapsack problem has not received much attention in literature as of yet. An application of the online fractional knapsack problem was introduced in [21], where the items are articles that are presented in a newsfeed to busy readers. The scarce resource is time and the profit is the information gain. Upon arrival of an article, readers see its length and a cue about the content. They can discard the article right away based on this information, or start reading it. They can read it completely or discard it at any time. The information gain is assumed to be proportional to the reading time and the cue about its content. Discarded articles are not considered again. Moreover, [21] contains a framework that turns any α-competitive algorithm for the online knapsack problem into an $(\alpha + e)$-competitive algorithm for the online fractional knapsack problem. The approach is based on flipping a biased coin (w.r.t. α) and either executing the given algorithm or the famous secretary algorithm, where the first n/e items are discarded and then the first item that is better than the best item seen so far is chosen. Together with the result of [1], this led to the previously best known competitive ratio of 9.37 for the online fractional knapsack problem. In this paper, we cut the competitive ratio by more than half, namely to 4.39. To this end, we observed that the algorithm in [1] already contains a secretary algorithm on its own and that small and large items are treated separately due to the nature of the integral knapsack solution. Instead of treating it as a black box, relaxing the integrality constraints allows us to unify the small-item and large-item cases and output fractional values instead of binary ones, thereby achieving a better competitive ratio. Moreover, this also removes the randomization from the algorithm making it deterministic, and the expected value that determines the competitive ratio solely depends on the randomness of the input permutation.

1.1   Related Work

The study of the online (integral) knapsack problem was initiated by Marchetti-Spaccamela and Vercellis [28], who showed that there is no constant-competitive deterministic online algorithm. Moreover, Chakrabarty et al. [34] extended the same hardness result to randomized algorithms. Motivated by the difficulty of the adversarial model for this problem, a number of beyond worst-case scenarios have been investigated. The most popular of these is the random order model that has received increasing attention in the field of online algorithms. In this model, the online knapsack problem was first studied by [5], showing a $10e$-competitive algorithm. Kesselheim et al. [24] developed an 8.06-competitive randomized algorithm for the generalized assignment problem, which generalizes to a setting with multiple knapsacks of different capacities. Albers et al. [1] achieved the currently best known upper bound of 6.65. The online fractional variant under adversarial arrivals was first considered in [30]. There, the knapsack capacity is augmented by a factor $1 \leq R \leq 2$, and the algorithm can remove previously packed items. The plain version of the fractional knapsack problem in the random order model was studied in [21], presenting a 9.37-competitive randomized algorithm, which we improve upon. Recently, [31] considered a general version of the fractional online knapsack problem with multiple knapsacks and rate constraints. A further branch of research considers the infinitesimal assumption (or that the packing is allowed to be *fractional*), i.e., the profit of a single item is small compared to the profit of the optimal integral solution, under which Vaze [33] gave a 2e-competitive algorithm. In addition, a common approach combined with the infinitesimal assumption is to assume that the density (profit-size ratio) of each item is in a known range [8,9,34]. Another problem closely related to the random order model is the secretary problem [14,26], that is, a special case of the online knapsack problem in which the weights are uniform and equal to the weight constraint. One natural generalization of the latter is the k-secretary problem [2,11,25], where k elements need to be selected, as well as the matroid secretary problem [6,15], where elements of a weighted matroid arrive in random order, and in both of these, the goal is to maximize the combined value of the selected elements. Other variants of online knapsack presented in the literature include removable models, where removals can incur no cost or a cancellation cost [3,4,17,18,20], reservation costs [7], an expected capacity constraint [32], and resource buffering [19]. Furthermore, there have been alternative approaches to the random order model, such as *stochastic* versions of the online knapsack problem [12,13,16,27,29], a random order model with bursts of adversarial time steps [22], and the advice complexity model [10].

1.2 Our Contribution

Theorem 1. *There exists a 4.39-competitive deterministic algorithm for the online fractional knapsack problem in the random order model.*

We achieve this result by analyzing a natural variation of the algorithm from [1]. As in the original version for the online knapsack problem, the algorithm works in three phases – the sampling phase, the secretary phase, and the knapsack phase. The transition between the phases happens at iterations $\lfloor cn \rfloor$ and $\lfloor dn \rfloor$, respectively, where $0 < c \leq d \leq 1$ are optimized to achieve the best possible competitive ratio when combining our analyses of the second and third phase contributions. The algorithm can be considered as a blending of two algorithms that share the first phase, which is possible because no items are picked in the sampling phase; thus, there is no interference between both algorithms. After the first phase, i.e., after roughly 47.5% of all items, the secretary algorithm takes over and exclusively decides the items of the second phase. It does so by selecting all items that have a larger profit than the most profitable item seen in the sampling phase. Should picking the current item exceed the knapsack capacity, the item is picked to the largest extent possible to fill the knapsack. If there is still capacity left after the second phase, i.e., after roughly 60.1% of all items, we switch from the secretary to the knapsack algorithm, which then exclusively fills in a fraction of each arriving item according to the optimal (fractional) solution of all items revealed so far and the remaining capacity.

It is interesting to note that we do not distinguish between large and small items in our analysis, in comparison to [1]. Furthermore, in the analysis of the secretary algorithm, we only account for the probability of picking the item that has the largest contribution to the objective value of the optimal (fractional) solution. This is sufficient to cover the case where the optimal solution consists of a single item, a situation that the knapsack algorithm (or our analysis of it) cannot handle well. In fact, if we forced $c = d$ in the parameter optimization, we could not get a better competitive ratio than 6.63, which is still better than the competitive ratio from [21], but significantly worse than the best ratio from this paper. On the other hand, if we did not use the knapsack algorithm at all ($d = 1$), we obtained the well-known secretary algorithm, where the sampling phase ends after skipping a $c = 1/e$ fraction of all items (justifying the naming).

2   Preliminaries

Definition 1. (*OFKP*). *We are given a set I of n items, each item $i \in I$ has size $s_i \in \mathbb{Q}_{>0}$ and a profit (value) $v_i \in \mathbb{Q}_{\geq 0}$. The goal is to find a maximum profit fractional packing into a knapsack of size $W \in \mathbb{Q}_{>0}$, i.e., a solution $x \in \mathbb{Q}^n_{\geq 0}$ s.t. $\sum_{i \in I} s_i x_i \leq W$ and $\sum_{i \in I} v_i x_i$ is maximized. The items are revealed one by one in a round-wise fashion. In round $\ell \in [n]$, the algorithm sees item ℓ with its size and profit. It has to decide immediately and irrevocably the fraction x_ℓ of the current item in the final packing.*

We make the following two assumptions without loss of generality: (i) No item has size larger than the knapsack capacity[1], (ii) Items have distinct values[2]. Using assumption (ii), we obtain that there is a unique (optimum) solution x^* for the given set of items I. In fact, profit-to-weight ratio ties can be broken by taking the most valuable element which is unique by assumption (ii).

Next, we formalize the optimal (fractional) solution for a given subset $Q \subseteq I$. Let the *density* of an item be the ratio of its profit to its size. The optimal (fractional) solution has a clear structure: There exists a density threshold ρ_Q such that any item $i \in Q$ with $v_i/s_i > \rho_Q$ has $x_i = 1$ and any item $i \in Q$ with $v_i/s_i < \rho_Q$ has $x_i = 0$. Meaning that the $k-1$ densest items are packed integrally and the remaining space is filled by the maximum feasible fraction of the k-th densest item. Let $I(\ell)$ denote the subset of items I revealed up to round ℓ and define $x^{(\ell)}$ to be the optimal (fractional) solution for the item set $I(\ell)$. Let OPT be the total profit of the optimal (fractional) solution x^* for the item set I, i.e., $\text{OPT} = \sum_{i \in I} v_i x_i^*$. For convenience of notation, let OPT also denote the set of items that are part of the optimal (fractional) solution, i.e., each item i whose $x_i^* > 0$, and let $\mathcal{K}$ be its cardinality. Additionally, we label items in descending order of contribution to OPT such that $v_i x_i^* \geq v_{i+1} x_{i+1}^*$ for all $i \in [n-1]$.

---

[1] Items whose size exceeds the capacity of the knapsack can be cut at the knapsack capacity.

[2] It can be accomplished in polynomial time by fixing a random (but consistent) tie-breaking between elements of the same value, based for instance on the identifier of the element [5].

As in [1], we will use the following well-known fact to obtain lower or upper bounds on sums in closed form.

Fact 2. *Let f be a non-negative real-valued function and let $a, b \in \mathbb{Z}$.*

(A) If f is non-increasing, then $\int_a^{b+1} f(t)\,\mathrm{d}t \leq \sum_{t=a}^b f(t) \leq \int_{a-1}^b f(t)\,\mathrm{d}t$.
(B) If f is non-decreasing, then $\int_{a-1}^b f(t)\,\mathrm{d}t \leq \sum_{t=a}^b f(t) \leq \int_a^{b+1} f(t)\,\mathrm{d}t$.

2.1 Review of the Blended Approach

A standard approach used by packing algorithms in the online setting is the following. Algorithms have a sampling phase, during which all items are rejected, and a decision phase, where items may be accepted according to some decision rule developed according to the information gathered in the sampling phase. The novel idea presented in [1] is to combine two algorithms, $\mathcal{A}_1$ and $\mathcal{A}_2$, in a blended manner. The strategy is to make a better use of the entire instance by letting the two algorithms have a common sampling phase, and, subsequently, using the sampling phase of one algorithm as the decision phase of the other. Let $0 < c \leq d \leq 1$ denote some constant parameters to be specified later. Rounds $1, \ldots, \lfloor cn \rfloor$ define the common sampling phase. For rounds $\lfloor cn \rfloor + 1, \ldots, \lfloor dn \rfloor$, there is the $\mathcal{A}_1$ decision phase, while, $\mathcal{A}_2$ continues its sampling phase. From round $\lfloor dn \rfloor + 1$ the algorithm $\mathcal{A}_1$ stops and the $\mathcal{A}_2$ decision phase starts until the end of the stream. As in [1], we make the assumption $cn, dn \in \mathbb{N}$ for the analysis, which does not affect the competitive ratio substantially for n large enough. Clearly, combining the two algorithms without an initial random choice on whether to run $\mathcal{A}_1$ or $\mathcal{A}_2$ [5,21,24] comes at the cost of possibly having a non-empty knapsack when $\mathcal{A}_2$ starts its decision phase, thus, deteriorating $\mathcal{A}_2$ performance. However, we will see that with some sufficiently high probability the algorithm $\mathcal{A}_1$ does not pack any item.

In the online integral knapsack algorithm developed in [1], the algorithm $\mathcal{A}_{\mathscr{L}}$ deals with all items that consume more than $1/3$ of the knapsack capacity. Items whose size is at most $1/3$ of the knapsack capacity are packed by algorithm $\mathcal{A}_{\mathscr{S}}$. The algorithm $\mathcal{A}_{\mathscr{L}}$ exploits the connection with the 2-secretary problem because at most two large items can fit in the knapsack. The algorithm $\mathcal{A}_{\mathscr{S}}$ packs the current small item integrally with probability equal to its value in the *optimal fractional solution* formed by the items occurred so far. One possible reason for which $\mathcal{A}_{\mathscr{L}}$ and $\mathcal{A}_{\mathscr{S}}$ operate on different instances, i.e., the instances that consist of large and small items respectively, is that the optimal integral solution is not *monotone*. Namely, in the optimal integral solutions for the items seen until rounds ℓ, m for $\ell < m$ denoted by $x^{(\ell)}$ and $x^{(m)}$ respectively, there might be an item i such that $x_i^{(m)} = 1 > x_i^{(\ell)} = 0$. Notably, this comes at the cost of not packing large items using $\mathcal{A}_{\mathscr{L}}$, as the large items in the optimal integral solution that algorithm $\mathcal{A}_{\mathscr{L}}$ considers may be arbitrarily different from the large items in the optimal fractional solution that $\mathcal{A}_{\mathscr{S}}$ considers.

2.2   Application of the Blended Approach in the Fractional Case

In the fractional variant of the problem, we can overcome the limitation described above. The fractional relaxation allows us to make both algorithms sharing the same instance of the problem. The *secretary* algorithm $\mathcal{A}_S$ packs the most profitable items, whereas the *knapsack* algorithm $\mathcal{A}_K$ packs both small and large items (fractionally when needed) that are part of the optimal fractional solution OPT. In the analysis of $\mathcal{A}_S$, the connection between the fractional knapsack problem and the k-secretary problem can be extended beyond the 2-secretary problem, unlike the integral case, losing only the symmetry property of packing an item before another when there is an item taken fractionally in the optimal (fractional) solution. However, we will see that the connection with the 1-secretary problem is enough for our purpose. Next, we give a high-level description of the proof of Theorem 1 using Algorithm 1.

Proof. Let $\mathcal{A}$ be the algorithm obtained by combining $\mathcal{A}_S$ and $\mathcal{A}_K$. Moreover, define p_i, q_i to be some probabilities that we introduce later. By setting parameters c, d to 0.47521, 0.60138, we will show

$$\mathbb{E}[\mathcal{A}] \geq \mathbb{E}[\mathcal{A}_S] + \mathbb{E}[\mathcal{A}_K] \geq \sum_{i \in \text{OPT}} p_i \cdot v_i x_i^* + \frac{c}{d} \sum_{i \in \text{OPT}} q_i \cdot v_i x_i^*$$

$$\geq \left(\frac{1}{4.39} - o(1) \right) \cdot \text{OPT}.$$

$\square$

In the remainder of this paper, we analyze the performance of $\mathcal{A}_S$ and $\mathcal{A}_K$. The algorithm $\mathcal{A}_S$ and its analysis are similar to the algorithm $\mathcal{A}_{\mathscr{L}}$ and its analysis in [1] based on the SINGLE-REF algorithm [2] for the k-secretary problem. The algorithm $\mathcal{A}_K$ and its analysis extend the approach of [1] to consider the possibility of packing items fractionally. Namely, we make $\mathcal{A}_K$ taking the largest possible fraction of an item that is part of the optimal fractional solution seen at round ℓ according to the available knapsack capacity. The resulting competitive ratio obtained by combining lower bounds on the expected profit of algorithms $\mathcal{A}_S$ and $\mathcal{A}_K$ is analyzed in Sect. 5.

Algorithm 1: Online Fractional Knapsack $\mathcal{A}$

Input   : permutation π of item set I, knapsack capacity W, parameters $c, d \in (0, 1]$ with $c \leq d$.
            Item i appears in round $\pi(i)$ and reveals v_i and s_i.
Output : feasible fractional knapsack packing, i.e., $0 \leq x_i \leq 1$ for item $i \in [n]$.

Let ℓ be the current round and i be the online item of round ℓ;

Let v^* be the maximum profit seen up to round $\lfloor cn \rfloor$;

for *round* $\ell \in \{\lfloor cn \rfloor + 1, \ldots, \lfloor dn \rfloor\}$ **do** $\triangleright$ Algorithm $\mathcal{A}_S$

    if $v_i > v^*$ **then**

$$\text{Set } x_i = \frac{1}{s_i} \cdot \min \left\{ s_i, \ W - \sum_{j : \pi(j) < \ell} s_j x_j \right\};$$

    else

        Set $x_i = 0$;

for *round* $\ell \in \{\lfloor dn \rfloor + 1, \ldots, n\}$ **do** $\triangleright$ Algorithm $\mathcal{A}_K$

    $x^{(\ell)} :=$ optimal fractional knapsack packing on $I(\ell)$;

$$\text{Set } x_i = \frac{1}{s_i} \cdot \min \left\{ s_i x_i^{(\ell)}, \ W - \sum_{j : \pi(j) < \ell} s_j x_j \right\};$$

---

3   Secretary Algorithm $\mathcal{A}_S$

The following is an adaptation of SINGLE-REF developed for the k-secretary problem in [2] and applied to the knapsack setting in [1]. There is a useful connection between the online knapsack problem under random arrival order and the k-secretary problem. The latter is defined as an unweighted version of online knapsack in the random order model, or equivalently, items can be seen as W/k large. In contrast, in our problem items may be larger than W/k.

Algorithm. The algorithm $\mathcal{A}_S$ works as follows. During the initial sampling phase of $\lfloor cn \rfloor$ rounds, the algorithm rejects all items and identifies as *best sample* the encountered element with the highest value. In rounds $\ell \in \{\lfloor cn \rfloor + 1, \ldots, \lfloor dn \rfloor\}$, the algorithm takes the largest possible fraction (according to the remaining knapsack capacity) of the items whose individual profit beats the profit of the best sample. Consequently, the first item beating the best sample will be taken integrally. Then, the algorithm maximizes the fraction of each subsequently accepted item.

Analysis. Let p_i for $i \in [n]$ be the probability that $\mathcal{A}_S$ packs the i-th most profitable item as the first element. When considering the probability p_i, we do not specify what the algorithm will further do, i.e., after the first accepted item, any or no second item may be included and so on for subsequent items. In the following, we report the lower bounds for the probabilities p_i showed in [1]. Note that these lower bounds do not depend on the integral or fractional variant considered as the *first* item can be always packed integrally. Let us define a permutation $\sigma : [n] \to [n]$ such that $v_{\sigma(1)} > v_{\sigma(2)} > \ldots > v_{\sigma(n)}$.

Lemma 1 (*[1]*). *We have the following lower bounds for the probability p_i that item $\sigma(i)$, $i \in [n]$, is accepted by $\mathcal{A}_S$ as the first item*

$$p_i \geq p(i) - o(1), \quad \text{with } p(i) = c \ln \frac{d}{c} + c \sum_{k=1}^{i-1} \binom{i-1}{k} (-1)^k \frac{d^k - c^k}{k}.$$

By the lemma above, we have $p(1) = c \ln \frac{d}{c}$ for $i = 1$. Furthermore, the value of OPT is upper bounded by the sum of the profits of the $\mathcal{K}$ most profitable items because items are sorted in decreasing order of contribution to OPT, thus, $v_{\sigma(i)} \geq v_i x_i^*$. This yields

$$\mathbb{E}[\mathcal{A}_S] \geq \sum_{i=1}^{\mathcal{K}} p_i v_{\sigma(i)} \geq \sum_{i=1}^{\mathcal{K}} p_i v_i x_i^*,$$

concluding the analysis of algorithm $\mathcal{A}_S$.

4 Knapsack Algorithm $\mathcal{A}_K$

In this section, we present the algorithm $\mathcal{A}_K$ that leverages the structure of optimal fractional solutions restricted to the items seen so far. In contrast to the small-item algorithm developed in [1], which uses the optimal fractional solutions to obtain an integral packing via randomized rounding, our deterministic algorithm uses the value of the current item in the fractional solution to the extent of the remaining knapsack capacity.

Algorithm. The algorithm $\mathcal{A}_K$ works as follows. During the initial sampling phase of $\lfloor dn \rfloor$ rounds, the algorithm rejects all items. In each round $\ell \geq \lfloor dn \rfloor + 1$, the algorithm computes an optimal fractional solution $x^{(\ell)}$ for $I(\ell)$. We pack an $x_i^{(\ell)}$ fraction of the current item if there is enough space, otherwise we pick the largest possible fraction according to the remaining space. Thus, the algorithm determines the fraction of item i, denoted by x_i, as follows

$$x_i = \frac{1}{s_i} \cdot \min \left\{ s_i x_i^{(\ell)}, \ W - \sum_{j \in I(\ell-1)} s_j x_j \right\}.$$

Analysis. We study the performance of algorithm $\mathcal{A}_K$ assuming it has the entire knapsack at its disposal, i.e., a capacity of W, and afterwards show how this occurs with constant probability. In our proofs, we consider an arbitrary fixed element $i \in$ OPT and define $\delta \in (0, 1]$ to be a parameter representing the fraction of the knapsack capacity W that element i occupies in our online solution. For a fixed δ, the proofs of Lemma 2 and Lemma 3 almost immediately follow from the small-item case analysis in [1]; we reproduce their proofs for completeness and include few changes in Lemma 2 that have to be made in order to adapt them to our fractional setting. In the second part of the analysis, in Lemma 4, we make use of δ to exploit the possibility of packing items fractionally.

Lemma 2 ([1]). *Let $i \in \text{OPT}$ and $x_i(\ell)$ be the fraction of item i that is packed by $\mathcal{A}_K$ in round ℓ. For $\ell \geq dn + 1$, it holds that*

$$\Pr\left[x_i(\ell) \geq \min\left\{\frac{\delta W}{s_i}, x_i^*\right\}\right] \geq \frac{1}{n}\left(1 - \frac{1}{1 - \delta}\ln\frac{\ell}{dn}\right).$$

Proof. In a random permutation item i arrives in round ℓ with probability $1/n$. In round $\ell \geq dn + 1$, the algorithm packs i for an $x_i^{(\ell)}$ fraction provided that there is enough space. Note that the rank w.r.t. profit-to-weight ratio of item i in $I(\ell)$ is less than or equal to its rank in I. According to the structure of the optimal fractional solutions, this implies that $x_i^{(\ell)} \geq x_i^*$. Moreover, if the current resource consumption X is at most $(1 - \delta)W$, then the current item i can be packed up to a fraction $\delta W/s_i$. By treating x_i^* as a parameter, it is not required to analyze the resource consumption in each round in expectation over all items. The latter approach appears in [23], which relies on the fact that in any step k of the algorithm the choice of the random permutation up to this point can be modeled as a sequence of independent random experiments. Let X_k be the resource consumption in round $k < \ell$. By assumption, the knapsack is empty after round dn, thus $X = \sum_{k=dn+1}^{\ell-1} X_k$. Let Q be the set of k visible items in round k. The set Q can be seen as uniformly drawn from all k-item subsets and any item $j \in Q$ is the current item of round k with probability $1/k$. The algorithm packs any item j for at most an $x_j^{(k)}$ fraction, thus

$$\mathbb{E}[X_k] \leq \sum_{j \in Q} \Pr[j \text{ occurs in round } k]s_j x_j^{(k)} = \frac{1}{k}\sum_{j \in Q}s_j x_j^{(k)} \leq \frac{W}{k},$$

where the last inequality holds because $x^{(k)}$ is a feasible solution for the knapsack of size W. By linearity of expectation and the previous inequality, the expected resource consumption up to round ℓ is

$$\mathbb{E}[X] = \sum_{k=dn+1}^{\ell-1} \mathbb{E}[X_k] \leq \sum_{k=dn+1}^{\ell-1} \frac{W}{k} \leq W\ln\frac{\ell}{dn}.$$

Applying Markov's inequality yields

$$\Pr[X < (1 - \delta)W] = 1 - \Pr[X \geq (1 - \delta)W]$$
$$\geq 1 - \frac{\mathbb{E}[X]}{(1 - \delta)W} \geq 1 - \frac{1}{1 - \delta}\ln\frac{\ell}{dn}.$$

$\square$

In the next Lemma, we use Lemma 2 to lower bound the total probability that a fixed fraction of a specific item will be packed.

Lemma 3 ([1]). *Let $i \in \mathrm{OPT}$ and x_i be the fraction of item i that is packed by $\mathcal{A}_K$. It holds that*

$$\Pr\left[x_i \geq \min\left\{\frac{\delta W}{s_i}, x_i^*\right\}\right] \geq 1 - d + \frac{1}{1-\delta}\left[1 - d - \left(1 + \frac{1}{n}\right) \cdot \ln\frac{1}{d}\right].$$

Proof Summing up the probabilities from Lemma 2 over all rounds $\ell \geq dn + 1$ gives

$$\Pr\left[x_i \geq \min\left\{\frac{\delta W}{s_i}, 1\right\}\right] = \sum_{\ell=dn+1}^{n} \Pr\left[x_i(\ell) \geq \min\left\{\frac{\delta W}{s_i}, x_i^*\right\}\right]$$

$$\geq \sum_{\ell=dn+1}^{n} \frac{1}{n}\left(1 - \frac{1}{1-\delta}\ln\frac{\ell}{dn}\right)$$

$$= \frac{1}{n}\left(n - dn - \frac{1}{1-\delta}\sum_{\ell=dn+1}^{n}\ln\frac{\ell}{dn}\right)$$

$$= 1 - d - \frac{1}{(1-\delta)n}\sum_{\ell=dn+1}^{n}\ln\frac{\ell}{dn}$$

$$\geq 1 - d - \frac{1}{(1-\delta)n}\left(n \cdot \ln\frac{1}{d} - n + dn + \ln\frac{1}{d}\right).$$

The last inequality follows from $\sum_{\ell=dn+1}^{n}\ln\frac{\ell}{dn} = \left(\sum_{\ell=dn}^{n-1}\ln\frac{\ell}{dn}\right) + \ln\frac{1}{d}$ and, by Fact 2, $\sum_{\ell=dn}^{n-1}\ln\frac{\ell}{dn} \leq \int_{dn}^{n}\ln\frac{\ell}{dn}\,d\ell$, which evaluates to $n \cdot \left(d - 1 + \ln\frac{1}{d}\right)$. The claim follows by rearranging terms. □

Observe that the above lower bound on the probability becomes negative for any $\delta \geq 2 + \ln(d)/(1-d)$. Therefore, it is only safe to use it for those items i that have a *utilization*

$$\mu_i := \frac{s_i x_i^*}{W} < \bar{\mu} := 2 + \frac{\ln d}{1-d},$$

where the utilization of an item measures its space consumption in the optimal fractional solution w.r.t. the total capacity and $\bar{\mu}$ denotes the maximum utilization allowed by our lower bound. In the following Lemma, we compute a lower bound on the expected profit obtained packing item $i \in \mathrm{OPT}$ using $\mathcal{A}_K$, crucially relying on the ability of packing items fractionally.

Lemma 4. *Let $i \in \mathrm{OPT}$ and $\mathbb{E}_i[\mathcal{A}_K] = q_i \cdot v_i x_i^*$ be the expected profit obtained by $\mathcal{A}_K$ from the fraction of item i that is packed in the optimal (offline) fractional solution. We have $q_i \geq q(\mu_i) - o(1)$ where*

$$q(\mu_i) = \frac{1}{\mu_i}\left((1-d) \cdot \min\{\mu_i, \bar{\mu}\} - \left(1 - d - \ln\frac{1}{d}\right)\ln(1 - \min\{\mu_i, \bar{\mu}\})\right).$$

Proof. Our goal is to compute the expected profit obtained from item $i \in \mathrm{OPT}$ by summing up the lower bounds on the probability that item i is packed for a fraction $x_i \in (0, x_i^*]$. To this end, let N be an arbitrarily large parameter. We define $\delta_j = 1 - j/N$ for all $j \in \{0, \ldots, N\}$. It follows that $\delta_j > \delta_{j-1}$. We may choose N appropriately such that there is an index $k \in \{0, \ldots, N-1\}$ with $\delta_k = \mu_i$. Note that μ_i is a rational number since the input data is rational. Let us define $m_j = \frac{\delta_j W}{s_i}$ for $j \in \{k, \ldots, N\}$. Recall that we consider a discrete probability space over the permutations of $[n]$. We consider the discrete random variable X_i for the fraction of item i that is selected by $\mathcal{A}_K$. Let Ω_i denote the finite set of values that X_i can attain. We have

$$\mathbb{E}_i[\mathcal{A}_K] = v_i \cdot \sum_{x_i \in \Omega_i} x_i \cdot \Pr[X_i = x_i]$$

$$\geq v_i \cdot m_k \cdot \Pr[X_i \geq m_k] + v_i \cdot \sum_{j=k+1}^{N} m_j \cdot \Pr[m_j \leq X_i < m_{j-1}].$$

Observe that $s_i x_i^* \leq \delta_k W$. Substituting $m_j = \frac{\delta_j W}{s_i x_i^*} \cdot x_i^* = \frac{\delta_j}{\delta_k} \cdot x_i^*$ and using the fact that

$$\Pr[m_j \leq x_i < m_{j-1}] = \Pr[x_i \geq m_j] - \Pr[x_i \geq m_{j-1}],$$

yields

$$\mathbb{E}_i[\mathcal{A}_K] \geq v_i x_i^* \cdot \Pr[x_i \geq m_k] + v_i x_i^* \cdot \sum_{j=k+1}^{N} \frac{\delta_j}{\delta_k} \left(\Pr[x_i \geq m_j] - \Pr[x_i \geq m_{j-1}] \right).$$

We rearrange the sum as follows

$$\mathbb{E}_i[\mathcal{A}_K] \geq v_i x_i^* \cdot \left(\sum_{j=k}^{N-1} \frac{\delta_j - \delta_{j+1}}{\delta_k} \cdot \Pr[x_i \geq m_j] + \frac{\delta_N}{\delta_k} \cdot \Pr[x_i \geq m_N] \right).$$

By definition, we have $\delta_j - \delta_{j+1} = 1/N$, $\delta_N = 0$, and $\delta_k = \mu_i$. This implies that

$$q_i \geq \frac{1}{\mu_i N} \sum_{j=k}^{N-1} \Pr[x_i \geq m_j] \geq \frac{1}{\mu_i N} \sum_{j=k'}^{N-1} \Pr[x_i \geq m_j].$$

for any integer $k' \geq k$, which we will choose as the smallest index such that the lower bounds from Lemma 3 are positive, i.e., $k' := \max\left\{ k, \left\lceil N \cdot \frac{d-1-\ln d}{1-d} \right\rceil \right\}$. That is,

$$q_i \geq \frac{1}{\mu_i N} \sum_{j=k'}^{N-1} \left[(1-d) + \left(1 - d - \ln \frac{1}{d} - \frac{1}{n} \ln \frac{1}{d} \right) \cdot \frac{1}{1-\delta_j} \right]$$

$$= \frac{1}{\mu_i} \left[(1-d) \left(1 - \frac{k'}{N} \right) + \left(1 - d - \ln \frac{1}{d} - \frac{1}{n} \ln \frac{1}{d} \right) \sum_{j=k'}^{N-1} \frac{1}{j} \right].$$

It is easy to check that $\left(1 - d - \ln \frac{1}{d} - \frac{1}{n} \ln \frac{1}{d}\right) \leq 0$ for $d \in (0,1]$. Thus, using Fact 2, we deduce

$$\sum_{j=k'}^{N-1} \frac{1}{j} \leq \sum_{j=k'}^{N} \frac{1}{j} \leq \int_{k'-1}^{N} \frac{1}{t}\, dt = \ln\left(\frac{N}{k'-1}\right) = \ln\left(\frac{1}{1 - \left(1 - \frac{k'}{N}\right) - \frac{1}{N}}\right).$$

By plugging in the above upper bound and observing that $\delta_{k'} = 1 - k'/N$, we obtain

$$q_i \geq \frac{1}{\mu_i}\left[(1-d)\delta_{k'} + \left(1 - d - \ln\frac{1}{d} - \frac{1}{n}\ln\frac{1}{d}\right)\ln\left(\frac{1}{1 - \delta_{k'} - \frac{1}{N}}\right)\right].$$

Note that $\frac{1}{n}\ln\frac{1}{d} = o(1)$ and that we can achieve $\ln\left(\frac{1}{1-\delta_{k'}}\right) \leq \ln\left(\frac{1}{1-\delta_{k'}-\frac{1}{N}}\right) \leq \ln\left(\frac{1}{1-\delta_{k'}}\right)+\varepsilon$ for any $\varepsilon > 0$, which yields the desired lower bound for $\mu_i = \delta_k < \bar{\mu}$. Moreover, if $k' = \left\lceil N \cdot \frac{d-1-\ln d}{1-d}\right\rceil$, we can obtain $\bar{\mu} - \varepsilon \leq \delta_{k'} \leq \bar{\mu}$ for any $\varepsilon > 0$, which completes the proof. $\square$

This establishes the profit obtained by items in OPT using $\mathcal{A}_K$. However, algorithm $\mathcal{A}$ can only benefit from $\mathcal{A}_K$ if algorithm $\mathcal{A}_S$ has not filled the knapsack completely. As we do not have any control over the (expected) size of the items packed by $\mathcal{A}_S$, we now condition on the event that it starts with an empty knapsack:

Lemma 5 ([1]). *With a probability of at least c/d, no item is packed by $\mathcal{A}_S$.*

Let ξ denote the event of an empty knapsack after round dn. The following Lemma bounds the overall expected profit from $\mathcal{A}_K$'s packing for algorithm $\mathcal{A}$ by applying Lemma 5.

Lemma 6. *We have*

$$\mathbb{E}_{\mathcal{A}_K}[\mathcal{A}] \geq \frac{c}{d} \sum_{i \in \text{OPT}} q_i \cdot v_i x_i.$$

Proof. By Lemma 5, the probability of an empty knapsack after round dn is at least c/d. Thus, we obtain

$$\mathbb{E}_{\mathcal{A}_K}[\mathcal{A}] \geq \Pr[\xi] \cdot \mathbb{E}[\mathcal{A}_K|\xi] \geq \frac{c}{d}\sum_{i \in \text{OPT}} \mathbb{E}_i[\mathcal{A}_K] = \frac{c}{d}\sum_{i \in \text{OPT}} q_i \cdot v_i x_i,$$

by linearity of expectation and Lemma 4. $\square$

5 Competitive Ratio Analysis

We provide a unified analysis of the competitive ratio achieved by algorithm $\mathcal{A}$ which combines $\mathcal{A}_S$ and $\mathcal{A}_K$. We determine the best choice for the parameters c and d w.r.t. the bounds that we have proven before. Note that our bounds only make sense for $0 < c \leq d \leq 1$ and $\bar{\mu} > 0$. The latter yields an additional lower bound on d, which is the smaller of the two roots of $2 - 2d + \ln d$, say, $d_{\min} \approx 0.20319$. In the following, we only consider the expected contribution by items that are contained in OPT. That is,

$$\mathbb{E}_{\text{OPT}}[\mathcal{A}] \geq \sum_{i \in \text{OPT}} p_i \cdot v_i x_i^* + \frac{c}{d} \sum_{i \in \text{OPT}} q_i \cdot v_i x_i^* = \sum_{i \in \text{OPT}} \left(p_i + \frac{c}{d} q_i \right) \cdot v_i x_i^*.$$

Our strategy is as follows. We first reason about adversarial instances. We use these to determine the parameters c and d that yield the best competitive ratio that is permitted by the analysis of $\mathcal{A}_S$ and $\mathcal{A}_K$ above. Afterwards, we formally prove that we indeed covered the worst-case, i.e., we achieve the postulated competitive ratio with the particular values for c and d on all instances. We have to take care of the following two situations:

(i) OPT only contains a single item.
(ii) OPT contains many items with a small utilization.

In the former case, we have that OPT must contain the most profitable item, which the secretary algorithm packs with a probability of p_1. Moreover, we start $\mathcal{A}_K$ with an empty knapsack with probability of at least c/d and pack the most profitable item there with a probability of $q_1 \geq q(1)$. Hence, we obtain

$$\mathbb{E}_{\text{OPT}}[\mathcal{A}] \geq \left(p_1 + \frac{c}{d} q_1 \right) v_1 x_1^* = \left(p_1 + \frac{c}{d} q_1 \right) \text{OPT} \geq \left(p(1) + \frac{c}{d} q(1) \right) \cdot \text{OPT},$$

where we now and in the remainder of this section omit the $o(1)$ for better readability. Let z denote the competitive ratio that we want to show. By the consideration above, we have that

$$z \leq p(1) + \frac{c}{d} q(1). \tag{1}$$

For the second case, algorithm $\mathcal{A}_K$ gives a lower bound of $c/d \cdot q(\mu)$ on the fraction that we pack of each item, w.r.t. its utilization in the optimal fractional solution, regardless of the cardinality of OPT. Since $q(\mu)$ is decreasing for $\mu \in (0, 1)$ and any choice of $d \in (d_{\min}, 1)$, we have the following constraint for the competitive ratio

$$z \leq \frac{c}{d} \cdot q(0). \tag{2}$$

In fact, the optimum of the resulting optimization problem

$$\begin{aligned}
\max_{c, d, z} \quad & z \\
\text{s.t.} \quad & z \leq c \cdot \ln \frac{d}{c} + \frac{c}{d} \cdot \left(2 - 2d + \ln d - \left(1 - d - \ln \frac{1}{d} \right) \cdot \ln \left(\frac{\ln \frac{1}{d}}{1 - d} - 1 \right) \right) \\
& z \leq \frac{c}{d} \cdot (2 - 2d + \ln d) \\
& 0 < c \leq d \leq 1
\end{aligned}$$

is attained when the two upper bounds on z are equal, which yields c in dependence on d. Hence, z is determined by a univariate function in d. It has a local maximum in the interval from $d_{\min}$ to 1, which is also a global maximum in that interval. A numerical computation yields that the maximum is attained for some $d \approx 0.6013835675554252$. This yields $c \approx 0.4752190514489393$ and a competitive ratio of 4.383238341343964.

In the following, we will show that the competitive ratio for all other cases is not worse for the parameters above completing the proof of Theorem 1.

Lemma 7. *For all $\mathcal{K} \in [n]$, the algorithm $\mathcal{A}$ is z-competitive for*

$$z = \frac{c}{d}(2 - 2d - \log(1/d)) > \frac{1}{4.39},$$

$c = 0.47521$, and $d = 0.60138$.

Proof. By Lemmas 1 and 6, the profit obtained by items in OPT is at least $\sum_{i \in \text{OPT}} \left(p(i) + \frac{c}{d}q(\mu_i)\right) \cdot v_i x_i^*$. Clearly, the case $\mathcal{K} = 1$ follows from the above optimization problem. For the general case $\mathcal{K} \geq 2$, the minimum coefficient over all terms in the summation, for the given c and d, may be smaller than the desired competitive ratio. However, if their average is larger than z, then we can exploit this as follows. The idea is that we transport *excess* from items with a smaller index (i.e., larger contribution to OPT), to items with larger index that may have a deficit. That is, we can redistribute the excess at item 1 to any larger indexed item. Note also that for any μ_i, it holds that $c/d \cdot q(\mu_i) \leq z$. Therefore, we deduce the following sufficient condition for our lemma

$$p(1) + \frac{c}{d}\sum_{i=1}^{\mathcal{K}} q(\mu_i) \geq z \cdot \mathcal{K}.$$

In particular, substituting $q(\mu_i)$ with its definition in $\sum_{i=1}^{\mathcal{K}} q(\mu_i)$, we have

$$(1-d) \cdot \min\left\{1, \frac{\bar{\mu}}{\mu_1}\right\} + \left(1 - d - \ln\frac{1}{d}\right)\ln\left(\frac{1}{1 - \min\{\mu_1, \bar{\mu}\}}\right)$$
$$+ \sum_{i=2}^{\mathcal{K}}\left(1 - d + \frac{1}{\mu_i}\left(1 - d - \ln\frac{1}{d}\right)\ln\left(\frac{1}{1-\mu_i}\right)\right).$$

Assuming that $\mu_1 \leq \bar{\mu} \approx 0.72428$, we can rewrite the above expression as follows

$$\mathcal{K} \cdot (1-d) - \left(1 - d - \ln\frac{1}{d}\right)\sum_{i=1}^{\mathcal{K}}\frac{1}{\mu_i}\ln(1 - \mu_i).$$

Recalling that $\left(1 - d - \ln\frac{1}{d}\right) < 0$, for $d \in (0, 1]$, and that $\sum_{i=1}^{\mathcal{K}} \mu_i \leq 1$, we observe that $\sum_{i=1}^{\mathcal{K}}\frac{1}{\mu_i}\ln(1 - \mu_i)$ is a separable concave function over the polyhedral domain $\{\mu \in [0, \bar{\mu}]^{\mathcal{K}} : \mathbf{1}^T\mu \leq 1\}$, which is minimized at the boundary, i.e., choosing $\mu_1 = \bar{\mu}, \mu_2 = 1 - \bar{\mu}$ and $\mu_j \to 0$ for $j \geq 3$. Note that for μ_j tending to zero, the term $\ln(1 - \mu_j)/\mu_j$ converges to -1 from below. Meaning that $\sum_{i=1}^{\mathcal{K}} \frac{1}{\mu} \ln(1 - \mu) \geq \frac{1}{\bar{\mu}} \ln(1 - \bar{\mu}) + \frac{1}{1-\bar{\mu}} \ln(\bar{\mu}) - (\mathcal{K} - 2)$, resulting in the following condition

$$p(1) + \frac{c}{d}\left(\mathcal{K} - \mathcal{K}d - \left(1 - d - \ln\frac{1}{d}\right) \cdot \left(\frac{1}{\bar{\mu}}\ln(1 - \bar{\mu}) + \frac{1}{1-\bar{\mu}}\ln(\bar{\mu}) - \mathcal{K} + 2\right)\right)$$

$$= \underbrace{p(1) - \frac{c}{d}\left(1 - d - \ln\frac{1}{d}\right)\left(2 + \frac{1}{\bar{\mu}}\ln(1 - \bar{\mu}) + \frac{1}{1-\bar{\mu}}\ln(\bar{\mu})\right) + z\mathcal{K}}_{\approx 0.02949},$$

after plugging in the definition of z. Let us now consider the case in which $\mu_1 > \bar{\mu}$. Then, the term $\sum_{i=1}^{\mathcal{K}} q(\mu_i)$ can be rewritten as follows

$$(1 - d) \cdot \frac{\bar{\mu}}{\mu_1} + \frac{1}{\bar{\mu}}\left(1 - d - \ln\frac{1}{d}\right)\ln\left(\frac{1}{1-\bar{\mu}}\right)$$

$$+ (\mathcal{K} - 1)(1 - d) - \left(1 - d - \ln\frac{1}{d}\right)\sum_{i=2}^{\mathcal{K}}\frac{1}{\mu_i}\ln(1 - \mu_i),$$

where the sum is minimized for $\mu_2 = 1 - \mu_1$ and $\mu_j \to 0$ for $j > 2$. The resulting function in μ_1 is decreasing over $(\bar{\mu}, 1]$ for our choice of parameters. Thus, it is lower bounded by

$$p(1) + \frac{c}{d}\left((1 - d)\bar{\mu} + \left(1 - d - \ln\frac{1}{d}\right)\ln\left(\frac{1}{1-\bar{\mu}}\right)\right)$$

$$+ \frac{c}{d}(\mathcal{K} - 1)\left(2 - 2d - \ln\frac{1}{d}\right)$$

$$= \underbrace{p(1) + \frac{c}{d}\left((1 - d)\bar{\mu} + \left(1 - d - \ln\frac{1}{d}\right)\ln\left(\frac{1}{1-\bar{\mu}}\right) - 2 + 2d + \ln\frac{1}{d}\right) + z\mathcal{K}}_{\approx 4 \cdot 10^{-6}}.$$

Both conditions are met for our choice of c, d, concluding the proof.   □

6   Future Directions

Potential improvements and future directions are:

(i) Better bounds on the probability of picking the i-th most valuable item in the secretary algorithm. We use the corresponding lower bounds from [1]; however, the first one already suffices for the single-item case, and an improvement of the many-item case would require that the amortized probability of picking the k-th most-valuable item does not go to 0 as k gets large.

(ii) Close the probability gap in the analysis. With our choice of c and d, we account for a probability of about 11% that the secretary algorithm packs the most profitable item, and for a probability of about 79% that the third phase starts with an empty knapsack, which we use to condition the expected profit of the knapsack algorithm – leaving a gap of about 10% that is lost for our analysis.

(iii) Interleaving of secretary and knapsack algorithm. Instead of running two separate phases for completing the secretary and the knapsack algorithm, they could run in parallel and the decisions on the current item could be based on a combination of both opinions (e.g., min, max, or coin flip).

(iv) Consequences for the online knapsack problem. Our focus was to improve the competitive ratio for the online fractional knapsack problem. Additionally, it would be interesting to investigate whether our result leads to new insights about the integer problem, e.g., w.r.t. the analysis of randomized rounding or in other closely related settings.

References

1. Albers, S., Khan, A., Ladewig, L.: Improved online algorithms for knapsack and GAP in the random order model. Algorithmica, February 2021. https://doi.org/10.1007/s00453-021-00801-2
2. Albers, S., Ladewig, L.: New results for the k-secretary problem. Theor. Comput. Sci. **863**, 102–119 (2021). https://doi.org/10.1016/j.tcs.2021.02.022
3. Babaioff, M., Hartline, J., Kleinberg, R.: Selling banner ads: Online algorithms with buyback. In: Fourth Workshop on Ad Auctions (2008)
4. Babaioff, M., Hartline, J.D., Kleinberg, R.D.: Selling ad campaigns: online algorithms with cancellations. In: Proceedings of the 10th ACM Conference on Electronic Commerce, pp. 61–70 (2009)
5. Babaioff, M., Immorlica, N., Kempe, D., Kleinberg, R.: A knapsack secretary problem with applications. In: Charikar, M., Jansen, K., Reingold, O., Rolim, J.D.. P.. (eds.) APPROX/RANDOM -2007. LNCS, vol. 4627, pp. 16–28. Springer, Heidelberg (2007). https://doi.org/10.1007/978-3-540-74208-1_2
6. Babaioff, M., Immorlica, N., Kempe, D., Kleinberg, R.: Matroid secretary problems. J. ACM (JACM) **65**(6), 1–26 (2018)
7. Böckenhauer, H.J., Burjons, E., Hromkovič, J., Lotze, H., Rossmanith, P.: Online simple knapsack with reservation costs. In: Bläser, M., Monmege, B. (eds.) 38th International Symposium on Theoretical Aspects of Computer Science (STACS 2021). Leibniz International Proceedings in Informatics (LIPIcs), vol. 187, pp. 16:1–16:18. Schloss Dagstuhl - Leibniz-Zentrum für Informatik, Dagstuhl, Germany (2021). https://doi.org/10.4230/LIPIcs.STACS.2021.16
8. Buchbinder, N., Naor, J.: Online primal-dual algorithms for covering and packing problems. In: Brodal, G.S., Leonardi, S. (eds.) ESA 2005. LNCS, vol. 3669, pp. 689–701. Springer, Heidelberg (2005). https://doi.org/10.1007/11561071_61
9. Buchbinder, N., Naor, J.: Improved bounds for online routing and packing via a primal-dual approach. In: 2006 47th Annual IEEE Symposium on Foundations of Computer Science (FOCS 2006), pp. 293–304. IEEE (2006)
10. Böckenhauer, H.J., Komm, D., Královič, R., Rossmanith, P.: The online knapsack problem: advice and randomization. Theor. Comput. Sci. **527**, 61–72 (2014). https://doi.org/10.1016/j.tcs.2014.01.027

11. Chan, T.H.H., Chen, F., Jiang, S.H.C.: Revealing optimal thresholds for generalized secretary problem via continuous LP: impacts on online K-item auction and bipartite K-matching with random arrival order, pp. 1169–1188. https://doi.org/10.1137/1.9781611973730.78

12. Dean, B.C., Goemans, M.X., Vondrák, J.: Adaptivity and approximation for stochastic packing problems. In: SODA, vol. 5, pp. 395–404. Citeseer (2005)

13. Dean, B.C., Goemans, M.X., Vondrák, J.: Approximating the stochastic knapsack problem: the benefit of adaptivity. Math. Oper. Res. **33**(4), 945–964 (2008). https://doi.org/10.1287/moor.1080.0330

14. Dynkin, E.B.: The optimum choice of the instant for stopping a Markov process. Soviet Math. **4**, 627–629 (1963)

15. Feldman, M., Svensson, O., Zenklusen, R.: A simple o(log log (rank))-competitive algorithm for the matroid secretary problem. In: Proceedings of the Twenty-Sixth Annual ACM-SIAM Symposium on Discrete Algorithms, pp. 1189–1201. SIAM (2014)

16. Goerigk, M., Gupta, M., Ide, J., Schöbel, A., Sen, S.: The robust knapsack problem with queries. Comput. Oper. Res. **55**, 12–22 (2015). https://doi.org/10.1016/j.cor.2014.09.010

17. Han, X., Kawase, Y., Makino, K.: Online unweighted knapsack problem with removal cost. Algorithmica **70**(1), 76–91 (2014)

18. Han, X., Kawase, Y., Makino, K.: Randomized algorithms for online knapsack problems. Theor. Comput. Sci. **562**, 395–405 (2015)

19. Han, X., Kawase, Y., Makino, K., Yokomaku, H.: Online Knapsack Problems with a Resource Buffer. In: Lu, P., Zhang, G. (eds.) 30th International Symposium on Algorithms and Computation (ISAAC 2019). Leibniz International Proceedings in Informatics (LIPIcs), vol. 149, pp. 28:1–28:14. Schloss Dagstuhl-Leibniz-Zentrum fuer Informatik, Dagstuhl, Germany (2019). https://doi.org/10.4230/LIPIcs.ISAAC.2019.28

20. Iwama, K., Taketomi, S.: Removable Online Knapsack Problems. In: Widmayer, P., Eidenbenz, S., Triguero, F., Morales, R., Conejo, R., Hennessy, M. (eds.) Automata, Languages and Programming, pp. 293–305. Springer, Berlin Heidelberg, Berlin, Heidelberg (2002)

21. Karrenbauer, A., Kovalevskaya, E.: Reading articles online. In: Wu, W., Zhang, Z. (eds.) COCOA 2020. LNCS, vol. 12577, pp. 639–654. Springer, Cham (2020). https://doi.org/10.1007/978-3-030-64843-5_43

22. Kesselheim, T., Molinaro, M.: Knapsack secretary with bursty adversary. In: Czumaj, A., Dawar, A., Merelli, E. (eds.) 47th International Colloquium on Automata, Languages, and Programming (ICALP 2020). Leibniz International Proceedings in Informatics (LIPIcs), vol. 168, pp. 72:1–72:15. Schloss Dagstuhl-Leibniz-Zentrum für Informatik, Dagstuhl, Germany (2020). https://doi.org/10.4230/LIPIcs.ICALP.2020.72

23. Kesselheim, T., Radke, K., Tönnis, A., Vöcking, B.: An optimal online algorithm for weighted bipartite matching and extensions to combinatorial auctions. In: Bodlaender, H.L.., Italiano, Giuseppe F.. (eds.) ESA 2013. LNCS, vol. 8125, pp. 589–600. Springer, Heidelberg (2013). https://doi.org/10.1007/978-3-642-40450-4_50

24. Kesselheim, T., Tönnis, A., Radke, K., Vöcking, B.: Primal beats dual on online packing LPs in the random-order model. In: Proceedings of the Forty-Sixth Annual ACM Symposium on Theory of Computing, STOC 2014, pp. 303–312. Association for Computing Machinery, New York (2014). https://doi.org/10.1145/2591796.2591810

25. Kleinberg, R.: A multiple-choice secretary algorithm with applications to online auctions. In: Proceedings of the Sixteenth Annual ACM-SIAM Symposium on Discrete Algorithms, SODA 2005, pp. 630–631. Society for Industrial and Applied Mathematics, USA (2005)
26. Lindley, D.V.: Dynamic programming and decision theory. J. Roy. Stat. Soc. Ser. C (Appl. Stat.) **10**(1), 39–51 (1961)
27. Lueker, G.S.: Average-case analysis of off-line and on-line knapsack problems. J. Algorithms **29**(2), 277–305 (1998). https://doi.org/10.1006/jagm.1998.0954
28. Marchetti-Spaccamela, A., Vercellis, C.: Stochastic on-line knapsack problems. Math. Program. **68**(1), 73–104 (1995). https://doi.org/10.1007/BF01585758
29. Marchetti-Spaccamela, A., Vercellis, C.: Stochastic on-line knapsack problems. Math. Program. **68**(1), 73–104 (1995)
30. Noga, J., Sarbua, V.: An online partially fractional knapsack problem. In: Proceedings of the 8th International Symposium on Parallel Architectures, Algorithms and Networks, pp. 108–112. IEEE Computer Society, Los Alamitos, CA, USA, December 2005. https://doi.org/10.1109/ISPAN.2005.19
31. Sun, B., Zeynali, A., Li, T., Hajiesmaili, M., Wierman, A., Tsang, D.H.: Competitive algorithms for the online multiple knapsack problem with application to electric vehicle charging. Proc. ACM Meas. Anal. Comput. Syst. **4**(3), November 2020. https://doi.org/10.1145/3428336
32. Vaze, R.: Online knapsack problem under expected capacity constraint. In: IEEE INFOCOM 2018-IEEE Conference on Computer Communications, pp. 2159–2167. IEEE (2018)
33. Vaze, R., Coupechoux, M.: Online budgeted truthful matching. SIGMETRICS Perform. Eval. Rev. **44**(3), 3–6 (2017). https://doi.org/10.1145/3040230.3040232
34. Zhou, Y., Chakrabarty, D., Lukose, R.: Budget constrained bidding in keyword auctions and online knapsack problems. In: Papadimitriou, C., Zhang, Shuzhong (eds.) WINE 2008. LNCS, vol. 5385, pp. 566–576. Springer, Heidelberg (2008). https://doi.org/10.1007/978-3-540-92185-1_63

Fractionally Subadditive Maximization Under an Incremental Knapsack Constraint

Yann Disser[1], Max Klimm[2], and David Weckbecker[1]([✉])

[1] TU Darmstadt, Darmstadt, Germany
{disser,weckbecker}@mathematik.tu-darmstadt.de
[2] TU Berlin, Berlin, Germany
klimm@math.tu-berlin.de

Abstract. We consider the problem of maximizing a fractionally sub-additive function under a knapsack constraint that grows over time. An incremental solution to this problem is given by an order in which to include the elements of the ground set, and the competitive ratio of an incremental solution is defined by the worst ratio over all capacities relative to an optimum solution of the corresponding capacity. We present an algorithm that finds an incremental solution of competitive ratio at most $\max\{3.293\sqrt{M}, 2M\}$, under the assumption that the values of singleton sets are in the range $[1, M]$, and we give a lower bound of $\max\{2.449, M\}$ on the attainable competitive ratio. In addition, we establish that our framework captures potential-based flows between two vertices, and we give a tight bound of 2 for the incremental maximization of classical flows with unit capacities.

Keywords: Incremental maximization · Knapsack constraint · Fractionally subadditive · Competitive ratio · Potential-based flow

1 Introduction

We consider an incremental knapsack problem of the form

$$\max_{S \subseteq E} f(S) \quad \text{s.t.} \quad w(S) \le C, \tag{1}$$

with a monotone objective function $f: 2^E \to \mathbb{R}_{\ge 0}$ over a finite ground set E of size $m := |E|$, weights $w: E \to \mathbb{R}_{\ge 0}$ associated with the elements of the ground set, and a growing capacity bound $C > 0$. We denote an optimum solution to (1) for a fixed capacity C by S_C^* and let $f^*(C)$ be its value as a function of C.

The growing capacity bound models the increase of available resources over time, e.g., the spending budget of consumers with a steady income and the

We acknowledge funding by the DFG through a short time project and subproject A007 of the CRC/TRR 154, as well as through grant DI 2041/2.

J. Koenemann and B. Peis (Eds.): WAOA 2021, LNCS 12982, pp. 206–223, 2021.
https://doi.org/10.1007/978-3-030-92702-8_13

regular cashflow of corporations. In these scenarios, it is natural to fix an order in which elements of the ground set should be purchased. As an example, think of a consumer with a steady income of 1 per time unit. There are three items $E = \{c, s, t\}$, a compact camera c at a price of $w(c) = 1$, a system camera s at a price of $w(s) = 2$, and a telephoto lens t at a price of $w(t) = 2$. For a subset $S \subseteq E$, suppose that $f(S) = 1$, if S contains the compact camera, but not the system camera, $f(S) = 2$ if it contains the system camera, but not the telephoto lens, $f(S) = 3$ if it contains both the system camera and the telephoto lens, and $f(S) = 0$, otherwise. If the buyer chooses to buy the compact camera first, they receive a utility of 1, which is optimal for the budget of 1, i.e., $f^*(1) = 1$. However, if the buyer delays their purchase until they can afford the system camera, the value at time 1 is 0 but the value at time 2 is 2, which is optimal, i.e., $f^*(2) = 2$. As this simple example illustrates, there is a tradeoff between the attained values at different times and, in particular, there is no purchase order that yields the optimal value at all times.

In this paper, we address this tradeoff from a perspective of competitive analysis. An incremental solution to our incremental knapsack problem is given by an ordering $\pi = (e_{\pi(1)}, e_{\pi(2)}, \dots, e_{\pi(m)})$ of the ground set. For capacity $C > 0$, let $\pi(C)$ be the items contained in the maximal prefix of π that fits into the capacity C, i.e., $\pi(C) = \{e_{\pi(1)}, e_{\pi(2)}, \dots, e_{\pi(k)}\}$ for some $k \in \mathbb{N}$ such that $\sum_{i=1}^{k} w(e_{\pi(i)}) \leq C$ and either $k = m$ or $\sum_{i=1}^{k+1} w(e_{\pi(i)}) > C$. We say that the ordering π is ρ-competitive with $\rho \geq 1$ if

$$f^*(C) \leq \rho \cdot f(\pi(C)) \quad \text{for all } C > 0.$$

We call an ordering *competitive* if it is ρ-competitive for some constant $\rho \geq 1$. It is easy to see that without any further assumptions, there is no hope to obtain a competitive ordering π. For illustration, consider the example above where the value of the system camera is changed to $M \in \mathbb{R}_{\geq 0}$. Since any competitive ordering is forced to buy the compact camera first in order to be competitive for $C = 1$, no ordering can be better than M-competitive at time 2. To avoid this issue, we will assume in the following that the value of each singleton set falls in a bounded interval $[1, M]$ for some constant $M \geq 1$, i.e., $f(e) \in [1, M]$ for all $e \in E$. We call such valuations M-*bounded*. Observe that even for 1-bounded valuations, complementarities between the items may prevent a competitive ordering. For illustration, adapt our example such that every non-empty subset of items yields a value of 1 unless it contains both the system camera and the telephoto lens for which the value is $N > M$. Again, any competitive ordering has to put the compact camera first, since it has lowest weight, and then cannot be better than N-competitive for budget $C = 4$. To avoid this issue, we additionally require that f is fractionally subadditive. Fractional subadditivity is a generalization of submodularity and a standard assumption in the combinatorial auction literature (cf. Nisan [26], Lehmann et al. [19]). Formally, a function $f: 2^E \to \mathbb{R}_{\geq 0}$ is called *fractionally subadditive* if $f(A) \leq \sum_{i=1}^{k} \alpha_i f(B_i)$ for all $A, B_1, B_2, \dots, B_k \in 2^E$ and all $\alpha_1, \alpha_2, \dots, \alpha_k \in \mathbb{R}_{\geq 0}$ such that $\sum_{i \in \{1, \dots, k\}: e \in B_i} \alpha_i \geq 1$ for all $e \in A$. Observe that every fractional subadditive function f is also subadditive, i.e., $f(A \cup B) \leq f(A) + f(B)$, but not vice-versa.

To summarize, we consider incremental solutions to (1), assuming that

(MO) f is monotone, i.e., $f(A) \geq f(B)$ for $A \supseteq B$,
(MB) f is M-bounded, i.e., $f(e) \in [1, M]$ for all $e \in E$,
(FS) f is fractionally subadditive.

Before stating our results, we illustrate the applicability of our framework to different settings.

Example 1. Every *submodular objective* is fractionally subadditive. This means that, for example, our framework captures the maximum coverage problem, where we are given a weighted set of sets $E \subseteq 2^U$ over a universe U. Every element of U has a value $v \colon U \to \mathbb{R}_{\geq 0}$ associated with it, and $f(S) = v\left(\bigcup_{X \in S} X\right)$ for all $S \subseteq E$ where we write $v(X) := \sum_{x \in X} v(x)$ for a set $X \in 2^U$. In this context, the M-boundedness condition demands that $v(X) \in [1, M]$ for all $X \in E$. Further examples include maximization versions of clustering and location problems.

Example 2. An objective function $f \colon E \to \mathbb{R}$ is called *XOS* if it can be written as the pointwise maximum of linear functions, i.e., there are $k \in \mathbb{N}$ and values $v_{e,i} \in \mathbb{R}$ for all $e \in E$ and $i \in \{1, \ldots, k\}$ such that $f(S) = \max_{\{1,\ldots,k\}} \sum_{e \in S} v_{e,i}$ for all $S \subseteq E$. The set of factionally subadditive functions and the set of XOS functions coincide (Feige [9]). XOS functions are a popular way to encode the valuations of buyers in combinatorial auctions since they often give rise to a succinct representation (cf. Nisan [26] and Lehman et al. [19]).

Example 3. The *weighted rank function of an independence system* is fractionally subadditive (Amanatidis et al. [1]). An independence system is a tuple $(E, \mathcal{I})$, where $\mathcal{I} \subseteq 2^E$ is closed under taking subsets and $\emptyset \in \mathcal{I}$. For a given weight function $w' \colon E \to \mathbb{R}_{\geq 0}$, the weighted rank function of the independence system $(E, \mathcal{I})$ is given by $f(S) = \max\{\sum_{e \in I} w'(e) | I \in \mathcal{I} \cap 2^S\}$. This setting captures well-known problems such as weighted d-dimensional matching for any $d \in \mathbb{N}$, weighted set packing, weighted maximum independent set, and knapsack (i.e., a secondary knapsack problem on the same set ground set).

Example 4. It turns out that a *potential-based flow between two vertices s and t* along a set E of parallel edges, each with source s and sink t, gives rise to a fractionally subadditive flow function (Proposition 1). A potential-based flow is given by a classical flow $f \colon E \to \mathbb{R}_{\geq 0}$ together with vertex potentials p_s and p_t, coupled via $p_s - p_t = \beta_e \psi(f(e))$ for all $e \in E$. Here, $\beta_e \in \mathbb{R}$ are parameters and ψ is a continuous, strictly increasing, and odd potential loss function. Different choices of ψ allow to model gas flows, water flows, and electrical flows, see Groß et al. [11]. In our incremental framework, $w \colon E \to \mathbb{R}_{\geq 0}$ are interpreted as construction costs of pipes or cables and the objective is to maximize the flow from s to t in terms of the objective

$$f(S \subseteq E) = \max_{S' \subseteq S} \max_{p \in \mathbb{R}_{\geq 0}} \sum_{e \in S'} \psi^{-1}(p/\beta_e) \quad \text{s.t.} \quad \psi^{-1}(p/\beta_e) \leq \mu_e \text{ for all } e \in S', \quad (2)$$

where $p := p_s - p_t$ and μ_e are edge capacities. The value of the objective is the maximum value of a feasible potential-based s-t-flow. Note that we need to allow turning off edges (by considering $S' \subseteq S$) in order to make f monotone. The M-boundedness condition corresponds to the assumption that $\mu_e \in [1, M]$ and a ρ-competitive ordering determines a construction order of pipes or cables such that the flow that can be sent through the network is at least a $1/\rho$-fraction of the optimum at all times.

Our Results. Our main results are bounds on the best possible competitive ratio for incremental solutions to (1) for objectives satisfying (MO), (MB), and (FS). In other words, we bound the loss in solution quality that we have to accept when asking for incremental solutions that optimize for all capacities simultaneously. Note that, as customary in online optimization, we do not impose restrictions on the computational complexity of finding incremental solutions. We show the following.

Theorem 1. *For monotone, M-bounded, and fractionally subadditive objectives, the best possible ρ for which the knapsack problem (1) admits a ρ-competitive incremental solution satisfies $\rho \in [\max\{2.449, M\}, \max\{3.293\sqrt{M}, 2M\}]$.*

In particular, for $M \geq 2.71$, the best possible competitive ratio is between M and $2M$, while the bounds for 1-bounded objectives simplify as follows.

Corollary 1. *For monotone, 1-bounded, and fractionally subadditive objectives, the best ρ for which the knapsack problem (1) admits a ρ-competitive incremental solution satisfies $\rho \in [2.449, 3.293]$.*

Most interestingly, our upper bound uses a simultaneous capacity- and value-scaling approach. In each phase, we increase our capacity and value thresholds and pick the smallest capacity for which the optimum solution exceeds our thresholds. This solution is then assembled by adding one element at a time in a specific order. The order is chosen based on a primal-dual LP formulation that relies on fractional subadditivity.

In Sect. 2, we describe our algorithmic approach in detail and give a proof of the upper bound. In Sect. 3, we complement our result with two lower bounds. As an additional motivation, in Sect. 4, we show that our framework captures potential-based flows as described in Example 4. In this context, a 1-bounded objective corresponds to unit capacities. As a contrast, we also show that classical s-t-flows with unit capacities admit 2-competitive incremental solutions, and this is best-possible.

Related Work. Bernstein et al. [4] considered a closely related framework for incremental maximization. Their framework assumes a growing cardinality constraint, which is a special case of our problem in (1) when all elements $e \in E$ have unit weight $w(e) = 1$. A natural incremental approach for a growing cardinality constraint is the greedy algorithm that includes in each step the element that increases the objective the most. This algorithm is well known to yield a

$(e/(e-1))$-approximation for submodular objective functions [25]. Several generalizations of this result to broader classes of functions are known. Recently, Disser and Weckbecker [6] unified these results by giving a tight bound for the approximation ratio of the greedy algorithm for γ-α-augmentable objective functions[1], which interpolates between known results for weighted rank functions of independence systems of bounded rank quotient, α-augmentable functions, and functions of bounded submodularity ratio. Sviridenko [28] showed that for a submodular function under a knapsack constraint, the greedy algorithm yields a $(1-1/e)$-approximation when combined with a partial enumeration procedure. This approximation guarantee is best possible as shown by Feige [8]. Yoshida [30] generalized the result of Sviridenko to submodular functions of bounded curvature.

Another closely related setting is the robust maximization of a linear function under a knapsack constraint. Here, the capacity of the knapsack is revealed in an online fashion while packing, and we ask for a packing order that guarantees a good solution for every capacity. Megow and Mestre [23] considered this setting under the assumption that we have to stop packing once an item exceeds the knapsack capacity and presented a polynomial time algorithm that has an instance-sensitive near-optimal competitive ratio. Navarra and Pinotti [24] added the mild assumption that all items fit in the knapsack and devised competitive solutions for this model. Disser et al. [5] allowed to discard items that do not fit and showed tight competitive ratios for this case. Kawase et al. [17] studied a generalization of this model in which the objective is submodular and devised a randomised competitive algorithm for this case. Since these models allow to discard items, these competitive rations do not translate to our model. Kobayashi and Takazawa [18] studied randomized strategies for cardinality robustness in the knapsack problem. Other online versions of the knapsack problem assume that items are revealed over time, e.g., see Matchetti-Spaccamela and Vercellis [21]. Thielen et al. [29] combined both settings and assumed that items appear over time while the capacity grows.

For incremental minimization, Lin et al. [20] introduced a general framework, based on a problem-specific augmentation routine, that subsumes several earlier results. An incremental maximization problem that received particular attention is the so-called robust matching problem which was introduced by Hassin and Rubinstein [13]. Here, we ask for a weighted matching such that the heaviest k edges of the matching approximate a maximum weight matching of cardinality k, for all cardinalities k. Hassin and Rubinstein [13] gave tight bounds on the deterministic competitive ratio of this problem, and Matuschke et al. [22] gave bounds on the randomized competitive ratio. Fujita et al. [10] and Kakimura et al. [14] considered extensions of this problem to independence systems. A similar variant of the knapsack problem where the k most valuable items are compared to an optimum solution of cardinality k was studied by Kakimura et al. [15].

---

[1] A function is called γ-α-augmentable if for all set $X, Y \subseteq E$ there exists $y \in Y$ with $f(X \cup \{y\}) - f(X) \geq (\gamma f(X \cup Y) - \alpha f(X))/|Y|$.

Incremental optimization has also been considered from an offline perspective, i.e., without uncertainty in items or capacities. Kalinowski et al. [16] and Hartline and Sharp [12] considered incremental flow problems where the average flow over time needs to be maximized (in contrast to the worst flow over time). Anari et al. [2] and Orlin et al. [27] considered general robust submodular maximization problems.

The class of fractionally additive valuations was introduced by Nisan [26] and Lehman et al. [19] under the name of XOS-valuations as a compact way to represent the utilities of bidders in combinatorial auctions. In a combinatorial auction, a set of elements E is auctioned off to a set of n bidders who each have a private utility function $f_i : 2^E \to \mathbb{R}_{\geq 0}$. In this context, a natural question is to maximize social welfare, i.e., to partition E into sets $E_1, E_2, \ldots, E_n$ with the objective to maximize $\sum_{i=1}^{n} f_i(E_i)$. Dobzinski and Schapira [7] gave a $(1 - 1/e)$-approximation for this problem.

2 Upper Bound

In the following, we fix a ground set E, a monotone, M-bounded and fractionally subadditive objective $f : 2^E \to \mathbb{R}_{\geq 0}$, and weights $w : E \to \mathbb{R}_{\geq 0}$. We present a refined variant of the incremental algorithm introduced in [4]. On a high level, the idea is to consider optimum solutions of increasing sizes, and to add all elements in these optimum solutions one solution at a time. By carefully choosing the order in which we add elements of a single solution, we ensure that elements contributing the most to the objective are added first. In this way, we can guarantee that either the solution we have assembled most recently, or the solution we are currently assembling provides sufficient value to stay competitive. While the algorithm of [4] only scales the capacity, our algorithm $\mathrm{ALG_{scale}}$ simultaneously scales capacities and solution values. In addition, we use a more sophisticated order in which we assemble solutions, based on a primal-dual LP formulation. We now describe our approach in detail.

Let $\lambda \approx 3.2924$ be the unique real root of the equation

$$0 = \lambda^7 - 2\lambda^6 - 3\lambda^5 - 3\lambda^4 - 3\lambda^3 - 2\lambda^2 - \lambda - 1,$$

$\delta := \frac{\lambda^3}{\lambda^2 + 1} \approx 3.0143$ and $\rho := \max\{\lambda\sqrt{M}, 2M\}$. Algorithm $\mathrm{ALG_{scale}}$ operates in phases of increasing capacities $C_1, \ldots, C_N \in \mathbb{R}_{\geq 0}$ with $C_1 := \min_{e \in E} w(e)$,

$$C_{i+1} = \min\{C \geq \delta C_i \mid f(S_C^*) \geq \rho f(S_{C_i}^*)\}$$

for $i \in \{1, \ldots, N - 2\}$, and $C_N := \sum_{e \in E} w(e) \in (C_{N-1}, \delta C_{N-1}]$, where $N \in \mathbb{N}$ is chosen accordingly. In phase $i \in \{1, \ldots, N\}$, $\mathrm{ALG_{scale}}$ adds the elements of the set $S_{C_i}^*$ one at a time. Recall that S_C^* is the optimum solution to (1) of capacity C. We may assume that previously added elements are added again (without any benefit), since this only hurts the algorithm.

To specify the order in which the elements of $S^*_{C_i}$ are added, consider the following linear program (LP_X) parameterized by $X \subseteq E$ (cf. [9]):

$$\min \sum_{B \subseteq E} \alpha_B f(B)$$

$$\text{s.t.} \sum_{B \subseteq E : e \in B} \alpha_B \geq 1, \quad \forall e \in X,$$

$$\alpha_B \geq 0, \quad \forall B \subseteq E,$$

and its dual

$$\max \sum_{e \in X} \gamma_e$$

$$\text{s.t.} \sum_{e \in B} \gamma_e \leq f(B), \quad \forall B \subseteq E,$$

$$\gamma_e \geq 0, \quad \forall e \in E.$$

Fractional subadditivity of f translates to $f(X) \leq \sum_{B \subseteq E} \alpha_B f(B)$ for all $\alpha \in \mathbb{R}^{2^E}$ feasible for (LP_X). The solution $\alpha^* \in \mathbb{R}^{2^E}$ with $\alpha^*_X = 1$ and $\alpha^*_B = 0$ for $X \neq B \subseteq E$ is feasible and satisfies $f(X) = \sum_{B \subseteq E} \alpha^*_B f(B)$. Together this implies that α^* is an optimum solution to (LP_X). By strong duality, there exists an optimum dual solution $\gamma^*(X) \in \mathbb{R}^E$ with

$$f(X) = \sum_{e \in X} \gamma^*_e(X). \tag{3}$$

In phase 1, the algorithm $\mathrm{ALG}_{\mathrm{scale}}$ adds the unique element in $S^*_{C_1}$. In phase 2, $\mathrm{ALG}_{\mathrm{scale}}$ adds an element of largest weight in $S^*_{C_2}$ first and the other elements in an arbitrary order. And in phase $i \in \{3, 4, ..., N\}$, $\mathrm{ALG}_{\mathrm{scale}}$ adds the elements of $S^*_{C_i}$ in an order $(e_1, ..., e_{|S^*_{C_i}|})$ such that, for all $j \in \{1, ..., |S^*_{C_i}| - 1\}$,

$$\frac{\gamma^*_{e_j}(S^*_{C_i})}{w(e_j)} \geq \frac{\gamma^*_{e_{j+1}}(S^*_{C_i})}{w(e_{j+1})}. \tag{4}$$

For $C \in [0, C_i]$ and with $j := \max\{j \in \{1, ..., |S^*_{C_i}|\} \mid w(\{e_1, ..., e_j\}) \leq C\}$, we let $S^*_{C_i,C} := \{e_1, ..., e_j\}$ denote the prefix of $S^*_{C_i}$ of capacity C. Furthermore, by π^A, we refer to the the permutation of E that represents the order in which the algorithm $\mathrm{ALG}_{\mathrm{scale}}$ adds the elements of E.

Roughly, we show that this algorithm is competitive as follows: In the first phase $\mathrm{ALG}_{\mathrm{scale}}$ obviously performs optimally. In all other phases, the solution added in the previous phase is large enough to be competitive until the solution added currently has a larger value. From this point until the end of the phase, the current solution is competitive.

Remark 1. In the construction of our algorithm, we assume to have oracle access to an optimum solution S^*_C of a given capacity $C \in \mathbb{R}_{\geq 0}$. Finding such a solution may not be possible in polynomial time. In [3], Badanidiyuru et al. give a $(2+\epsilon)$-approximation algorithm and show that no polynomial time algorithm can have an approximation ratio of less than 2, unless P = NP. Our algorithm $\text{ALG}_{\text{scale}}$ can use an α-approximation instead of an optimum solution, for a loss of factor α in its competitive ratio.

We first show that the dual variables $\gamma_e^*(X)$ associate a contribution to the overall objective to each element $x \in X$, and that this association is consistent under taking subsets of X.

Lemma 1. *Let $X \subseteq Y \subseteq E$. Then,*

$$f(X) \geq \sum_{e \in X} \gamma_e^*(Y).$$

Proof. Since $\gamma^*(Y)$ is a feasible solution for the dual of (LP_X), it is also a feasible solution for the dual of (LP_Y). Thus, since $\gamma^*(X)$ is an optimum solution of (LP_X),

$$\sum_{e \in X} \gamma_e^*(Y) \leq \sum_{e \in X} \gamma_e^*(X) \overset{(3)}{=} f(X). \qquad \square$$

The following lemma establishes that the order in which we add the elements of each optimum solution are decreasing in density, in an approximate sense.

Lemma 2. *Let $C, C' \in \mathbb{R}_{\geq 0}$ with $C \leq C' \leq w(E)$. Then*

$$f^*(C') \leq \frac{C'}{C}(f(S_{C',C}^*) + M).$$

Proof. If $S_{C'}^* = S_{C',C}^*$, the statement holds trivially. Suppose $|S_{C'}^*| > |S_{C',C}^*|$. Let $j := |S_{C',C}^*|$, and let $S_{C'}^* = \{e_1, ..., e_{|S_{C'}^*|}\}$ such that (4) holds. Note that, by definition, $S_{C',C}^* = \{e_1, ..., e_j\}$ and

$$w(\{e_1, ..., e_j\}) \leq C < w(\{\{e_1, ..., e_{j+1}\}). \tag{5}$$

We have

$$
\begin{aligned}
f^*(C') &\overset{(3)}{=} \sum_{i=1}^{|S_{C'}^*|} w(e_i) \frac{\gamma_{e_i}^*(S_{C'}^*)}{w(e_i)} \\
&\overset{(4)}{\leq} \left(\sum_{i=1}^{j+1} \gamma_{e_i}^*(S_{C'}^*)\right) + \frac{\sum_{i=1}^{j+1} w(e_i)}{w(\{e_1, ..., e_{j+1}\})} \sum_{i=j+2}^{|S_{C'}^*|} w(e_i) \frac{\gamma_{e_{j+1}}^*(S_{C'}^*)}{w(e_{j+1})} \\
&\overset{(4)}{\leq} \left(\sum_{i=1}^{j+1} \gamma_{e_i}^*(S_{C'}^*)\right) + \frac{\left(\sum_{i=1}^{j+1} \gamma_{e_i}^*(S_{C'}^*)\right)}{w(\{e_1, ..., e_{j+1}\})} \sum_{i=j+2}^{|S_{C'}^*|} w(e_i) \\
&\overset{(5)}{<} \left(\sum_{i=1}^{j+1} \gamma_{e_i}^*(S_{C'}^*)\right) + \frac{\left(\sum_{i=1}^{j+1} \gamma_{e_i}^*(S_{C'}^*)\right)}{C}(C' - C)
\end{aligned}
$$

$$= \frac{C'}{C}\left(\left(\sum_{i=1}^{j}\gamma_{e_i}^*(S_{C'}^*)\right)+\gamma_{e_{j+1}}^*(S_{C'}^*)\right)$$

$$\overset{\text{Lemma 1}}{\leq} \frac{C'}{C}(f(\{e_1,...,e_j\})+f(\{e_{j+1}\}))$$

$$\leq \frac{C'}{C}(f(S_{C',C}^*)+M).$$

$\square$

Since every set $S \subseteq E$ with $w(S) \leq C$ satisfies $f(S) \leq f^*(C)$, and since we have $w(S_{C',C}^*) \leq C$, we immediately obtain the following.

Corollary 2. *Let $C, C' \in \mathbb{R}_{\geq 0}$ with $C \leq C' \leq w(E)$. Then*

$$f^*(C') \leq \frac{C'}{C}(f^*(C)+M).$$

With this, we are now ready to show the upper bound of our main result.

Theorem 2. *The incremental solution computed by* $\text{ALG}_{\text{scale}}$ *is ρ-competitive for* $\rho = \max\{\lambda\sqrt{M}, 2M\} \approx \max\{3.2924\sqrt{M}, 2M\}$.

Proof. We have to show that, for all sizes $C \in \mathbb{R}_{\geq 0}$, we have $f^*(C) \leq \rho f(\pi^A(C))$. We will do this by analyzing the different phases of the algorithm. Observe that, for all $i \in \{2, ..., N-1\}$, we have

$$f^*(C_i) \overset{\text{Def. of ALG}_{\text{scale}}}{\geq} \rho f^*(C_{i-1})$$
$$\geq \rho^{i-1}f^*(C_1)$$
$$\overset{\text{(MB)}}{\geq} \rho^{i-1}$$
$$\overset{\text{Def. of }\rho}{\geq} (\lambda\sqrt{M})^{i-1}. \tag{6}$$

In phase 1, we have $C \in (0, C_1]$. Since C_1 is the minimum weight of all elements and we start by adding $S_{C_1}^*$, i.e., the optimum solution of size C_1, the value $\pi^A(C)$ is optimal.

Consider phase 2, and suppose $C \in (C_1, C_2)$. If $C_2 > \delta C_1$ holds, then C_2 is the smallest value such that $f^*(C_2) \geq \rho f^*(C_1)$, i.e., by monotonicity of f, we have $f(\pi^A(C)) \geq f(\pi^A(C_1)) = f^*(C_1) > \frac{1}{\rho}f^*(C)$. Now assume $C_2 = \delta C_1$. If $C \in (C_1, 3C_1)$, i.e., any solution of size C cannot contain more than two elements, or if $C \in (C_1, C_2)$ and $S_{C_2}^*$ contains at most 2 elements, by fractional subadditivity and M-boundedness of f, we have $f^*(C) \leq |S_{C_2}^*|M \leq 2M$ and thus, by (6), $f(\pi^A(C)) \geq f^*(C_1) \geq 1 \geq \frac{1}{2M}f^*(C) \geq \frac{1}{\rho}f^*(C)$. Now suppose $C \in [3C_1, C_2)$ and that $S_{C_2}^*$ contains at least 3 elements. The prefix $\pi^A(C_1+C_2)$ contains all elements from $S_{C_1}^* \cup S_{C_2}^*$, the prefix $\pi^A(C_2) = \pi^A(C_1 + C_2 - C_1)$ contains at least all but one element of $S_{C_2}^*$, and the prefix $\pi^A(C_2 - C_1)$ contains at least all but 2 elements from $S_{C_2}^*$ because C_1 is the weight of any element is at least C_1. Since $C \geq 3C_1 > (\delta - 1)C_1 = C_2 - C_1$, $\pi^A(C)$ contains at least all but 2 elements from $S^*_{C_2}$. Recall that in phase 2 the algorithm adds the element with the highest objective value first. Therefore, and because $|S^*_{C_2}| \geq 3$, we have $f(\pi^A(C)) \geq \frac{1}{3}f(S^*_{C_2}) \geq \frac{1}{\rho}f^*(C)$.

Consider phase 2 and suppose $C \in [C_2, C_1 + C_2]$. We have

$$f^*(C_1 + C_2) \leq f^*(C_2) + M \tag{7}$$

because f is subadditive and because C_1 is the minimum weight of all elements. Furthermore, we have

$$f(\pi^A(C_2)) \geq f^*(C_2) - M \geq \rho - M \geq M \tag{8}$$

where the first inequality follows from subadditivity of f and the fact that the prefix $\pi^A(C_2)$ contains at least all but one element from $S^*_{C_2}$. Combining (7) and (8), we obtain $f(\pi^A(C_2)) \geq f^*(C_1 + C_2) - 2M \geq f^*(C_1 + C_2) - 2f(\pi^A(C_2))$, i.e., by monotonicity,

$$f^*(C) \leq f^*(C_1 + C_2) \leq 3f(\pi^A(C_2)) \leq \rho f(\pi^A(C_2)) \leq \rho f(\pi^A(C)).$$

Now consider phase $i \in \{3, ..., N\}$ and $C \in (\sum_{j=1}^{i-1} C_j, \sum_{j=1}^{i} C_j]$. Note that, for $j \in \{1, ..., i\}$, $C_i \geq \delta^{i-j}C_j$ and hence

$$\sum_{j=1}^{i-1} \frac{C_j}{C_i} \leq \sum_{j=1}^{i-1} \frac{1}{\delta^{i-j}} < \sum_{j=1}^{\infty} \frac{1}{\delta^i} = \frac{1}{\delta - 1} < 1,$$

i.e., we have $\sum_{j=1}^{i-1} C_j \leq C_i \leq \sum_{j=1}^{i} C_j$. If $i = N$ and $\sum_{j=1}^{i-1} C_j \geq C_i$, we have nothing left to show. Thus, suppose that $\sum_{j=1}^{N-1} C_j \leq C_N$.

Suppose $C \in (\sum_{j=1}^{i-1} C_j, C_i)$. We will show that in this case, the value of the solution C_{i-1}, which is already added by the algorithm, is large enough to guarantee competitiveness. If $C_i > \delta C_{i-1}$ holds, then C_i is the smallest integer such that $f^*(C_i) \geq \rho f^*(C_{i-1})$, i.e., we have

$$f(\pi^A(C)) \overset{(MO)}{\geq} f(\pi^A(C_{i-1})) \geq f^*(C_{i-1}) > \frac{1}{\rho}f^*(C).$$

For the case that $C_i = \delta C_{i-1}$, we distinguish between two different cases:

<u>Case 1</u> ($i = 3$): Let $c := (\frac{1}{\lambda\sqrt{M}} + \frac{1}{\lambda^2})\delta C_2$. Note that $(\frac{1}{\lambda} + \frac{1}{\lambda^2})\delta = \frac{\lambda^2}{\lambda+1} - 1 - \frac{1}{\delta}$ by definition of λ and δ and thus

$$
\begin{aligned}
c &\leq \left(\frac{1}{\lambda} + \frac{1}{\lambda^2}\right)\delta C_2 \\
&= \left(\frac{\lambda^2}{\lambda+1} - 1 - \frac{1}{\delta}\right)C_2 \\
&\leq \frac{\lambda^2}{\frac{\lambda}{\sqrt{M}} + 1}C_2 - C_2 - C_1
\end{aligned}
$$

$$= \frac{\lambda^2 M}{\lambda\sqrt{M} + M}C_2 - C_2 - C_1. \tag{9}$$

We will show that $\pi^{A}(C_1 + C_2)$ is competitive up to size $C_1 + C_2 + c$, and that $\pi^{A}(C_1 + C_2 + c)$ is competitive up to size C_3. We have

$$f^*(C_1 + C_2 + c) \overset{\text{Corollary 2}}{\leq} \frac{C_1 + C_2 + c}{C_2}(f^*(C_2) + M)$$

$$\overset{(9)}{\leq} \frac{C_1 + C_2 + \left(\frac{\lambda^2 M}{\lambda\sqrt{M}+M}C_2 - C_2 - C_1\right)}{C_2}(f^*(C_2) + M)$$

$$= \frac{\lambda^2 M}{\lambda\sqrt{M} + M}\left(1 + \frac{M}{f^*(C_2)}\right)f^*(C_2)$$

$$\overset{(6)}{\leq} \frac{\lambda^2 M}{\lambda\sqrt{M} + M}\left(1 + \frac{M}{\lambda\sqrt{M}}\right)f^*(C_2)$$

$$= \lambda\sqrt{M}\frac{\lambda\sqrt{M}}{\lambda\sqrt{M} + M}\left(\frac{\lambda\sqrt{M} + M}{\lambda\sqrt{M}}\right)f^*(C_2)$$

$$\leq \rho f^*(C_2)$$

$$\leq \rho f(\pi^{A}(C_1 + C_2)),$$

where the last inequality follows from the fact that the algorithm starts by packing $S^*_{C_1}$ and $S^*_{C_2}$ before any other elements and needs capacity $C_1 + C_2$ to assemble both sets, i.e., $S^*_{C_2} \subseteq \pi^{A}(C_1 + C_2)$.

Since $\text{ALG}_{\text{scale}}$ adds the elements from $S^*_{C_3}$ after those from $S^*_{C_1}$ and $S^*_{C_2}$, we have $S^*_{C_3,c} \subseteq \pi^{A}(C_1 + C_2 + c)$, and thus

$$f(\pi^{A}(C_1 + C_2 + c)) \overset{\text{(MO)}}{\geq} f(S^*_{C_3,c})$$

$$\overset{\text{Lemma 2}}{\geq} \frac{c}{C_3}f^*(C_3) - M$$

$$\overset{C_3 = \delta C_2}{=} \left(\left(\frac{1}{\lambda\sqrt{M}} + \frac{1}{\lambda^2}\right) - \frac{M}{f^*(C_3)}\right)f^*(C_3)$$

$$\overset{(6)}{\geq} \left(\frac{1}{\lambda\sqrt{M}} + \frac{1}{\lambda^2} - \frac{M}{\lambda^2 M}\right)f^*(C_3)$$

$$= \frac{1}{\lambda\sqrt{M}}f^*(C_3)$$

$$\geq \frac{1}{\rho}f^*(C_3).$$

This, together with monotonicity of f, implies $f^*(C) \leq \rho f(\pi^{A}(C))$ for all $C \in (C_1 + C_2, C_3]$.

$\underline{\text{Case 2}}$ ($i \geq 4$): Recall that $C \in (\sum_{j=1}^{i-1} C_j, C_i)$. We have

$$f^*(C) \overset{\text{(MO)}}{\leq} f^*(C_i)$$

$$\text{Corollary 2} \atop \leq \frac{C_i}{C_{i-1}}(f^*(C_{i-1}) + M)$$

$$\stackrel{C_i = \delta C_{i-1}}{=} \delta\left(1 + \frac{M}{f^*(C_{i-1})}\right)f^*(C_{i-1})$$

$$\stackrel{(6)}{\leq} \delta\left(1 + \frac{M}{\lambda^2 M}\right)f^*(C_{i-1})$$

$$= \frac{\lambda^3}{\lambda^2 + 1}\left(1 + \frac{1}{\lambda^2}\right)f^*(C_{i-1})$$

$$= \lambda f^*(C_{i-1})$$

$$\leq \rho f(\pi^{\mathrm{A}}(C)).$$

Thus, also in this case, we found $f^*(C) \leq \rho f(\pi^{\mathrm{A}}(C))$ for all $C \in \left(\sum_{j=1}^{i-1} C_j, C_i\right)$.

Now, consider $C \in \left[C_i, \sum_{j=1}^{i} C_j\right]$. Up to this budget, the algorithm had a capacity of $C - \sum_{j=1}^{i-1} C_j > C - C_i \geq 0$ to pack elements from $S^*_{C_i}$, i.e., $S^*_{C_i, C - \sum_{j=1}^{i-1} C_j} \subseteq \pi^{\mathrm{A}}(C)$. We will show that the value of this set is large enough to guarantee competitiveness in this case. We have

$$f(\pi^{\mathrm{A}}(C))$$

$$\stackrel{\text{(MO)}}{\geq} f(S^*_{C_i, C - \sum_{j=1}^{i-1} C_j})$$

$$\stackrel{\text{Lemma 2}}{\geq} \frac{C - \sum_{j=1}^{i-1} C_j}{C_i} f^*(C_i) - M$$

$$\stackrel{\text{Corollary 2}}{\geq} \frac{C - \sum_{j=1}^{i-1} C_j}{C_i}\left(\frac{C_i}{C} f^*(C) - M\right) - M$$

$$= \left(\frac{C - \sum_{j=1}^{i-1} C_j}{C} - \frac{C - \sum_{j=1}^{i-1} C_j}{C_i} \cdot \frac{M}{f^*(C)} - \frac{M}{f^*(C)}\right)f^*(C)$$

$$\stackrel{C_i \leq C \leq \sum_{j=1}^{i} C_i}{\geq} \left(1 - \sum_{j=1}^{i-1}\frac{C_j}{C_i} - 1 \cdot \frac{M}{f^*(C)} - \frac{M}{f^*(C)}\right)f^*(C)$$

$$\stackrel{(6), C_{j+1} \geq \delta C_j}{\geq} \left(1 - \sum_{j=1}^{i-1}\frac{1}{\delta^{i-j}} - \frac{2M}{\rho^{i-1}}\right)f^*(C)$$

$$\geq \left(1 - \sum_{j=1}^{\infty}\frac{1}{\delta j} - \frac{2M}{\rho^{i-1}}\right)f^*(C)$$

$$= \left(1 - \left(\frac{1}{1 - \delta^{-1}} - 1\right) - \frac{2M}{\lambda^2 M}\right)f^*(C)$$

$$\geq 0.319 \cdot f^*(C)$$

$$\geq \frac{1}{\rho}f^*(C).$$

$\square$

For 1-bounded objectives, Theorem (2) immediately yields the following.

Corollary 3. *If $M = 1$, the incremental solution computed by $\mathrm{ALG}_{\mathrm{scale}}$ is 3.2924-competitive.*

3 Lower Bound

In this section, we show the second part of Theorem 1, i.e., we give a lower bound on the competitive ratio for the incremental knapsack problem (1) with monotone, M-bounded, and fractionally subadditive objectives, and we show a lower bound for the special case with $M = 1$.

Theorem 3. *For monotone, M-bounded, and fractionally subadditive objectives, the knapsack problem (1) does not admit a ρ-competitive incremental solution for $\rho < M$.*

Proof. Consider the set $E = \{e_1, e_2\}$ with weights $w(e_i) = i$ for $i \in \{1, 2\}$ and values $v(e_1) = 1$ and $v(e_2) = M$. We define $f(X \subseteq E) := \sum_{e \in X} v(e)$ to be the objective. It is easy to see that f is monotone, M-bounded and modular and thus fractionally subadditive.

Consider some competitive algorithm with competitive ratio $\rho \geq 0$ for the knapsack problem (1). In order to be competitive for capacity 1, the algorithm has to add element e_1 first. Thus, the solution of the algorithm of size 2 cannot contain element e_2, i.e., the value of the solution of capacity 2 given by the algorithm has value 1. The optimal solution of capacity 2 has value M, and thus $\rho \geq M$.   □

For $M \in [1, \sqrt{6})$, we obtain a stronger lower bound.

Theorem 4. *For monotone, 1-bounded, and fractionally subadditive objectives, the knapsack problem (1) does not admit a ρ-competitive incremental solution for $\rho < \sqrt{6}$.*

Proof. Consider the set $E = E_1 \cup E_2 \cup E_3 = \{e_1\} \cup \{e_2, e_3, e_4\} \cup \{e_5, ..., e_{10}\}$, the weights

$$w(e) = \begin{cases} 101, & \text{for } e \in E_1, \\ 102, & \text{for } e \in E_2, \\ 103, & \text{else,} \end{cases}$$

and the objective $f \colon E \to \mathbb{R}_{\geq 0}$ with

$$f(X \subseteq E) = \max\{|X \cap E_1|, \frac{\sqrt{6}}{3}|X \cap E_2|, |X \cap E_3|, |X \cap \{e_2\}|, |X \cap \{e_3\}|, |X \cap \{e_4\}|\}.$$

It is easy to see that $f(e) = 1$ for all $e \in E$ and that f is monotone. Furthermore, f is fractionally subadditive because it is an XOS-valuation since all terms in the maximum of the definition of f are modular set functions on E.

Suppose there was an algorithm ALG with competitive ratio $\rho < \sqrt{6}$. Let $\pi: \{1, ..., 10\} \to \{1, ..., 10\}$ be a permutation that represents the order in which ALG adds the elements of E. To be competitive for capacity $C = 101$, we must have $\pi(1) = 1$. For capacity $C = 306$, the optimum solution E_2 has value $\sqrt{6}$. To be competitive for this capacity, ALG has to achieve an objective value of $\frac{\sqrt{6}}{\rho} > 1$. This is not possible by adding one element of E_2 and one of E_3. Furthermore, adding two elements of E_3 is not possible with a capacity of C, since the elements in E_3 have weight 103, i.e., we must have $\{\pi(2), \pi(3)\} \subset \{2, 3, 4\}$. Without loss of generality, we can assume $(\pi(2), \pi(3)) = (2, 3)$. We distinguish between two different cases:

Case 1: $\pi(4) = 4$. We consider the capacity $C = 6 \cdot 103 = 618$. The solution of capacity at most C given by ALG can contain at most two elements of E_3. Thus, the value of the solution of ALG of capacity $C = 618$ is $\sqrt{6}$ which is not ρ-competitive since the optimum solution E_3 of capacity $C = 618$ has value 6 and $\rho\sqrt{6} < 6$.

Case 2: $\pi(4) \in \{5, ..., 10\}$. We consider the capacity $C = 5 \cdot 103 = 515$. The solution of capacity at most C given by ALG can contain at most two elements of E_3. Thus, the value of the solution of ALG of capacity $C = 515$ is 2, which is not ρ-competitive since the optimum solution E_3 of capacity $k = 515$ has value 5, and we have $2\rho < 5$.

This is a contradiction to the fact that ALG has competitive ratio $\rho < \sqrt{6}$.

□

4   Application to Flows

In this section we show that our algorithm $\text{ALG}_{\text{scale}}$ can be used to solve problems as given in Example 4, where we are given a graph $G = (V, E)$ consisting of two nodes s and t with a collection of edges between them, and want to dertermine an order in which to build the edges while maintaining a potential-based flow between s and t that is as large as possible. As mentioned in Example 4, the corresponding objective (2) is monotone and M-bounded for $\mu_e \in [1, M]$. Now we show fractional subadditivity.

Proposition 1. *The function*

$$f(S \subseteq E) = \max_{S' \subseteq S} \max_{p \in \mathbb{R}_{\geq 0}} \sum_{e \in S'} \psi^{-1}(p/\beta_e) \quad s.t. \quad \psi^{-1}(p/\beta_e) \leq \mu_e \text{ for all } e \in S'$$

is fractionally subadditive.

Proof. For $e \in E$, let

$$p_e := \beta_e \psi(\mu_e)$$

be the maximum potential difference between s and t such that the flow along e induced by the potential difference p_e is still feasible, i.e., does not violate the capacity constraint μ_e. For $e, e' \in E$, we define $x_e(p_{e'})$ to be the flow value along e induced by a potential difference of $p_{e'}$ between s and t if this flow is feasible and 0 otherwise. For $S \subseteq E$, we have

$$f(S) = \max_{S' \subseteq S} \max_{p \in \mathbb{R}_{\geq 0}} \sum_{e \in S'} \psi^{-1}(p/\beta_e) \quad \text{s.t.} \quad \psi^{-1}(p/\beta_e) \leq \mu_e \text{ for all } e \in S'$$

$$= \max_{e' \in E} \sum_{e \in S} x_e(p_{e'}),$$

i.e., f is an XOS-function and thus fractionally subadditive (see Example (2)). $\square$

We now turn to an incremental variant of classical s-t-flows. In the problem INCREMENTAL MAXIMUM s-t-FLOW we are given a graph $G = (V, E)$ and consider the objective

$$f(X \subseteq E) = \max\{F \mid \text{there exists an } s\text{-}t\text{-flow of value } F \text{ in } (V, X)\}.$$

We restrict ourselves to the case where all edges have unit weight and unit capacity.

To solve this problem, we describe the algorithm QUICKEST-INCREMENT from [16]: The algorithm starts by adding the shortest path and then iteratively adds the smallest set of edges that increase the maximum flow value by at least 1. Let $r \in \mathbb{N}$ be the number of iterations until QUICKEST-INCREMENT terminates. For $i \in \{0, 1, ..., r\}$, let λ_i be the size of the set added in iteration i, i.e., λ_0 is the length of the shortest s-t-path, λ_1 the size of the set added in iteration 1, and so on. For $k \in \{1, ..., |E|\}$, we denote the solution of size k of the algorithm by S_k^A.

With $X_{\max} \in \mathbb{R}_{\geq 0}$ defined as the maximal possible s-t-flow value in the underlying graph, for $j \in \{1, ..., \lfloor X_{\max} \rfloor\}$, we denote by c_j the minimum number of edges required to achieve a flow value of at least j.

In [16] the authors relate the values λ_i and c_j in the following way.

Lemma 3. *In the unit capacity case, we have $\lambda_i \leq c_j/(j - i)$ for all $i, j \in \mathbb{N}$ with $0 \leq i < j \leq r$.*

Using this estimate, we can find a bound on the competitive ratio of QUICKEST-INCREMENT.

Theorem 5. *The algorithm QUICKEST-INCREMENT has a competitive ratio of at most 2 for the problem INCREMENTAL MAXIMUM s-t-FLOW with unit capacities.*

Proof. Note that, since we consider unit capacities, we have $X_{\max} - 1 = r \in \mathbb{N}$ because QUICKEST-INCREMENT increases the value of the solution by exactly 1 in each iteration.

Consider some size $k \in \{1, ..., |E|\}$. If $k < c_1$, we have $f(S_k^*) = 0$, i.e., every solution is competitive. If $k \geq c_1$, let $j := f(S_k^*)$. Note that $f(S_{c_j}^*) = j = f(S_k^*)$ and therefore $k \geq c_j$. By Lemma 3, we have

$$\sum_{i=0}^{\lceil \frac{j}{2} \rceil - 1} \lambda_i \le \sum_{i=0}^{\lceil \frac{j}{2} \rceil - 1} \frac{c_j}{j - i}$$

$$= c_j \sum_{i=0}^{\lceil \frac{j}{2} \rceil - 1} \frac{1}{j - i}$$

$$\le c_j \sum_{i=0}^{\lceil \frac{j}{2} \rceil - 1} \frac{1}{j - \lceil \frac{j}{2} \rceil + 1}$$

$$= c_j \lceil \frac{j}{2} \rceil \frac{1}{\lfloor \frac{j}{2} \rfloor + 1} \le c_j \tag{10}$$

This implies $f(S_k^A) \ge f(S_{c_j}^A) \overset{(10)}{\ge} \lceil \frac{j}{2} \rceil \ge \frac{1}{2} j = \frac{1}{2} f(S_k^*)$.   □

As it turns out, there exists no algorithm with a better competitive ratio than QUICKEST-INCREMENT.

Theorem 6. *There is no algorithm with a competitive ratio of less than 2 for the problem* INCREMENTAL MAXIMUM s-t-FLOW *with unit capacities.*

Proof. Consider the graph $G = (V, E)$ with

$$V := \{s, t, u_1, u_2, u_3, v_1, v_2, v_3\},$$
$$E := \{(s, u_1), (s, v_1), (u_1, u_2), (v_1, v_2), (u_3, t), (v_3, t), (u_1, v_3)\},$$

with unit capacities (cf. Figure 1).

We consider the problem INCREMENTAL MAXIMUM s-t-FLOW and an algorithm ALG which finds an approximation for this problem. In order to be competitive for size $k = 3$, ALG has to add the blocking path s, u_1, v_3, t first. But this implies that the solution of size $k = 8$ does not contain both paths, s, u_1, u_2, u_3, t and s, v_1, v_2, v_3, t, i.e., the optimal solution of size $k = 8$ has flow value 2 while the solution of ALG has flow value 1. Thus, the competitive ratio of ALG cannot be smaller than 2.   □

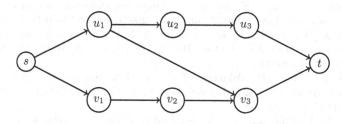

Fig. 1. A lower bound instance with best possible competitive ratio 2 for the problem INCREMENTAL MAXIMUM s-t-FLOW

References

1. Amanatidis, G., Birmpas, G., Markakis, E.: Coverage, matching, and beyond: new results on budgeted mechanism design. In: Proceedings of the 12 International Conference on Web and Internet Economics (WINE), pp. 414–428 (2016)
2. Anari, N., Haghtalab, N., Naor, S., Pokutta, S., Singh, M., Torrico, A.: Structured robust submodular maximization: Offline and online algorithms. In: Proceedings of the 22nd International Conference on Artificial Intelligence and Statistics (AISTATS), vol. 89, pp. 3128–3137 (2019)
3. Badanidiyuru, A., Dobzinski, S., Oren, S.: Optimization with demand oracles. In: Proceedings of the 13th ACM Conference on Electronic Commerce, EC 2012, pp. 110–127. Association for Computing Machinery (2012). https://doi.org/10.1145/2229012.2229025
4. Bernstein, A., Disser, Y., Groß, M., Himburg, S.: General bounds for incremental maximization. Math. Program. (2020). https://doi.org/10.1007/s10107-020-01576-0
5. Disser, Y., Klimm, M., Megow, N., Stiller, S.: Packing a knapsack of unknown capacity. SIAM J. Discret. Math. **31**(3), 1477–1497 (2017)
6. Disser, Y., Weckbecker, D.: Unified greedy approximability beyond submodular maximization. arXiv:2011.00962 (2020)
7. Dobzinski, S., Schapira, M.: An improved approximation algorithm for combinatorial auctions with submodular bidders. In: Proceedings of the Seventeenth Annual ACM-SIAM Symposium on Discrete Algorithm, SODA 2006, pp. 1064–1073. Society for Industrial and Applied Mathematics (2006)
8. Feige, U.: A threshold of ln n for approximating set cover. J. ACM **45**(4), 634–652 (1998)
9. Feige, U.: On maximizing welfare when utility functions are subadditive. SIAM J. Comput. **39**(1), 122–142 (2009). https://doi.org/10.1137/070680977
10. Fujita, R., Kobayashi, Y., Makino, K.: Robust matchings and matroid intersections. SIAM J. Discret. Math. **27**(3), 1234–1256 (2013)
11. Groß, M., Pfetsch, M.E., Schewe, L., Schmidt, M., Skutella, M.: Algorithmic results for potential-based flows: easy and hard cases. Networks **73**(3), 306–324 (2019). https://doi.org/10.1002/net.21865
12. Hartline, J., Sharp, A.: Incremental flow. Networks **50**(1), 77–85 (2007)
13. Hassin, R., Rubinstein, S.: Robust matchings. SIAM J. Discret. Math. **15**(4), 530–537 (2002)
14. Kakimura, N., Makino, K.: Robust independence systems. SIAM J. Discret. Math. **27**(3), 1257–1273 (2013)
15. Kakimura, N., Makino, K., Seimi, K.: Computing knapsack solutions with cardinality robustness. Jpn. J. Ind. Appl. Math. **29**(3), 469–483 (2012)
16. Kalinowski, T., Matsypura, D., Savelsbergh, M.W.: Incremental network design with maximum flows. Eur. J. Oper. Res. **242**(1), 51–62 (2015). https://doi.org/10.1016/j.ejor.2014.10.003
17. Kawase, Y., Sumita, H., Fukunaga, T.: Submodular maximization with uncertain knapsack capacity. SIAM J. Discret. Math. **33**(3), 1121–1145 (2019). https://doi.org/10.1137/18M1174428
18. Kobayashi, Y., Takazawa, K.: Randomized strategies for cardinality robustness in the knapsack problem. Theor. Comput. Sci. **699**, 53–62 (2017)
19. Lehmann, B., Lehmann, D., Nisan, N.: Combinatorial auctions with decreasing marginal utilities. Games Econ. Behav. **55**(2), 270–296 (2006). https://doi.org/10.1016/j.geb.2005.02.006

20. Lin, G., Nagarajan, C., Rajaraman, R., Williamson, D.P.: A general approach for incremental approximation and hierarchical clustering. SIAM J. Comput. **39**(8), 3633–3669 (2010)
21. Marchetti-Spaccamela, A., Vercellis, C.: Stochastic on-line knapsack problems. Math. Program. **68**, 73–104 (1995)
22. Matuschke, J., Skutella, M., Soto, J.A.: Robust randomized matchings. Math. Oper. Res. **43**(2), 675–692 (2018)
23. Megow, N., Mestre, J.: Instance-sensitive robustness guarantees for sequencing with unknown packing and covering constraints. In: Proceedings of the 4th Conference on Innovations in Theoretical Computer Science (ITCS), pp. 495–504 (2013)
24. Navarra, A., Pinotti, C.M.: Online knapsack of unknown capacity: how to optimize energy consumption in smartphones. Theor. Comput. Sci. **697**, 98–109 (2017)
25. Nemhauser, G.L., Wolsey, L.A., Fisher, M.L.: An analysis of approximations for maximizing submodular set functions. I. Math. Program. **14**, 265–294 (1978). https://doi.org/10.1007/BF01588971
26. Nisan, N.: Bidding and allocation in combinatorial auctions. In: Proceedings of the 2nd ACM Conference on Electronic Commerce (EC), pp. 1–12 (2000). https://doi.org/10.1145/352871.352872
27. Orlin, J.B., Schulz, A.S., Udwani, R.: Robust monotone submodular function maximization. Math. Program. **172**(1–2), 505–537 (2018)
28. Sviridenko, M.: A note on maximizing a submodular set function subject to a knapsack constraint. Oper. Res. Lett. **32**, 41–43 (2004). https://doi.org/10.1016/s0167-6377(03)00062-2
29. Thielen, C., Tiedemann, M., Westphal, S.: The online knapsack problem with incremental capacity. Math. Methods Oper. Res. **83**(2), 207–242 (2016)
30. Yoshida, Y.: Maximizing a monotone submodular function with a bounded curvature under a knapsack constraint. SIAM J. Discret. Math. **33**(3), 1452–1471 (2019)

Improved Analysis of Online Balanced Clustering

Marcin Bienkowski[1]($\boxtimes$)(iD), Martin Böhm[1](iD), Martin Koutecký[2](iD),
Thomas Rothvoß[3], Jiří Sgall[2](iD), and Pavel Veselý[2](iD)

[1] Institute of Computer Science, University of Wrocław, Wrocław, Poland
`marcin.bienkowski@cs.uni.wroc.pl`
[2] Computer Science Institute of Charles University, Faculty of Mathematics
and Physics, Prague, Czechia
[3] University of Washington, Seattle, USA

Abstract. In the online balanced graph repartitioning problem, one has
to maintain a clustering of n nodes into ℓ clusters, each having $k = n/\ell$
nodes. During runtime, an online algorithm is given a stream of communication requests between pairs of nodes: an inter-cluster communication
costs one unit, while the intra-cluster communication is free. An algorithm can change the clustering, paying unit cost for each moved node.

This natural problem admits a simple $O(\ell^2 \cdot k^2)$-competitive algorithm
COMP, whose performance is far apart from the best known lower bound
of $\Omega(\ell \cdot k)$. One of open questions is whether the dependency on ℓ can
be made linear; this question is of practical importance as in the typical
datacenter application where virtual machines are clustered on physical
servers, ℓ is of several orders of magnitude larger than k. We answer this
question affirmatively, proving that a simple modification of COMP is
$(\ell \cdot 2^{O(k)})$-competitive.

On the technical level, we achieve our bound by translating the problem to a system of linear integer equations and using Graver bases to
show the existence of a "small" solution.

Keywords: Clustering · Graph partitioning · Balanced partitioning ·
Graver basis · Online algorithms · Competitive analysis

1 Introduction

We study the *online balanced graph repartitioning problem*, introduced by Avin
et al. [3]. In this problem, an algorithm has to maintain a time-varying partition
of n nodes into ℓ clusters, each having $k = n/\ell$ nodes. An algorithm is given
an online stream of communication requests, each involving a pair of nodes.

Supported by Polish National Science Centre grant 2016/22/E/ST6/00499, by
Center for Foundations of Modern Computer Science (Charles University project
UNCE/SCI/004), by the project 19-27871X of GA ČR, and by NSF CAREER grant
1651861.

J. Koenemann and B. Peis (Eds.): WAOA 2021, LNCS 12982, pp. 224–233, 2021.
https://doi.org/10.1007/978-3-030-92702-8_14

A communication between a pair of nodes from the same cluster is free, while inter-cluster communication incurs unit cost. In response, an algorithm may change the mapping of nodes to clusters, also paying a unit cost for changing a cluster of a single node. After remapping, each cluster has to contain k nodes again.

We focus on an online scenario, where an (online) algorithm has to make irrevocable remapping decisions after each communication request without the knowledge of the future. The problem can be seen as a dynamic (and online) counterpart of a so-called ℓ-balanced graph partitioning problem [1], where the goal is to partition the graph into ℓ equal-size parts to minimize the total weight of edges in the cut. In particular, $\ell = 2$ corresponds to the well-studied graph bisection problem [9,13].

A main practical motivation originates from server virtualization in datacenters. There, nodes correspond to n virtual machines run on ℓ physical ones. Each physical machine (a cluster) has the capacity for accommodating k virtual machines. Communication requests are induced by distributed applications running in the datacenter. While communication within a physical machine is practically free, the inter-cluster communication (between different physical machines) generates considerable load and affects the overall running time (see, e.g., [5]). Due to the current capabilities of modern architectures, migrating a virtual machine to another physical machine (remapping of a node) is possible, but it incurs a certain load.[1]

To evaluate the performance of an online algorithm ALG, we use a standard notion of competitive ratio [4] which is the supremum over all possible inputs of ALG-to-OPT cost, where OPT denotes the optimum offline solution for the problem.

1.1 Component-Based Approach

Our contribution is related to the following natural algorithm (henceforth called COMP) proposed by Avin et al. [3]. We define it in detail in Sect. 3; here, we give its informal description. COMP operates in phases, in each phase keeping track of components of nodes that communicated in this phase (i.e., connected components of a graph whose edges are communication requests). COMP always keeps all nodes of a given component in a single cluster; we call this property *component invariant*. When components are modified, to maintain the invariant, some nodes may have to change their clusters; the associated cost can be trivially bounded by n. As this changes the mapping of nodes to clusters, we

---

[1] In the reality, the network load generated by a single request is much smaller than the load of migrating a whole virtual machine. This has been captured by some papers, which assigned cost $\alpha \geq 1$ to the latter event (α is then a parameter of the problem). However, this additional difficulty can be resolved by standard rent-or-buy approaches (reacting only to every α-th request between a given pair of nodes). Therefore, and also to keep the description simple, in this paper, we assume that $\alpha = 1$.

call it *remapping event*. If such remapping does not exist[2], the current phase terminates, and it is possible to show that OPT paid at least 1 in this phase. The number of remapping events is equal to the number of times connected components are modified, which can be upper-bounded by $n - 1$. Thus, the overall cost of COMP in a single phase is $O(n^2) = O(\ell^2 \cdot k^2)$, while that of OPT is at least 1. This shows that COMP is $O(\ell^2 \cdot k^2)$-competitive.

1.2 Related Work

Perhaps surprisingly, no better algorithm than COMP (even randomized one) is known for the general case. Some improvements were, however, given for specific values of k and ℓ.

The dependency of the competitive ratio on k is at least linear for deterministic algorithms: a lower bound of $\Omega(k)$ follows by the reduction from online paging [15] and holds already for $\ell = 2$ clusters [2]. For $k = 2$, $O(1)$-competitive (deterministic) algorithms are known: a 7-competitive algorithm was given by Avin et al. [2,3] and was later improved to a 6-competitive one by Pacut et al. [12]. However, already for $k \geq 3$, the competitive ratio of any deterministic algorithm cannot be better than $\Omega(k \cdot \ell)$ [11,12]. For the special case of $k = 3$, Pacut et al. [12] showed that a variant of COMP is $O(\ell)$-competitive. The lack of progress towards improving the $O(\ell^2 \cdot k^2)$ upper bound for the general case motivated the research of simplified variants.

Henzinger et al. [8] initiated the study of a so-called *learning variant*, where there exists a fixed partitioning (unknown to an algorithm) of nodes into clusters, and the communication requests are *consistent* with this mapping, i.e., all requests are given between same-cluster node pairs. Hence, the implicit goal of an algorithm is to learn such static mapping. The deterministic lower bound of $\Omega(k \cdot \ell)$ also holds for this variant, and, furthermore, there exists a deterministic algorithm that asymptotically matches this bound [11,12].

Another strand of research focused on a resource-augmented scenario, where each cluster of an online algorithm is able to accommodate $(1 + \epsilon) \cdot k$ nodes (but the online algorithm is still compared to OPT, whose clusters have to keep k nodes each). For $\epsilon > 1$, the first deterministic algorithm was given by Avin et al. [2]; the achieved ratio was $O(k \cdot \log k)$ (for an arbitrary ℓ). Suprisingly, the ratio remains $\Omega(k)$ even for large ϵ (as long as the algorithm cannot keep all nodes in a single cluster) [2].

When these two simplifications are combined (i.e., the learning variant is studied in a resource-augmented scenario), asymptotically optimal results are due to Henzinger et al. [7,8]. They show that for any fixed $\epsilon > 0$, the deterministic ratio is $\Theta(\ell \cdot \log k)$ [7,8] and the randomized ratio is $\Theta(\log \ell + \log k)$ [7]. Furthermore, for $\epsilon > 1$, their deterministic algorithm is $O(\log k)$-competitive [7].

---

[2] Deciding whether such remapping exists is NP-hard. As typical for online algorithms, however, our focus is on studying the disadvantage of not knowing the future rather than on computational complexity.

1.3 Our Contribution

We focus on the general variant of the balanced graph repartitioning problem. We study a variant of COMP (see Subsect. 1.1) in which *each remapping event is handled in a way minimizing the number of affected clusters.*

We show that the number of nodes remapped this way is $2^{O(k)}$ (in comparison to the trivial bound of $n = \ell \cdot k$). The resulting bound on the competitive ratio is then $(\ell \cdot k) \cdot 2^{O(k)} = \ell \cdot 2^{O(k)}$, i.e., we replaced the quadratic dependency on ℓ in the competitive ratio by the linear one. We note that the resulting algorithm retains the $O(\ell^2 \cdot k^2)$-competitiveness guarantee of the original COMP algorithm as well.

Given the lower bound of $\Omega(\ell \cdot k)$ [11,12], the resulting strategy is optimal for a constant k. We also note that for the datacenter application described earlier, k is of several orders of magnitude smaller than ℓ.

We achieve our bound by translating the remapping event to a system of linear integer equations so that the size of the solution (sum of values of variables) is directly related to the number of affected clusters. Then, we use algebraic tools such as Graver bases to argue that these equations admit a "small" solution.[3]

2 Preliminaries

An offline part of the input is a set of $n = \ell \cdot k$ nodes and an initial valid mapping of these nodes into ℓ clusters. We call a mapping *valid* if each cluster contains exactly k nodes.

An online part of the input is a set of requests, each being a pair of nodes (u, v). The request incurs cost 1 if u and v are mapped to different clusters. After each request, an online algorithm may modify the current node mapping to a new valid one, paying 1 for each node that changes its cluster.

For an input I and an algorithm ALG, ALG(I) denotes its cost on input I, whereas OPT(I) denotes the optimal cost of an offline solution. ALG is γ-competitive if there exists a constant β such that for any input I, it holds that $\text{ALG}(I) \le \gamma \cdot \text{OPT}(I) + \beta$. While β has to be independent of the online part of the sequence, it may depend on the offline part, i.e., be a function of parameters ℓ and k.

3 Better Analysis for COMP

Algorithm COMP [3] splits input into phases. In each phase, COMP maintains an auxiliary partition $\mathcal{R}$ of the set of nodes into *components*; initially, each node is in its own singleton component. COMP maintains the following *component invariant*: for each component $S \in \mathcal{R}$, all its nodes are inside the same cluster.

Assume a request (u, v) arrives. Two cases are possible.

---

[3] One could also bound the size of a solution along the lines of Schrijver [14, Corollary 17.1b], which boils down to a determinant bound, same as our proof. In general, Graver basis elements may be much smaller, but in out specific case, the resulting bound would be asymptotically the same.

- If u and v are within the same component of $\mathcal{R}$, then by the component invariant, they are in the same cluster. COMP serves this request without paying anything, without changing $\mathcal{R}$, and without remapping nodes.
- If u and v are in different components S_a and S_b, COMP merges these components into $S_{ab} = S_a \uplus S_b$ by removing S_a and S_b from $\mathcal{R}$, and adding S_{ab} to $\mathcal{R}$. (By $A \uplus B$ we denote the *disjoint union* of sets A and B.)

  If components S_a and S_b were in different clusters, the resulting component S_{ab} now spans two clusters, which violates the component invariant. To restore it, COMP verifies whether there exists a valid mapping of nodes into clusters preserving the component invariant (also for the new component S_{ab}). In such case, a *remapping event* occurs: among all such mappings, COMP chooses one minimizing the total number of affected clusters. (The original variant of COMP [3] simply chose any such mapping.) If, however, no such valid mapping exists, COMP resets $\mathcal{R}$ to the initial partition in which each node is in its own component and starts a new phase.

Lemma 1. *Assume each remapping event affects at most $f(\ell, k)$ clusters (for some function f). Then COMP is $O(\ell \cdot k^2 \cdot f(\ell, k))$-competitive.*

Proof. Fix any input I and split it into phases according to COMP. Each phase (except possibly the last one) terminates because there is no valid node mapping that would preserve the component invariant. That is, for any fixed valid mapping of nodes to clusters, the phase contains an inter-cluster request. Thus, during any phase, either OPT pays at least 1 for node remapping, or its mapping is fixed within phase, and then it pays at least 1 for serving requests.

On the other hand, within each phase, COMP modifies family $\mathcal{R}$ of components at most $n-1 = \ell \cdot k - 1$ times since the number of components decreases by one with each modification. Each time it happens, it pays 1 for the request, and then, if the remapping event is triggered, it pays additionally at most $k \cdot f(\ell, k)$ as it remaps at most k nodes from each affected cluster. Hence, the overall cost in a single phase is $(\ell \cdot k - 1) \cdot (1 + k \cdot f(\ell, k)) = O(\ell \cdot k^2 \cdot f(\ell, k))$.

The cost of COMP in the last phase is universally bounded by $O(\ell \cdot k^2 \cdot f(\ell, k))$ and in the remaining phases, the COMP-to-OPT cost ratio is at most $O(\ell \cdot k^2 \cdot f(\ell, k))$, which concludes the proof. $\qquad\square$

The trivial upper bound on $f(\ell, k)$ is ℓ, which together with Lemma 1 yields the already known bound of $O(\ell^2 \cdot k^2)$ [3]. In the remaining part, we show that if COMP tries to minimize the number of affected clusters for each remapping event, then $f(\ell, k)$ can be upper-bounded by $2^{O(k)}$. Thus, our analysis beats the simple approach when the number of clusters ℓ is much larger than the cluster capacity k. This ratio is also optimal for constant k as the lower bound on the competitive ratio is $\Omega(\ell \cdot k)$ [11,12].

3.1 Analyzing a Remapping Event

Recall that we want to analyze a remapping event, where we have a given valid mapping of nodes to clusters and a family of components $\mathcal{R}$ satisfying the component invariant. COMP merges two components S_a and S_b into one, and the resulting component $S_{ab} = S_a \uplus S_b$ spans two clusters. As we assume that it is possible to remap nodes to satisfy the component invariant, the size of S_{ab} is at most k.

We first express our setup in the form of a system of linear integer equations. The definition below assumes that component invariant holds, i.e., each component is entirely contained in some cluster.

Definition 1 (Cluster configuration). *A configuration of a cluster C is a vector $\mathbf{c} = (c_1, \ldots, c_k)$ where $c_i \geq 0$ is the number of components of size i in C.*

We denote the set of all possible cluster configurations by $\mathcal{C}$, i.e., $\mathcal{C}$ contains all possible k-dimensional vectors $\mathbf{c} = (c_1, \ldots, c_k)$ such that $\sum_{i=1}^{k} i \cdot c_i = k$. For succinctness, we use $\mathrm{nd}(\mathbf{c}) = \sum_{i=1}^{k} i \cdot c_i$. The number of configurations is equal to the partition function $\pi(k)$, which denotes the number of possibilities how k can be expressed as a sum of a multiset of non-negative integers. From the known bounds on π, we get $|\mathcal{C}| = \pi(k) \leq 2^{O(\sqrt{k})}$ [6].

We take two clusters that contained components S_a and S_b, respectively, and we virtually replace them by a *pseudo-cluster* $\hat{C}$ that contains all their components (including S_{ab} and excluding S_a and S_b). Let $\hat{\mathbf{c}}$ be the configuration of this pseudo-cluster; note that $\mathrm{nd}(\hat{\mathbf{c}}) = 2k$, and hence $\hat{\mathbf{c}} \notin \mathcal{C}$. We define the extended set of configurations $\mathcal{C}^{\mathrm{ext}} = \mathcal{C} \uplus \{\hat{\mathbf{c}}\}$.

Let $\mathbf{x}$ be a $|\mathcal{C}^{\mathrm{ext}}|$-dimensional vector, indexed by possible configurations from $\mathcal{C}^{\mathrm{ext}}$, such that, for any configuration $\mathbf{c} \in \mathcal{C}^{\mathrm{ext}}$, $x_{\mathbf{c}}$ is the number of clusters with configuration $\mathbf{c}$ before remapping event. Note that $\mathbf{x} \geq \mathbf{0}$, $x_{\hat{\mathbf{c}}} = 1$ and $\|\mathbf{x}\|_1 = \sum_{\mathbf{c} \in \mathcal{C}^{\mathrm{ext}}} x_{\mathbf{c}} = \ell - 1$. Let

$$\mathbf{u} = \sum_{\mathbf{c} \in \mathcal{C}^{\mathrm{ext}}} x_{\mathbf{c}} \cdot \mathbf{c}. \tag{1}$$

That is, $\mathbf{u} = (u_1, u_2, \ldots, u_k)$, where u_i is the total number of components of size i (in all clusters). Clearly, $\mathrm{nd}(\mathbf{u}) = (\ell - 2) \cdot k + 2 \cdot k = \ell \cdot k$. We rewrite (1) as

$$\mathbf{u} = A\mathbf{x}, \tag{2}$$

where A is the matrix with k rows and $|\mathcal{C}^{\mathrm{ext}}|$ columns. Its columns are equal to vectors of configurations from $\mathcal{C}^{\mathrm{ext}}$.

As $\mathbf{x}$ describes the current state of the clusters, in the following, we focus on finding an appropriate vector $\mathbf{y}$ describing a *target* state of the clusters, i.e., their state after the remapping event takes place.

Definition 2. *An integer vector $\mathbf{y}$ is a valid target vector if it is $|\mathcal{C}^{\mathrm{ext}}|$-dimensional and satisfies $\mathbf{y} \geq \mathbf{0}$, $A\mathbf{y} = \mathbf{u}$ and $y_{\hat{\mathbf{c}}} = 0$.*

Lemma 2. *For any valid target vector $\mathbf{y}$, it holds that $\|\mathbf{y}\|_1 = \ell$.*

Proof. Let $\ell' = \|\mathbf{y}\|_1$. Then $\mathbf{y} = \sum_{i=1}^{\ell'} \mathbf{y}^i$, where $\mathbf{y}^i$ is equal to 1 for some configuration $\mathbf{c} \neq \hat{\mathbf{c}}$ and 0 everywhere else. As $\mathbf{u} = A\mathbf{y} = \sum_{i=1}^{\ell'} A\mathbf{y}^i$, we obtain

$\ell \cdot k = \mathrm{nd}(\mathbf{u}) = \sum_{i=1}^{\ell'} \mathrm{nd}(A\mathbf{y}^i)$. For any i, vector $A\mathbf{y}^i$ is a single column of A corresponding to a configuration $\mathbf{c} \neq \hat{\mathbf{c}}$, and thus $\mathrm{nd}(A\mathbf{y}^i) = k$. This implies that $\ell' = \ell$, which concludes the proof.   □

Lemma 3. *There exist a valid target vector* $\mathbf{y}$.

Proof. After the remapping event takes place, it is possible to map nodes to different clusters so that the component invariant holds (after merging S_a and S_b into S_{ab}), i.e., each component is entirely contained in some cluster (not in the pseudo-cluster). Thus, each cluster has a well-defined configuration in $\mathcal{C}$, and $y_\mathbf{c}$ is simply the number of clusters with configuration $\mathbf{c}$ after remapping.   □

Lemma 4. *Fix a valid target vector* $\mathbf{y}$. *Then, there exists a node remapping that affects* $(1/2) \cdot \|\mathbf{x} - \mathbf{y}\|_1 + 1/2$ *clusters.*

Proof. We define vector $\tilde{\mathbf{x}}$, such that $\tilde{x}_\mathbf{c} = x_\mathbf{c}$ for $\mathbf{c} \neq \hat{\mathbf{c}}$, and $\tilde{x}_{\hat{\mathbf{c}}} = 2$. Hence $\|\tilde{\mathbf{x}}\|_1 = \ell$. By Lemma 2, $\|\mathbf{y}\|_1 = \ell$ as well. For $\mathbf{c} \in \mathcal{C}$, the value of $\tilde{x}_\mathbf{c}$ denotes how many clusters have configuration $\mathbf{c}$, with $\tilde{x}_{\hat{\mathbf{c}}} = 2$ simply denoting that there are two clusters whose configuration is not equal to any configuration from $\mathcal{C}$.

Now, for any configuration $\mathbf{c} \in \mathcal{C}^{\mathrm{ext}}$, we fix $\min\{\tilde{x}_\mathbf{c}, y_\mathbf{c}\}$ clusters with configuration $\mathbf{c}$. Our remapping does not touch these clusters and there exists a straightforward node remapping which involves only the remaining clusters. Their number is equal to

$$(1/2) \cdot \sum_{\mathbf{c} \in \mathcal{C}^{\mathrm{ext}}} |\tilde{x}_\mathbf{c} - y_\mathbf{c}| = 1 + (1/2) \cdot \sum_{\mathbf{c} \in \mathcal{C}} |\tilde{x}_\mathbf{c} - y_\mathbf{c}| = 1 + (1/2) \cdot \sum_{\mathbf{c} \in \mathcal{C}} |x_\mathbf{c} - y_\mathbf{c}|$$

$$= (1/2) + (1/2) \cdot \sum_{\mathbf{c} \in \mathcal{C}^{\mathrm{ext}}} |x_\mathbf{c} - y_\mathbf{c}|$$

$$= (1/2) + (1/2) \cdot \|\mathbf{x} - \mathbf{y}\|_1,$$

and thus the lemma follows.   □

We note that a valid target vector $\mathbf{y}$ guaranteed by Lemma 3 may be completely different from vector $\mathbf{x}$ describing the current clustering, and thus it is possible that $\|\mathbf{x} - \mathbf{y}\|_1 = \Omega(\ell)$. We however show that on the basis of $\mathbf{y}$, we may find a valid target vector $\mathbf{y}'$, such that $\|\mathbf{x} - \mathbf{y}'\|_1$ is small, i.e., at most $2^{O(k)}$.

3.2   Using Graver Basis

Lemma 3 guarantees the existence of vector $\mathbf{z} = \mathbf{x} - \mathbf{y}$, encoding the reorganization of the clusters. We already know that $A\mathbf{z} = \mathbf{0}$ must hold, but there are other useful properties as well; for instance, if $z_\mathbf{c} > 0$ for a configuration $\mathbf{c}$, then $z_\mathbf{c} \leq x_\mathbf{c}$ (i.e. the corresponding reorganization does not try to remove more clusters of configuration $\mathbf{c}$ than $x_\mathbf{c}$). Our goal is to find another vector $\mathbf{w}$ that also encodes the reorganization and $\|\mathbf{w}\|_1$ is small.

The necessary condition for $\mathbf{w}$ is that it satisfies $A\mathbf{w} = \mathbf{0}$, and thus we study properties of matrix A, defined by (2), in particular its Graver basis. For an introduction to Graver bases, we refer the interested reader to a book by Onn [10].

Definition 3 (Sign-compatibility). *Given two vectors **a** and **b** of the same length, we say that they are* sign-compatible *if for each coordinate i the sign of a_i is the same as the sign of b_i (i.e., $a_i \cdot b_i \geq 0$). We say that $\mathbf{a} \sqsubseteq \mathbf{b}$ if $\mathbf{a}$ and $\mathbf{b}$ are sign-compatible and $|a_i| \leq |b_i|$ for every coordinate i. Note that $\sqsubseteq$ imposes a partial order.*

Definition 4 (Graver basis). *Given an integer matrix A, its* Graver basis $\mathcal{G}(A)$ *is the set of $\sqsubseteq$-minimal elements of the lattice $\mathcal{L}^*(A) = \{\mathbf{h} \mid A\mathbf{h} = \mathbf{0}, \mathbf{h} \in \mathbb{Z}^n, \mathbf{h} \neq \mathbf{0}\}$.*

Lemma 5 (Lemma 3.2 of [10]). *Any vector $\mathbf{h} \in \mathcal{L}^*(A)$ is a sign-compatible sum $\mathbf{h} = \sum_i \mathbf{g}^i$ of Graver basis elements $\mathbf{g}^i \in \mathcal{G}(A)$, with some elements possibly appearing with repetitions.*

Having the tools above, we may now prove the existence of a remapping involving small number of clusters.

Lemma 6. *If there exists a valid target vector $\mathbf{y}$, then there exists a valid target vector $\mathbf{y}'$, such that $\mathbf{x} - \mathbf{y}' \in \mathcal{G}(A)$.*

Proof. By (2) and the lemma assumption, $A\mathbf{x} = A\mathbf{y}$. Let $\mathbf{z} = \mathbf{x} - \mathbf{y}$. Then, $z_{\hat{\mathbf{c}}} = x_{\hat{\mathbf{c}}} - y_{\hat{\mathbf{c}}} = 1$ and $A\mathbf{z} = \mathbf{0}$, and thus $\mathbf{z} \in \mathcal{L}^*(A)$.

Using Lemma 5, we may express $\mathbf{z}$ as $\mathbf{z} = \sum_i \mathbf{g}^i$, where $\mathbf{g}^i \in \mathcal{G}(A) \subseteq \mathcal{L}^*(A)$ for all i and all $\mathbf{g}^i$ are sign-compatible with $\mathbf{z}$. As $z_{\hat{\mathbf{c}}} = 1$, the sign-compatibility means that there exists $\mathbf{g}^j$ appearing in the sum with $g_{\hat{\mathbf{c}}}^j = 1$. We set $\mathbf{w} = \mathbf{g}^j$.

Let $\mathbf{y}' = \mathbf{x} - \mathbf{w}$. Clearly $\mathbf{x} - \mathbf{y}' = \mathbf{w} \in \mathcal{G}(A)$. It remains to show that $\mathbf{y}'$ is a valid target vector. We have $A\mathbf{y}' = A\mathbf{x} - A\mathbf{w} = \mathbf{u} - \mathbf{0} = \mathbf{u}$ and $y_{\hat{\mathbf{c}}}' = x_{\hat{\mathbf{c}}} - w_{\hat{\mathbf{c}}} = 1 - 1 = 0$. To show that $\mathbf{y}' \geq \mathbf{0}$, we consider two cases. If $z_{\mathbf{c}} \geq 0$, then by sign-compatibility $0 \leq w_{\mathbf{c}} \leq z_{\mathbf{c}}$, and thus $y_{\mathbf{c}}' = x_{\mathbf{c}} - w_{\mathbf{c}} \geq x_{\mathbf{c}} - z_{\mathbf{c}} = y_{\mathbf{c}} \geq 0$. On the other hand, if $z_{\mathbf{c}} < 0$, then again by sign-compatibility, $w_{\mathbf{c}} \leq 0$, and thus $y_{\mathbf{c}}' = x_{\mathbf{c}} - w_{\mathbf{c}} \geq x_{\mathbf{c}} \geq 0$. $\qquad\square$

To complete our argument, it remains to bound $\|\mathbf{g}\|_1$, where $\mathbf{g}$ is an arbitrary element of $\mathcal{G}(A)$. We start with a known bound for ℓ^∞-norm of any element of the Graver basis.

Lemma 7 (Lemma 3.2 of [10]). *Let q be the number of columns of integer matrix M. Let $\Delta(M)$ denote the maximum absolute value of the determinant of a square sub-matrix of M. Then $\|\mathbf{g}\|_\infty \leq q \cdot \Delta(M)$ for any $\mathbf{g} \in \mathcal{G}(M)$.*

Lemma 8. *For any $\mathbf{g} \in \mathcal{G}(A)$, it holds that $\|\mathbf{g}\|_1 \leq 2^{O(k)}$.*

Proof. We start by showing that $\Delta(A) \leq e^k$. Let $\tilde{A}$ be the matrix A with each row multiplied by its index. By the definition of a configuration, the column sums of $\tilde{A}$ are equal to k, with the exception of the column corresponding to the configuration $\hat{\mathbf{c}}$ whose sum is equal to $2k$. As all entries of $\tilde{A}$ are non-negative, the same holds for ℓ_1-norms of columns of $\tilde{A}$.

Fix any square sub-matrix B of A and let j be the number of its columns (rows). Let $\tilde{B}$ be the corresponding sub-matrix of $\tilde{A}$. Since $\tilde{B}$ is obtained from B

by multiplying its rows by j distinct positive integers, $|\det(\tilde{B})| \geq j! \cdot |\det(B)|$. (This relation holds with equality if and only if B contains (a part of) the first j rows of A.)

It therefore remains to upper-bound $|\det(\tilde{B})|$. Hadamard's bound on determinant states that the absolute value of a determinant is at most the product of lengths (ℓ_2-norms) of its column vectors, which are in turn bounded by the ℓ_1-norms of columns of $\tilde{B}$. These are not greater than ℓ_1-norms of the corresponding columns of $\tilde{A}$ and thus $|\det(\tilde{B})| \leq 2 \cdot k^j$.

Combining the above bounds and using $j \leq k$, we obtain

$$|\det(B)| \leq \frac{|\det(\tilde{B})|}{j!} \leq \frac{2 \cdot k^j}{j!} \leq \frac{2 \cdot k^k}{k!} \leq 2 \cdot e^{k-1} \leq e^k.$$

As B was chosen as an arbitrary square sub-matrix of A, $\Delta(A) \leq e^k$. Our matrix A has $|\mathcal{C}^{\text{ext}}|$ columns, and hence Lemma 7 implies that $\|\mathbf{w}\|_\infty \leq |\mathcal{C}^{\text{ext}}| \cdot 2^{O(k)}$. As $\mathbf{w}$ is $|\mathcal{C}^{\text{ext}}|$-dimensional, $\|\mathbf{w}\|_1 \leq |\mathcal{C}^{\text{ext}}| \cdot \|\mathbf{w}\|_\infty \leq |\mathcal{C}^{\text{ext}}|^2 \cdot e^k$. Finally using $|\mathcal{C}^{\text{ext}}| \leq 2^{O(\sqrt{k})}$, we obtain $\|\mathbf{w}\|_1 \leq 2^{O(k)}$. $\square$

Corollary 1. *The remapping event of* COMP *affects at most* $2^{O(k)}$ *clusters.*

Proof. Combining Lemma 3 with Lemma 6 yields the existence of a valid target vector $\mathbf{y}'$ satisfying $\|\mathbf{x} - \mathbf{y}'\|_1 \in \mathcal{G}(A)$. By Lemma 8, $\|\mathbf{x} - \mathbf{y}'\|_1 \leq 2^{O(k)}$. Thus, plugging $\mathbf{y}'$ to Lemma 4 yields the corollary. $\square$

3.3   Competitive Ratio

Combining Lemma 1 with Corollary 1 immediately yields the desired bound on the competitive ratio of COMP.

Theorem 1. *The variant of* COMP *in which each remapping event is handled in a way minimizing the number of affected clusters is* $(\ell \cdot 2^{O(k)})$-*competitive.*

References

1. Andreev, K., Räcke, H.: Balanced graph partitioning. Theory Comput. Syst. **39**(6), 929–939 (2006). https://doi.org/10.1007/s00224-006-1350-7
2. Avin, C., Bienkowski, M., Loukas, A., Pacut, M., Schmid, S.: Dynamic balanced graph partitioning. SIAM J. Disc. Math. **34**(3), 1791–1812 (2020). https://doi.org/10.1137/17M1158513
3. Avin, C., Loukas, A., Pacut, M., Schmid, S.: Online balanced repartitioning. In: Proceedings 30th International Symposium on Distributed Computing (DISC), pp. 243–256 (2016). https://doi.org/10.1007/978-3-662-53426-7_18
4. Borodin, A., El-Yaniv, R.: Online Computation and Competitive Analysis. Cambridge University Press, Cambridge (1998)
5. Chowdhury, M., Zaharia, M., Ma, J., Jordan, M.I., Stoica, I.: Managing data transfers in computer clusters with Orchestra. In: ACM SIGCOMM, pp. 98–109 (2011). https://doi.org/10.1145/2018436.2018448

6. Erdős, P.: On an elementary proof of some asymptotic formulas in the theory of partitions. Ann. Math. **43**(3), 437–450 (1942). https://doi.org/10.2307/1968802
7. Henzinger, M., Neumann, S., Räcke, H., Schmid, S.: Tight bounds for online graph partitioning. In: Proceedings of 32nd ACM-SIAM Symposium on Discrete Algorithms (SODA), pp. 2799–2818 (2021). https://doi.org/10.1137/1.9781611976465.166
8. Henzinger, M., Neumann, S., Schmid, S.: Efficient distributed workload (re-)embedding. In: 2019 SIGMETRICS/Performance Joint International Conference on Measurement and Modeling of Computer Systems, pp. 43–44 (2019). https://doi.org/10.1145/3309697.3331503
9. Krauthgamer, R., Feige, U.: A polylogarithmic approximation of the minimum bisection. SIAM Rev. **48**(1), 99–130 (2006). https://doi.org/10.1137/050640904
10. Onn, S.: Nonlinear discrete optimization. Zurich Lectures in Advanced Mathematics, European Mathematical Society (2010)
11. Pacut, M., Parham, M., Schmid, S.: Brief announcement: deterministic lower bound for dynamic balanced graph partitioning. In: Proceedings 39th ACM Symposium on Principles of Distributed Computing (PODC), pp. 461–463 (2020). https://doi.org/10.1145/3382734.3405696
12. Pacut, M., Parham, M., Schmid, S.: Optimal online balanced graph partitioning. In: Proceedings 40th IEEE International Conference on Computer Communications (INFOCOM), pp. 1–9 (2021). https://doi.org/10.1109/INFOCOM42981.2021.9488824
13. Räcke, H.: Optimal hierarchical decompositions for congestion minimization in networks. In: Proceedings of 40th ACM Symposium on Theory of Computing (STOC), pp. 255–264 (2008). https://doi.org/10.1145/1374376.1374415
14. Schrijver, A.: Theory of Linear and Integer Programming. John Wiley & Sons, Hoboken (1998)
15. Sleator, D.D., Tarjan, R.E.: Amortized efficiency of list update and paging rules. Commun. ACM **28**(2), 202–208 (1985). https://doi.org/10.1145/2786.2793

Precedence-Constrained Covering Problems with Multiplicity Constraints

Stavros G. Kolliopoulos[1]([✉]) and Antonis Skarlatos[2][iD]

[1] Department of Informatics and Telecommunications, National and Kapodistrian University of Athens, Athens, Greece
sgk@di.uoa.gr
[2] Department of Computer Sciences, University of Salzburg, Salzburg, Austria
antonis.skarlatos@plus.ac.at

Abstract. We study the approximability of covering problems when the set of items chosen to satisfy the covering constraints must form an ideal of a given partial order. We examine the general case with multiplicity constraints, where item i can be chosen up to d_i times. For the basic Precedence-Constrained Knapsack problem (PCKP) we answer an open question of McCormick et al. [10] and show the existence of approximation algorithms with strongly-polynomial bounds. PCKP is a special case, with a single covering constraint, of a Precedence-Constrained Covering Integer Program (PCCP). For a general PCCP where the number of covering constraints is $m \geq 1$, we show that an algorithm of Pritchard and Chakrabarty [11] for Covering Integer Programs can be extended to yield an f-approximation, where f is the maximum number of variables with nonzero coefficients in a covering constraint. This is nearly-optimal under standard complexity-theoretic assumptions and surprisingly matches the bound achieved for the problem without precedence constraints.

Keywords: Covering integer programs · Precedence constraints

1 Introduction

Covering problems form a core area of discrete optimization, which has been intensively investigated from the perspectives of algorithms, polyhedral combinatorics, and computational complexity. The paradigmatic covering problem is a *covering integer program (CIP)* which is an integer program of the form

$$\min\{c^T x \mid Ax \geq D, \ 0 \leq x \leq d, \ x \in \mathbb{Z}^n\}$$

where c is the *cost* vector, A is an $m \times n$ nonnegative integer matrix, $D \in \mathbb{Z}_+^m$ is the *demand* vector and $d \in \mathbb{Z}_+^n$ contains the *multiplicity upper bounds* on the variables. The constraints $Ax \geq D$ are the *covering constraints* and $0 \leq x \leq d$ are the *multiplicity constraints*. In this paper we study the *Precedence-Constrained Covering Problem (PCCP)*, the substantial generalization of CIPs with added precedence constraints. Given a partial order $\preceq$ on the set $[n]$, a feasible solution must meet the covering, the multiplicity constraints and moreover for every $i \in [n]$

© Springer Nature Switzerland AG 2021
J. Koenemann and B. Peis (Eds.): WAOA 2021, LNCS 12982, pp. 234–251, 2021.
https://doi.org/10.1007/978-3-030-92702-8_15

$$x_j - x_i \geq 0, \ \forall j \preceq i. \tag{1}$$

Inequalities (1) imply that the support S of a feasible x must be an *ideal* of the partial order, i.e., if $i \in S$, all elements j with $j \preceq i$, must also be in S. Our paper focuses on the effect of the multiplicity constraints on the approximability of precedence-constrained covering problems.

Our first set of results concerns the fundamental Precedence-Constrained Knapsack problem. *Knapsack* is one of the most well-studied problems in combinatorial optimization. The input consists of a set U of items where item i has a nonnegative weight u_i and a cost c_i. In the covering variant that we consider in this paper a scalar demand D is specified and one wants to select a subset of the items with total weight at least D and minimum total cost. Knapsack is weakly NP-hard and is well known to have an FPTAS [6]. LP-based approximation algorithms have also been studied for this problem [1,2]. The natural LP relaxation has an unbounded integrality gap. Carr et al. introduced the Knapsack Cover (KC) valid inequalities [2], which fall into the family of cover inequalities introduced by Wolsey [14], and used them to obtain a 2-appoximation algorithm that invokes the ellipsoid as a subroutine. Carnes and Shmoys obtained a primal-dual 2-approximation algorithm which again exploits the KC inequalities [1]. McCormick et al. [10] observe that the 2-approximation of [1] can be extended to the case of a general d vector, as opposed to $d = 1$, i.e., the all-ones case.

The *Precedence-Constrained Knapsack problem (PCKP)* is the special case of PCCP with exactly one covering constraint. Thus PCKP is the generalization of Knapsack where the input items form a partially ordered set $\mathcal{P} = (U, \preceq)$ and there are d_i copies available of item i. A number of x_i copies can be chosen for i only if for all $j \prec i$, $x_j \geq x_i$. In addition to being an interesting problem in itself, PCKP, especially in its packing version, arises as a substructure in many complex integer programming problems, including production scheduling and open pit mining (see, e.g., [5,13]).

Standard Knapsack is the special case of PCKP where the input partial order is an antichain. The addition of precedence constraints makes the problem quite challenging. McCormick et al. [10] studied the special case of PCKP where $d = 1$, denoted (0,1)-PCKP. They showed that the natural LP relaxation for PCKP even after the addition of the KC inequalities has an $\Omega(|U|)$ integrality gap. In the same paper they proposed a new valid formulation (cf. (P_{PCKP}) below) and gave a $w(\mathcal{P})$-approximation via a primal-dual algorithm. Here $w(\mathcal{P})$ denotes the width of the input partial order, i.e., the size of the largest antichain. Importantly the algorithm of [10] is not Knapsack-"oblivious": with a different analysis they show that it achieves a 2-approximation for Knapsack, i.e., when the entire U is an antichain. The $w(\mathcal{P})$-approximation is also obtained for the special case of PCCP were all upper bounds d_i are equal to 1 [10]. In terms of negative results, very little is known. McCormick et al. show that there is no PTAS for (0,1)-PCKP unless NP $\subseteq \cap_{\varepsilon > 0}$BPTIME($2^{n^\varepsilon}$). For general multiplicity upper bounds a pseudopolynomial algorithm is given in [10] that achieves a $w(\mathcal{P})\Delta$-approximation where $\Delta = \max_i\{d_i\}$. The algorithm works for PCCP and not simply for PCKP. To our knowledge this is the best-known result for a general d vector. The authors in [10] posed as an open problem the existence of an algorithm with strongly-polynomial bounds. In this paper we show among other results that such algorithms exist both for PCKP and for the PCCP generalization.

Our Contribution for PCKP. We first point out in Sect. 3 that the linear relaxation of the valid formulation (P_{PCKP}) introduced in [10] has an integrality gap of $\Omega(w(\mathcal{P}))$ even for (0,1)-PCKP. Our main results are in Sect. 4. We first give a primal-dual algorithm that achieves an $O(w^2(\mathcal{P}))$-approximation when all nonzero d_is have the same value Δ. The relaxation of [10] and the associated primal-dual algorithm focus on the minimal items outside a given ideal and selects one such item, say i. In order to estimate the desirability of item i for inclusion the weight of the successors of i is "projected" on i, which yields the effective weight $\overline{u}_i$. We take a different approach and allow our algorithm to select any item outside the current ideal A, not necessarily a minimal one. Our notion of effective weight $\underline{u}_i$ in our relaxation (P_{PCKP1}) "collects" instead the weights of the predecessors of i outside A. Selecting the predecessors of i to the amount of x_i each when i is selected x_i times is necessary for a feasible solution including i. This is not in general the case for the successors of i. On the other hand our definition of $\underline{u}_i$ leads to an extra $w(\mathcal{P})$ factor in the approximation ratio, compared with the 0–1 case, but allows the handling of non-unit upper bounds.

Using a different and simpler relaxation (P_{PCKP2}) we give an $O(n^*)$-approximation for the case of general d, where n^* denotes the minimum number of nonzero variables in an optimal solution. In both relaxations we propose the multiplicity constraints come into play only through the coefficients of valid inequalities of the Knapsack Cover (KC) variety; it is up to the primal-dual algorithm to ensure they are respected. Moreover the feasible region of the second one is even oblivious to the precedence constraints. The associated polytope is defined solely by the standard KC inequalities. The partial order matters only in the objective function where we use a modified cost for each item. It is interesting that these approaches are sufficient to establish the stated upper bounds on the approximation ratio. It is easy to see that both our algorithms achieve a 2-approximation for Knapsack.

Previous Work for CIPs. For CIPs without precedence constraints, there are two lines of approach that parameterize the approximation ratio as a function either of the numbers of rows or the number of columns in the constraint matrix A. An $O(\log \alpha)$-approximation algorithm was given by Kolliopoulos and Young [8] where α is the maximum number of nonzero entries in a column of A. Because Set Cover is a special case, this is best possible unless $\mathsf{P} = \mathsf{NP}$ [12]. When the parameter of interest is the row sparsity f, i.e., the maximum number of nonzero entries in a row of A, no $(f-1-\varepsilon)$-approximation is possible for any fixed $\varepsilon > 0$, unless $\mathsf{P} = \mathsf{NP}$ [4] and an $(f-\varepsilon)$-approximation would violate the Unique Games Conjecture [7]. A near-optimal f-appproximation was achieved by Pritchard and Chakrabarty [11] and using different methods by Koufogiannakis and Young [9].

Our Contribution for PCCP. We show in Sect. 5 that there is an f-approximation for PCCP. Using the hardness results for CIPs stated above no $(f - 1 - \varepsilon)$-approximation is possible for any fixed $\varepsilon > 0$, unless $\mathsf{P} = \mathsf{NP}$ [4] and an $(f - \varepsilon)$-approximation would violate the Unique Games Conjecture [7]. It is fairly surprising that the near-optimal f-approximation for CIPs without precedence constraints holds when an arbitrary partial order is also part of the input. No such relation is known for approximations that are parameterized by the number of rows. Our algorithm is a rather simple extension of the LP-rounding algorithm of Pritchard and Chakrabarty [11] for CIPs. Their algorithm has a monotonicity property according to which the relative order of fractional variable values with the same upper bound is maintained in the integer values. We show that this monotonicity can be exploited to ensure that the integer solution respects also the precedence constraints.

2 Preliminaries

Consider a partially ordered set $\mathcal{P} = (U, \preceq)$, where U is a set of elements and $\preceq$ is an order relation on U. Two elements $i, j \in U$ are *comparable* if $i \preceq j$ or $j \preceq i$ holds and *incomparable* otherwise. A subset $A \subseteq U$ of elements is called a *chain* if for every pair of elements $i, j \in A$ it holds that i, j are comparable. Likewise, a subset $A \subseteq U$ of elements is called an *antichain* if for every pair of elements $i, j \in A$ it holds that i, j are incomparable. The *size* of the maximum antichain will be denoted as $w(\mathcal{P})$. A vector $x \in \mathbb{R}^{|U|}$ *respects the precedence constraints* if $x_i \geq x_j$ when $i \preceq j$.

A subset $A \subseteq U$ is called an *ideal* if for every $i \in A$ it is true that all the elements j with the property $j \preceq i$ are also part of A. In other words, for an ideal A, $i \in A$ and $j \preceq i$ implies $j \in A$. A set $A \subseteq U$ is *closed* under $\mathcal{P}$ if A is an ideal. For a partial order $\mathcal{P}$ the set of all the ideals of $\mathcal{P}$ will be denoted as $\mathcal{L}(\mathcal{P})$. For an ideal $A \in \mathcal{L}(\mathcal{P})$ we define $\mathcal{P}(A) = (U \setminus A, \preceq)$ to be the partial order $\mathcal{P}$ restricted to the elements that do not belong to A and $\min \mathcal{P}(A) = \{i \in U \setminus A \mid \nexists j \in U \setminus A \text{ such that } j \prec i\}$ to be the set of minimal items of the partial order $\mathcal{P}(A)$.

The Knapsack Cover inequalities introduced in [2] motivate the following definitions. For any $A \subseteq U$, $D(A) := \max\left\{D - \sum_{i \in A} u_i d_i, 0\right\}$, and for $i \notin A$, $u_i(A) := \min\{u_i, D(A)\}$. The *Knapsack Cover (KC) inequality* corresponding to a set A is defined as

$$\sum_{i \in U \setminus A} u_i(A) x_i \geq D(A).$$

For any $A \subseteq U$, the KC inequality for A is satisfied by any feasible integer solution for PCKP. It is therefore a *valid inequality*. Given an integer program IP with solution set $S \subseteq \mathbb{Z}^n$, a *valid relaxation* of IP is a linear (or integer) program whose feasible region contains S.

3 Integrality Gap

The following definitions are given in [10]. For an ideal $A \in \mathcal{L}(\mathcal{P})$ and an element $j \in U \setminus A$, $X_j(A) := \{i \in U \setminus A \mid i \preceq j \wedge i \in \min \mathcal{P}(A)\}$. For $i \in \min \mathcal{P}(A)$ its *effective weight* is defined as $\bar{u}_i(A) := \min\{D(A), u_i + \sum_{j:\, i \prec j} u_j(A) / |X_j(A)|\}$. The following integer program was introduced in [10] and was shown to be valid for PCKP.

$$\text{minimize} \sum_{i \in U} c_i x_i$$

$$\text{subject to} \sum_{i \in \min \mathcal{P}(A)} \bar{u}_i(A) x_i \geq D(A),\ \forall A \in \mathcal{L}(\mathcal{P})$$

$$x \in \{0,1\}^{|U|} \qquad (P_{PCKP})$$

The primal-dual algorithm of [10] is based on the linear relaxation of (P_{PCKP}) and its dual which follows.

$$\text{maximize} \sum_{A \in \mathcal{L}(\mathcal{P})} y(A) D(A)$$

$$\text{subject to} \sum_{A \in \mathcal{L}(\mathcal{P}):\, i \in \min \mathcal{P}(A)} \bar{u}_i(A) y(A) \leq c_i,\ \forall i \in U$$

$$y(A) \geq 0,\ \forall A \in \mathcal{L}(\mathcal{P}) \qquad (D_{PCKP})$$

Theorem 1. *There is an infinite family of instances of (0,1)-PCKP for which the integrality gap of the linear relaxation of (P_{PCKP}) is at least $w(\mathcal{P})/4$. For any positive integer k the family contains an instance with $w(P) > k$.*

Proof. Consider a partial order with $2n$ elements consisting of the n chains $\{e_i^1, e_i^2\}$, $i \in [n]$, with $\{e_i^1\}_{i \in [n]}$ being the minimal elements. The weights and the costs of the bottom layer are defined as $u_i^1 = c_i^1 = 1$, $\forall i \in [n]$. The weights and the costs of the top layer are defined as $u_i^2 = c_i^2 = \frac{n}{2}$, $\forall i \in [n]$. Let us consider a PCKP instance with $D = \frac{n}{2}$. The optimal integral solution of (P_{PCKP}) for this instance cannot have cost less than $\frac{n}{2}$. Suppose to the contrary that there exists a feasible integer solution x with cost less than $\frac{n}{2}$. Let S denote the support of x. S cannot contain any element from the top layer and also must contain less than $\frac{n}{2}$ elements from the bottom layer. By construction S is an ideal thus the corresponding constraint for $A = S$ must be satisfied. Each element of A has weight equal to one and $|A| < \frac{n}{2}$ therefore $D(A) > 0$. However the left side of the constraint sums to zero as we have picked elements only from the set S. Therefore x cannot be feasible.

Let us now build a solution for the linear relaxation of (P_{PCKP}). Set $x_i^1 = \frac{2}{n}$, $x_i^2 = 0$, $\forall i \in [n]$. We show that x is feasible. Constraints corresponding to sets that intersect the top layer do not matter. We are interested only in ideals, and if A contains an element e_i^2 from the top layer, it must also contain the corresponding element e_i^1. But then $D - \sum_{i \in A} u_i d_i < 0$ and the corresponding constraint will be trivially satisfied, because $D(A) = 0$. For the same reason, A cannot contain $\frac{n}{2}$ or more elements from the bottom layer. Therefore the only constraints that we should check are the ones whose corresponding subset A contains less than $\frac{n}{2}$ elements from the bottom layer.

Consider a subset $A \subseteq U$, with $|A| < \frac{n}{2}$, that contains elements only from the bottom layer. The corresponding constraint will be the following:

$$\sum_{1 \leq i \leq n:\, i \notin A} \overline{u_i^1}(A) \cdot x_i^1 + \sum_{1 \leq i \leq n:\, i \in A} \overline{u_i^2}(A) \cdot x_i^2 \geq D(A)$$

However from the definition of $\overline{u}(A)$, it holds that $\overline{u_i^1}(A) = \overline{u_i^2}(A) = D(A)$, $\forall i \in [n]$, and the corresponding constraint is equivalent to

$$\sum_{1 \leq i \leq n:\, i \notin A} x_i^1 + \sum_{1 \leq i \leq n:\, i \in A} x_i^2 \geq 1 \Leftrightarrow \sum_{1 \leq i \leq n:\, i \notin A} \frac{2}{n} + \sum_{1 \leq i \leq n:\, i \in A} 0 \geq 1 \Leftrightarrow$$

$$(n - |A|) \cdot \frac{2}{n} \geq 1 \Leftrightarrow |A| \leq \frac{n}{2}$$

Since $|A| \leq \frac{n}{2}$, the corresponding constraint will be satisfied. Therefore the solution x is a feasible one. The cost of x equals $\sum_{i=1}^{n} x_i^1 + \sum_{i=1}^{n} x_i^2 \cdot \frac{n}{2} = n \cdot \frac{2}{n} = 2$. The gap between the optimal solution and the optimal LP solution is at least $\frac{\frac{n}{2}}{2} = \frac{n}{4}$. As the width $w(\mathcal{P})$ of the partial order is equal to n the theorem follows. $\square$

For the proof above to work it is not essential that $w(\mathcal{P}) = \Theta(|U|)$. The instance may be augmented with a chain of arbitrary size that has e_1^2 as a minimal element. Thus $|U|$ can become arbitrarily larger than $w(\mathcal{P})$ while the above arguments continue to hold.

4 PCKP with Multiplicity Constraints

In this section we present two polynomial-time algorithms with strongly polynomial approximation bounds for PCKP. The analysis of the first algorithm assumes that all variables x_i, $i \in U$, have the same multiplicity bound $d_i = \Delta \in \mathbb{Z}_{>0}$. The presentation assumes throughout a d_i bound for each variable. We point out explicitly the places in the proof where we need the common-value assumption.

Formally PCKP with general multiplicity constraints can be described as the following integer program.

$$\text{minimize} \sum_{i \in U} c_i x_i$$

$$\text{subject to} \sum_{i \in U} u_i x_i \geq D$$

$$x_i - x_j \geq 0, \ \forall i \preceq j$$

$$x_i \leq d_i, \ \forall i \in U$$

$$x \in \mathbb{Z}_+^{|U|}$$

The primal-dual algorithm of [10] uses the linear relaxation of (P_{PCKP}) and focuses on the minimal elements outside a given ideal A. The weight of non-mininal elements is projected on their minimal predecessors. Our primal-dual algorithm is allowed to pick at each iteration any element and not only minimal elements. In order to keep the solution closed under the partial order, when an element is chosen, the algorithm is forced to pick all the unchosen predecessors. We need the following definitions.

$$\text{UP}(i) = \{j \in U : i \preceq j\}$$

$$\text{DW}(i) = \{j \in U : j \preceq i\}$$

$$\underline{U}(x) = \sum_{i \in U} u_i \cdot \max_{j \in \text{UP}(i)} x_j$$

$$\underline{c}_i = \sum_{j \in \text{DW}(i)} c_j$$

$$\underline{u}_i(A) = \min \left\{ \sum_{j \in \text{DW}(i) \setminus A} u_j, \ D(A) \right\}$$

The new linear relaxation that we propose follows. Its validity is easy to see.

$$\text{minimize} \sum_{i \in U} \underline{c}_i x_i$$

$$\text{subject to} \sum_{i \in U \setminus A} \underline{u}_i(A) x_i \geq D(A), \ \forall A \in \mathcal{L}(\mathcal{P})$$

$$x_i \geq 0, \ \forall i \in U \qquad (P_{PCKP1})$$

The dual of (P_{PCKP1}) is the following LP:

$$\text{maximize} \sum_{A \in \mathcal{L}(\mathcal{P})} y(A) D(A)$$

$$\text{subject to} \sum_{A \in \mathcal{L}(\mathcal{P}): i \in U \setminus A} \underline{u}_i(A) y(A) \leq \underline{c}_i, \ \forall i \in U$$

$$y(A) \geq 0, \ \forall A \in \mathcal{L}(\mathcal{P}) \qquad (D_{PCKP1})$$

There are no explicit multiplicity constraints in the formulation, but the algorithm that we will develop will respect them. The following auxiliary lemma is a straightforward consequence of Dilworth's Theorem.

Lemma 1. *Consider a finite partially ordered set $\mathcal{P} = (U, \preceq)$ where each $i \in U$ has an associated value v_i. Let α be an upper bound on the sum of values of the elements in a chain of $\mathcal{P}$. Then the total sum of values of all the elements of $\mathcal{P}$ is at most $\alpha \cdot w(\mathcal{P})$.*

Proof. From Dilworth's Theorem [3], there is a chain decomposition $\{C_1, \ldots, C_{w(\mathcal{P})}\}$ of size $w(\mathcal{P})$. The total sum of values of all the elements of $\mathcal{P}$ can be bounded in the following way: $\sum_{i \in U} v_i = \sum_{j=1}^{w(\mathcal{P})} \sum_{i \in C_j} v_i \leq \sum_{j=1}^{w(\mathcal{P})} \alpha = \alpha \cdot w(\mathcal{P})$.   □

We will often convert a nonnegative integer vector x which respects the precedence constraints, to another vector $\tau(x)$, where the corresponding components of $\tau(x)$ are defined as follows:

$$\forall i \in U : \tau(x)_i = x_i - \max_{j \in \mathrm{UP}(i) \setminus \{i\}} x_j$$

Observe that $\tau(x)$ is a nonnegative integer vector as well, due to the fact that x respects the precedence constraints. The intuition behind $\tau(x)$ is the following. Assume we choose two elements i, j, such that $i \preceq j$, x_i and x_j times respectively, where $x_i \geq x_j$. The weight and the cost of i have been included into the corresponding value and cost of j in the new definitions of $\underline{u}$ and $\underline{c}$. Hence in the analysis it will be sufficient to count no more than $x_i - x_j$ extra occurrences of the element i.

Given a vector $y \in \mathbb{R}^{|U|}$ denote by $B_i(y) = \sum_{j \in \mathrm{UP}(i)} y_j$ the total sum of the y-values of the elements of $\mathrm{UP}(i)$ and by $C_i(y)$ the maximum sum of the y-values of elements over all chains in $\mathrm{UP}(i)$. Lemmas 2–4 present some useful relationships between x and $\tau(x)$, with respect to the underlying partial order.

Lemma 2. *For all elements $i \in U$, it holds that $B_i(\tau(x)) \geq x_i$.*

Proof. We will use strong induction on a topological ordering of the elements of the partial order $\mathcal{P}$. The ordering starts from the maximal elements.

- Base Case: Element i is a maximal element of $\mathcal{P}$. Then $\mathrm{UP}(i) = \{i\}$ and $\tau(x)_i = x_i$. Hence $B_i(\tau(x)) = x_i$.
- Inductive Hypothesis: Assume that for $p \geq 0$, the sum $B_i(\tau(x))$ corresponding to element i in the p-th position of the topological ordering, is at least x_i.
- Inductive Step: We prove that for $p \geq 0$, the corresponding sum $B_i(\tau(x))$ of element i in the $(p+1)$-th position of the topological ordering, is at least x_i. Let $k = \arg\max_{j \in \mathrm{UP}(i) \setminus \{i\}} x_j$. From the inductive hypothesis, the corresponding value $B_k(\tau(x))$ is at least x_k. For the sum $B_i(\tau(x))$, it holds that $B_i(\tau(x)) \geq \tau(x)_i + B_k(\tau(x)) \geq \tau(x)_i + x_k$. Therefore as $\tau(x)_i = x_i - x_k$, it holds that $B_i(\tau(x)) \geq x_i$.   □

Lemma 3. *For all elements $i \in U$, it holds that $C_i(\tau(x)) \le x_i$.*

Proof. We will use strong induction in a topological ordering of the elements of the partial order $\mathcal{P}$. The ordering starts from the maximal elements.

- Base Case: Element i is a maximal element of $\mathcal{P}$. Then $\mathrm{UP}(i) = \{i\}$ and $\tau(x)_i = x_i$. Hence $C_i(\tau(x)) = x_i$.
- Inductive Hypothesis: Assume that for $p \ge 0$, the sum $C_i(\tau(x))$ corresponding to element i in the p-th position of the topological ordering, is at most x_i.
- Inductive Step: We prove that for any $p \ge 0$, the corresponding sum $C_i(\tau(x))$ of element i in the $(p+1)$-th position of the topological ordering, is at most x_i. Let $k = \arg\max_{j \in \mathrm{UP}(i) \setminus \{i\}} x_j$ and $k' = \arg\max_{j \in \mathrm{UP}(i) \setminus \{i\}} C_j(\tau(x))$. From the inductive hypothesis, the corresponding value $C_{k'}(\tau(x))$ is at most $x_{k'}$. For the sum $C_i(\tau(x))$, it holds that $C_i(\tau(x)) = \tau(x)_i + C_{k'}(\tau(x)) \le \tau(x)_i + x_{k'}$. Therefore as $\tau(x)_i = x_i - x_k$ and $x_{k'} \le x_k$, it holds that $C_i(\tau(x)) \le x_i$. □

Lemma 4. *For all elements $i \in U$, it holds that $B_i(\tau(x)) \le w(\mathcal{P}) \cdot x_i$.*

Proof. According to Lemma 1, $B_i(\tau(x)) \le w(\mathcal{P}) \cdot C_i(\tau(x))$. From Lemma 3, $B_i(\tau(x)) \le w(\mathcal{P}) \cdot x_i$. □

If x is a solution of interest for PCKP, in the analysis of the primal-dual algorithm that relies on the (P_{PCKP1}) formulation, the truncated vector $\tau(x)$ can be used instead. The next lemma relates the cost of x with the cost of $\tau(x)$ under $\underline{c}$. The lemma shows that the truncated vector $\tau(x)$ is feasible for the new formulation, which is crucial in order to compare the real optimum with the optimum of (P_{PCKP1}) in Lemma 6.

Lemma 5. *Let $\hat{x}$ be a nonnegative integer vector which respects the precedence constraints. Then inequalities (2) hold. Assuming that A is an ideal in $\mathcal{L}(\mathcal{P})$ and $\hat{x}$ is also a feasible solution of PCKP, inequality (3) holds as well.*

$$\sum_{i \in U} c_i \cdot \hat{x}_i \le \sum_{i \in U} \underline{c}_i \cdot \tau(\hat{x})_i \le w(\mathcal{P}) \cdot \sum_{i \in U} c_i \cdot \hat{x}_i \tag{2}$$

$$\sum_{i \in U \setminus A} \underline{u}_i(A) \cdot \tau(\hat{x})_i \ge D(A) \tag{3}$$

Proof. We have that $\sum_{i \in U} \underline{c}_i \tau(\hat{x})_i$ equals

$$\sum_{i \in U} \sum_{j \in \mathrm{DW}(i)} c_j \tau(\hat{x})_i = \sum_{j \in U} c_j \sum_{i \in \mathrm{UP}(j)} \tau(\hat{x})_i = \sum_{j \in U} c_j B_j(\tau(\hat{x})) \ge \sum_{j \in U} c_j \hat{x}_j.$$

The inequality above follows from Lemma 2. Using Lemma 4 we obtain that

$$\sum_{i \in U} \underline{c}_i \tau(\hat{x})_i = \sum_{j \in U} c_j B_j(\tau(\hat{x})) \le w(\mathcal{P}) \sum_{j \in U} c_j \cdot \hat{x}_j.$$

If there exists an $i \in U \setminus A$, such that $\underline{u}_i(A) \cdot \tau(\hat{x})_i \ge D(A)$, inequality (3) trivially holds. Otherwise for all $i \in U \setminus A : \underline{u}_i(A) \cdot \tau(\hat{x})_i < D(A)$, and as $\tau(\hat{x})$

is an integer nonnegative vector, either $\underline{u}_i(A) < D(A)$ or $\tau(\hat{x})_i = 0$. Hence it is true that

$$\underline{u}_i(A) \cdot \tau(\hat{x})_i = \sum_{j \in \mathrm{DW}(i) \setminus A} u_j \cdot \tau(\hat{x})_i.$$

The sum can be evaluated as follows:

$$\sum_{i \in U \setminus A} \underline{u}_i(A)\tau(\hat{x})_i = \sum_{i \in U \setminus A} \sum_{j \in \mathrm{DW}(i) \setminus A} u_j\tau(\hat{x})_i = \sum_{j \in U \setminus A} u_j \cdot \left(\sum_{i \in \mathrm{UP}(j) \setminus A} \tau(\hat{x})_i \right)$$

$$\overset{A \in \mathcal{L}(\mathcal{P})}{=} \sum_{j \in U \setminus A} u_j \cdot \left(\sum_{i \in \mathrm{UP}(j)} \tau(\hat{x})_i \right) \overset{\text{Lemma 2}}{\geq} \sum_{j \in U \setminus A} u_j\hat{x}_j \overset{\text{feasibility of } \hat{x}}{\geq}$$

$$\max\left\{ D - \sum_{i \in A} u_i d_i, \ 0 \right\} = D(A).$$

The proof of the lemma is complete. $\square$

The objective function of the new formulation overestimates the cost of i as the cost of every element in $\mathrm{DW}(i)$ is counted in $\underline{c}_i$. Nevertheless based on Lemma 5, for every integer feasible solution $\hat{x}$ of PCKP, there exists another integer feasible solution $\tau(\hat{x})$ of (P_{PCKP1}) with cost that is larger only by a factor of $w(\mathcal{P})$. Let OPT denote the value of the PCKP optimum and $\underline{OPT}$ the value of the optimum of (P_{PCKP1}) .

Lemma 6. *For any feasible instance $\mathcal{I}$ of PCKP it holds that $\underline{OPT}(\mathcal{I}) \leq w(\mathcal{P})\cdot OPT(\mathcal{I})$.*

We now proceed to the first main theorem of this section.

Theorem 2. *There exists an $O(w^2(\mathcal{P}))$-approximation primal-dual polynomial-time algorithm for PCKP when all the multiplicity upper bounds d_i, $i \in U$, are equal to the same positive integer Δ.*

Proof. Consider the primal-dual algorithm in Fig. 1 which is running over the (D_{PCKP1}) linear program and outputs the primal solution $\hat{x}$.

If there exists a feasible integer solution, the algorithm will find one because of the assumption that all the d_is are equal. In every iteration of the while-loop the computation of $\underline{U}(x)$ credits the variables for the items in $\underline{S}$ with a value of Δ. If the while-loop exits because $\underline{S} = U$, then $\sum_{i \in U} u_i\Delta < D$ and the instance is infeasible. The output vector $\hat{x}$ is closed under the partial order, $\hat{x}_i \leq d_i$ for all $i \in U$, and at the end of the algorithm

$$\sum_{i \in U} u_i\hat{x}_i = \sum_{i \in U} u_i \cdot \max_{j \in \mathrm{UP}(i)} x_j = \underline{U}(x) \geq D$$

Therefore $\hat{x}$ covers the demand as well. Thus $\hat{x}$ is feasible for PCKP.

$S = \emptyset,\ \underline{S} = \emptyset,\ y = 0$

```
while (U(x) < D && |S| < |U|) {
    Increase y(S) until a dual constraint of (D_PCKP1)
    becomes tight for element i
    S = S ∪ {i}
    S = S ∪ DW(i)
    x_i = d_i
}


// pruning step
Let l be the last element inserted in S.
Set x_l to the minimum integer value such that U(x) ≥ D.
```

$\forall i \in U : \hat{x}_i = \max_{j \in \mathrm{UP}(i)} x_j.$

Fig. 1. Algorithm for Theorem 2.

Based on Lemma 5, the cost of $\hat{x}$ can be bounded by the corresponding cost of $\tau(\hat{x})$ in terms of (P_{PCKP1}) and also $\tau(\hat{x})$ is an integer feasible solution of (P_{PCKP1}). However the feasibility of $\tau(\hat{x})$ is not necessary for the analysis of the cost. Observe that if $\tau(\hat{x})_i > 0$, then the corresponding dual constraint of element i is tight. To prove it, suppose to the contrary that the corresponding constraint is not tight. Then $\hat{x}_i = \max_{j \in \mathrm{UP}(i) \setminus \{i\}} x_j$, and $\tau(\hat{x})_i$ would be equal to zero. The cost of $\hat{x}$ can be evaluated as follows:

$$\sum_{i \in U} c_i \hat{x}_i \leq \sum_{i \in S} \underline{c}_i \tau(\hat{x})_i = \sum_{i \in S} \tau(\hat{x})_i \cdot \Big(\sum_{A \in \mathcal{L}(\mathcal{P}):\ i \in U \setminus A} \underline{u}_i(A) \cdot y(A) \Big)$$

$$= \sum_{A \in \mathcal{L}(\mathcal{P})} y(A) \cdot \Big(\sum_{i \in S \setminus A} \tau(\hat{x})_i \underline{u}_i(A) \Big)$$

$$= \sum_{A \in \mathcal{L}(\mathcal{P})} y(A) \cdot \Big(\sum_{i \in S \setminus A \wedge i \neq l} \tau(\hat{x})_i \underline{u}_i(A) + \tau(\hat{x})_l \underline{u}_l(A) \Big)$$

$$\leq \sum_{A \in \mathcal{L}(\mathcal{P})} y(A) \cdot \Big(\Big(\sum_{i \in S \setminus A \wedge i \neq l} \tau(\hat{x})_i \cdot \Big(\sum_{j \in \mathrm{DW}(i) \setminus A} u_j \Big) \Big) + \tau(\hat{x})_l \underline{u}_l(A) \Big) \quad (4)$$

$$< \sum_{A \in \mathcal{L}(\mathcal{P})} y(A)\big(w(\mathcal{P})D(A) + 2D(A)\big) = (w(\mathcal{P}) + 2) \sum_{A \in \mathcal{L}(\mathcal{P})} y(A)D(A)$$

$$\leq (w(\mathcal{P}) + 2) \cdot \underline{OPT} \leq O(w^2(\mathcal{P})) \cdot OPT.$$

The last inequality follows from Lemma 6. In order to explain (4), let us fix an ideal $A \in \mathcal{L}(\mathcal{P})$ such that $y(A) > 0$. Notice that if we form a sequence out of the elements of $\underline{S}$ in the order in which they were added by the algorithm, such an A corresponds to a consecutive prefix of the sequence. Let z be equal to the final vector x except that its l-th component is set to zero. Construct a vector $\hat{z}$ in the following way: $\forall i \in U : \hat{z}_i = \max_{j \in \mathrm{UP}(i)} z_j$. Note that z, $\hat{z}$, $\tau(\hat{z})$

intuitively correspond to the vectors $x, \hat{x}, \tau(\hat{x})$ that would result if the algorithm terminated without executing the last iteration of the while-loop. We have that $\underline{U}(z) = \sum_{i \in U} u_i \cdot \hat{z}_i < D$.

Claim 1. *We have that* $(\tau(\hat{x})_l - 1) \cdot \underline{u}_l(A) < D(A)$.

Proof of Claim. Since l is a maximal element in $\underline{S}$, we have that $\tau(\hat{x})_l = \hat{x}_l = x_l$. We emphasize that x_l above is the final value of the variable, after the pruning step. By definition, $\underline{u}_l(A) = \min\left\{ \sum_{j \in \mathrm{DW}(l) \backslash A} u_j, \ D(A) \right\}$. Clearly $0 < \hat{x}_l \leq d_l$.

Case 1: $\underline{u}_l(A) = \sum_{j \in \mathrm{DW}(l) \backslash A} u_j$. Let $\underline{S}_0$ be the set $\underline{S}$ before the insertion of l into S. Define $F := \mathrm{DW}(l) \setminus (\underline{S}_0 \cup \{l\})$. Before l is added to S the items of F contribute zero to $\underline{U}(x)$. In the final solution, for all $i \in F$, $\hat{x}_i = \hat{x}_l$ and for all $i \in \mathrm{DW}(l) \setminus F$, $\hat{x}_i \geq \hat{x}_l$. Decreasing $\hat{x}_l$ by one leads to infeasibility. Therefore

$$Q := \sum_{i \in \underline{S}_0 \backslash A} u_i \hat{x}_i + \sum_{i \in F} u_i (\hat{x}_i - 1) + u_l(\tau(\hat{x})_l - 1) < D(A).$$

Since $(\tau(\hat{x})_l - 1) \sum_{j \in \mathrm{DW}(l) \backslash A} u_j \leq Q$, the claim follows.

Case 2: $\underline{u}_l(A) = D(A)$. In this case $\sum_{j \in \mathrm{DW}(l) \backslash A} u_j \geq D(A)$. Therefore $\hat{x}_l = 1$ and $\tau(\hat{x})_l - 1 = 0$. The proof of the claim is complete.

For $i \notin \mathrm{DW}(l)$, $\tau(\hat{x})_i = \tau(\hat{z})_i$. For $i \in \mathrm{DW}(l)$, $\hat{x}_i = \max\{\hat{z}_i, x_l\}$. We examine the two cases. If $\arg\max\{\hat{z}_i, x_l\} = \hat{z}_i$, then $\tau(\hat{x})_i \leq \tau(\hat{z})_i$ (observe that the inequality becomes strict iff $\max_{j \in UP(i) \backslash i} \hat{z}_j < x_l$). If $\arg\max\{\hat{z}_i, x_l\} = x_l$, then $\tau(\hat{x})_i = 0$, if $i \neq l$, and $\tau(\hat{x})_i = x_l$, if $i = l$. Therefore for any $j \in U$

$$\sum_{i \in \mathrm{UP}(j) \wedge i \neq l} \tau(\hat{x})_i \leq \sum_{i \in \mathrm{UP}(j)} \tau(\hat{z})_i \tag{5}$$

We obtain that

$$\left(\sum_{i \in S \backslash A \wedge i \neq l} \tau(\hat{x})_i \cdot \left(\sum_{j \in \mathrm{DW}(i) \backslash A} u_j \right) \right) + \tau(\hat{x})_l \cdot \underline{u}_l(A)$$

$$= \left(\sum_{j \in U \backslash A} u_j \cdot \left(\sum_{i \in \mathrm{UP}(j) \wedge i \neq l} \tau(\hat{x})_i \right) \right) + \tau(\hat{x})_l \cdot \underline{u}_l(A) \tag{6}$$

$$\leq \left(\sum_{j \in U \backslash A} u_j \cdot \left(\sum_{i \in \mathrm{UP}(j)} \tau(\hat{z})_i \right) \right) + \tau(\hat{x})_l \cdot \underline{u}_l(A) \tag{7}$$

$$\leq w(\mathcal{P}) \cdot \left(\sum_{j \in U \backslash A} u_j \cdot \hat{z}_j \right) + \tau(\hat{x})_l \cdot \underline{u}_l(A) \tag{8}$$

$$< w(\mathcal{P})D(A) + 2D(A)$$

For the equality above consider the items indexed by i in the rhs. If the indexing was over

$$(i \in \mathrm{UP}(j) \setminus A) \wedge i \neq l$$

the rhs would be trivially a rearrangement of the terms of the lhs (recall that $\tau(\hat{x})_i > 0$ implies that $i \in S$). But as A is an ideal, for $j \notin A$, all the successors of j will be outside of A. Hence indeed the rhs is a rearrangement of the lhs and the equality follows. Inequality 6 follows from (5). Inequality 7 follows from Lemma 4. Inequality 8 holds because (i) $\underline{u}_l(A) \le D(A)$ and Claim 1 imply that $\tau(\hat{x})_l \cdot \underline{u}_l(A) \le 2D(A)$ and (ii) A is an ideal and all the multiplicity bounds are equal and as a result

$$\sum_{i \in U} u_i \cdot \hat{z}_i = \sum_{i \in U \setminus A} u_i \cdot \hat{z}_i + \sum_{i \in A} u_i \cdot d_i < D.$$

The proof of Theorem 2 is complete. □

Consider a different valid formulation which is similar to (P_{PCKP1}) but uses the actual weights of the elements.

$$\text{minimize} \sum_{i \in U} \underline{c}_i x_i$$

$$\text{subject to} \sum_{i \in U \setminus A} u_i(A) x_i \ge D(A), \ \forall A \subseteq U$$

$$x_i \ge 0, \ \forall i \in U \qquad (P_{PCKP2})$$

and its dual is the following one:

$$\text{maximize} \sum_{A \subseteq U} y(A) D(A)$$

$$\text{subject to} \sum_{A \subseteq U: \, i \in U \setminus A} u_i(A) y(A) \le \underline{c}_i, \ \forall i \in U$$

$$y(A) \ge 0, \ \forall A \subseteq U \qquad (D_{PCKP2})$$

Let x^* be an optimal solution of PCKP with minimum support of size n^*. The vector x^* is also feasible for (P_{PCKP1}) with cost at most n^* times larger than before. The cost of each element i can be included in at most n^* elements above it, and also every element above i can be selected at most x_i^* times. Let $\underline{OPT'}$ denote the optimum of (P_{PCKP2}).

Lemma 7. *For any feasible instance $\mathcal{I}$ of PCKP it holds that $\underline{OPT'}(\mathcal{I}) \le n^* \cdot OPT(\mathcal{I})$, where n^* is the minimum number of items in an optimal solution.*

Theorem 3. *There exists a $2n^*$-approximation primal-dual polynomial-time algorithm for PCKP with general multiplicity constraints, where n^* is the minimum number of items in an optimal solution.*

Proof. Consider the primal-dual algorithm in Fig. 2 which is running over the (D_{PCKP2}) linear program. In contrast to the algorithm of Theorem 2 the dual variable that is raised in each iteration does not necessarily correspond to an ideal.

```
S = ∅, y = 0

while (U(x) < D && |S| < |U|) {
    Increase y(S) until a dual constraint of (D_{PCKP2})
    becomes tight for element i
    S = S ∪ {i}
    x_i = d_i
}

// pruning step
Let l be the last element inserted in S.
Set x_l equal to the minimum integer value such that U(x) ≥ D
```

$\forall i \in U : \hat{x}_i = \max_{j \in \mathrm{UP}(i)} x_j$

Fig. 2. Algorithm for Theorem 3.

If there exists a feasible solution, the algorithm will find one. Assume to the contrary that $\underline{U}(x) < D$ and the loop exits without finding a solution. Then $S = U$. For every $i \in S$, $x_i = d_i$ and $\underline{U}(x) = \sum_{i \in U} u_i \max_{j \in \mathrm{UP}(i)} x_j = \sum_{i \in U} u_i d_i$. Therefore the instance is infeasible.

The produced integer solution $\hat{x}$ is closed under the partial order, $\hat{x}_i \leq d_i$ for all $i \in U$ and as

$$\sum_{i \in U} u_i \hat{x}_i = \sum_{i \in U} u_i \cdot \max_{j \in \mathrm{UP}(i)} x_j = \underline{U}(x) \geq D$$

the solution $\hat{x}$ covers the demand as well. Therefore $\hat{x}$ is feasible for PCKP. Based on Lemma 5, the cost of $\hat{x}$ can be bounded by the corresponding cost of $\tau(\hat{x})$ in terms of (P_{PCKP2}). Observe that if $\tau(\hat{x})_i > 0$, then the corresponding dual constraint of element i is tight. The cost of $\hat{x}$ can be evaluated as follows:

$$\sum_{i \in U} c_i \hat{x}_i \leq \sum_{i \in S} \underline{c}_i \tau(\hat{x})_i \sum_{i \in S} \tau(\hat{x})_i \cdot \sum_{A \subseteq U: i \in U \setminus A} u_i(A) \cdot y(A)$$

$$= \sum_{A \subseteq U} y(A) \sum_{i \in S \setminus A} \tau(\hat{x})_i \cdot u_i(A)$$

$$= \sum_{A \subseteq U} y(A) \cdot \left(\sum_{i \in S \setminus A \wedge i \neq l} u_i(A) \tau(\hat{x})_i + u_l(A) \tau(\hat{x})_l \right) \tag{9}$$

$$\leq 2 \sum_{A \subseteq U} y(A) D(A) \leq 2\underline{OPT}' \leq 2n^* \cdot OPT$$

The last inequality follows from Lemma 7. For (9), let us fix a subset $A \subseteq U$ such that $y(A) > 0$. We have that

$$\sum_{i \in S \setminus A \wedge i \neq l} u_i(A) \tau(\hat{x})_i + u_l(A) \tau(\hat{x})_l \leq \sum_{i \in S \setminus A \wedge i \neq l} u_i \hat{x}_i + u_l(A) \hat{x}_l \leq 2D(A).$$

The last inequality holds because $\sum_{i \in S \setminus A \,\wedge\, i \neq l} u_i \hat{x}_i + u_l(A)(\hat{x}_l - 1) < D(A)$ and $u_l(A) \leq D(A)$. $\hfill\square$

5   PCCP with General Multiplicity Constraints

Consider the following precedence-constrained CIP where N and M are finite sets.

$$\text{minimize} \sum_{i \in N} c_i x_i$$

$$\text{subject to} \sum_{i \in N} u_i^t x_i \geq D^t, \quad \forall t \in M \tag{10}$$

$$x_{i_1} - x_{i_2} \geq 0, \ \forall i_1 \preceq i_2 \tag{11}$$

$$0 \leq x_i \leq d_i, \ \forall i \in N \tag{12}$$

$$x \in \mathbb{Z}^n \qquad (P_{PCCP}) \tag{13}$$

Because of the precedence constraints, we can safely assume that for each $i_1 \preceq i_2$: $d_{i_1} \geq d_{i_2}$.

Define the floor of a vector to be taken componentwise, so that $\lfloor y \rfloor_i = \lfloor y_i \rfloor$. The following two definitions are from [11]. A linear constraint of the form $a^T x \geq b$ is called ρ-*roundable* for some scalar $\rho > 1$, if for all nonnegative real x, $a^T x \geq b$ implies that $a^T \lfloor \rho x \rfloor \geq b$. Two constraints $a^T x \geq b$ and $(a')^T x \geq b'$ are $\mathbb{Z}_+$-*equivalent* if they admit the same set of integer solutions, i.e., for all nonnegative integral x

$$a^T x \geq b \Leftrightarrow (a')^T x \geq b'.$$

Pritchard and Chakrabarty [11] provided an efficient algorithm that transforms any covering constraint to an equivalent f-roundable one, where f is the maximum number of nonzero elements in a row of the $|M| \times |N|$ matrix $[u_i^t]$.

Proposition 1 ([11, Proposition 9]).   *In polynomial-time one can transform all covering constraints* (10) *of* (P_{PCCP}) *to* $\mathbb{Z}_+$-*equivalent ones that are* f-*roundable.*

By adding KC inequalities to the CIP [2,8] we obtain the following valid relaxation.

$$\text{minimize} \sum_{i \in N} c_i x_i$$

$$\text{subject to} \sum_{i \in N \setminus A} u_i^t(A) x_i \geq D^t(A), \ \forall A \subseteq N, \ \forall t \in M \tag{14}$$

$$x_{i_1} - x_{i_2} \geq 0, \ \forall i_1 \preceq i_2 \tag{15}$$

$$0 \leq x_i \leq d_i, \ \forall i \in N \qquad (P_{PCCPKC}) \tag{16}$$

We now describe a relaxed separation oracle (see [2,8] for details) that we will use to find in polynomial-time a feasible fractional solution to (P_{PCCPKC}) with some desirable properties. For a candidate solution x, there are two cases. Either x respects the precedence constraints or it does not.

- x *does not respect the precedence constraints:* In polynomial time, a pair of elements i_1, i_2 such that $i_1 \preceq i_2$ and $x_{i_1} - x_{i_2} < 0$ can be found. This is the violated inequality returned by the separation oracle.
- x *respects the precedence constraints:* Let $A = \{i \in N \mid x_i \geq \frac{d_i}{f}\}$. For every $t \in M$, the corresponding inequality

$$\sum_{i \in N \setminus A} u_i^t(A) \cdot x_i \geq D^t(A)$$

is checked. If for at least one t the inequality is violated, then this is the inequality returned by the separation oracle. Otherwise, the ellipsoid method terminates.

Let us denote by $\bar{x}$ the final solution returned by the ellipsoid method. Define an integer solution $\hat{x}$ in the following way:

$$\forall i \in N : \hat{x}_i = \min\{d_i, \lfloor f \cdot \bar{x}_i \rfloor\}.$$

To prove our main theorem Theorem 5, the feasibility of $\hat{x}$ is necessary. The next theorem from [11] ensures that if the covering constraints are f-roundable and the corresponding KC inequalities have been included, then $\hat{x}$ satisfies all the demands.

Theorem 4 [11, Theorem 1]). *If all covering constraints* (10) *of* (P_{PCCPKC}) *are* f*-roundable, the integer vector* $\hat{x}$ *defined above satisfies all demands, that is* $\forall t \in M : \sum_{i \in N} u_i^t \cdot \hat{x}_i \geq D^t$ *and moreover* $\sum_{i \in N} c_i \hat{x}_i \leq f \sum_{i \in N} c_i \bar{x}_i.$

Theorem 5. *There exists an* f*-approximation rounding polynomial-time algorithm for PCCP with general multiplicity constraints.*

Proof. It is clear from the construction that $\hat{x}$ satisfies the multiplicity constraints (12). Also as $\bar{x}$ respects the precedence constraints (15) and $f \geq 0$, it holds that:

$$\forall i_1 \preceq i_2 : \bar{x}_{i_1} - \bar{x}_{i_2} \geq 0 \Leftrightarrow \forall i_1 \preceq i_2 : \bar{x}_{i_1} \geq \bar{x}_{i_2} \Leftrightarrow$$
$$\forall i_1 \preceq i_2 : f \cdot \bar{x}_{i_1} \geq f \cdot \bar{x}_{i_2} \Rightarrow \forall i_1 \preceq i_2 : \lfloor f \cdot \bar{x}_{i_1} \rfloor \geq \lfloor f \cdot \bar{x}_{i_2} \rfloor$$

From this and the fact that for a pair of elements $i_1 \preceq i_2 : d_{i_1} \geq d_{i_2}$, we can conclude that $\forall i_1 \preceq i_2 : \hat{x}_{i_1} - \hat{x}_{i_2} \geq 0$, and the integer solution $\hat{x}$ respects also the precedence constraints (15). Moreover based on Proposition 1 and Theorem 4, the solution $\hat{x}$ satisfies all the demands. Therefore $\hat{x}$ is a feasible solution of (P_{PCCP}) with cost at most f times larger than the cost of $\bar{x}$. As the cost of $\bar{x}$ is a lower bound on the optimal cost, the theorem follows. $\qquad\square$

6   Discussion

In this work we obtained the first approximation algorithms with strongly polynomial bounds for PCKP and PCCP. Our f-approximation for PCCP is near-
optimal under standard complexity-theoretic assumptions. It is an open question
of immediate interest, whether there exists an $O(w(\mathcal{P}))$-approximation for PCKP
with general multiplicity constraints. Moreover there is a need for improved hardness results for the problem as there is a considerable gap between the lower and
the upper bounds. Finally, the approximability of PCCP as a function of the
number of covering constraints remains as a challenging open question.

Acknowledgements. The authors thank the anonymous reviewers for comments that
improved the presentation.

References

1. Carnes, T., Shmoys, D.: Primal-dual schema for capacitated covering problems.
   In: Lodi, A., Panconesi, A., Rinaldi, G. (eds.) IPCO 2008. LNCS, vol. 5035, pp.
   288–302. Springer, Heidelberg (2008). https://doi.org/10.1007/978-3-540-68891-
   4_20
2. Carr, R.D., Fleischer, L., Leung, V.J., Phillips, C.A.: Strengthening integrality
   gaps for capacitated network design and covering problems. In: SODA, pp. 106–
   115 (2000)
3. Dilworth, R.P.: A decomposition theorem for partially ordered sets. In: Gessel, I.,
   Rota, G.C. (eds.) Classic Papers in Combinatorics, pp. 139–144. Springer, Heidelberg (2009). https://doi.org/10.1007/978-0-8176-4842-8_10
4. Dinur, I., Guruswami, V., Khot, S., Regev, O.: A new multilayered PCP and the
   hardness of hypergraph vertex cover. SIAM J. Comput. **34**(5), 1129–1146 (2005).
   https://doi.org/10.1137/S0097539704443057
5. Espinoza, D., Goycoolea, M., Moreno, E.: The precedence constrained knapsack
   problem: Separating maximally violated inequalities. Disc. Appl. Math. **194**, 65–80
   (2015). https://doi.org/10.1016/j.dam.2015.05.020
6. Kellerer, H., Pferschy, U., Pisinger, D.: Knapsack problems. In: Du, D.Z., Parda-
   los, P.M. (eds.) Handbook of Combinatorial Optimization, pp. 299–428. Springer,
   Heidelberg (2004)
7. Khot, S., Regev, O.: Vertex cover might be hard to approximate to within 2-epsilon.
   J. Comput. Syst. Sci. **74**(3), 335–349 (2008). https://doi.org/10.1016/j.jcss.2007.
   06.019
8. Kolliopoulos, S.G., Young, N.E.: Approximation algorithms for covering/packing
   integer programs. J. Comput. Syst. Sci. **71**, 495–505 (2005)
9. Koufogiannakis, C., Young, N.E.: Greedy Δ-approximation algorithm for covering with arbitrary constraints and submodular cost. Algorithmica **66**(1), 113–152
   (2013). https://doi.org/10.1007/s00453-012-9629-3
10. McCormick, S.T., Peis, B., Verschae, J., Wierz, A.: Primal-dual algorithms for
    precedence constrained covering problems. Algorithmica **78**(3), 771–787 (2017)
11. Pritchard, D., Chakrabarty, D.: Approximability of sparse integer programs. Algo-
    rithmica **61**(1), 75–93 (2011)

12. Trevisan, L.: Non-approximability results for optimization problems on bounded degree instances. In: Proceedings of the Thirty-Third Annual ACM Symposium on Theory of Computing, pp. 453–461 (2001)
13. Woeginger, G.J.: On the approximability of average completion time scheduling under precedence constraints. Disc. Appl. Math. **131**(1), 237–252 (2003)
14. Wolsey, L.: Facets for a linear inequality in 0–1 variables. Math. Program. **8**, 168–175 (1975)

Contention Resolution, Matrix Scaling and Fair Allocation

Nikhil Bansal[1(✉)] and Ilan Reuven Cohen[2(✉)]

[1] TU Eindhoven and CWI, Amsterdam, The Netherlands
n.bansal@tue.nl
[2] Bar-Ilan University, Ramat Gan, Israel
ilan-reuven.cohen@biu.ac.il

Abstract. A contention resolution (CR) scheme is a basic algorithmic primitive, that deals with how to allocate items among a random set S of competing players, while maintaining various properties. We consider the most basic setting where a *single* item must be allocated to some player in S. Here, in a seminal work, Feige and Vondrak (2006) designed a *fair* CR scheme when the set S is chosen from a product distribution. We explore whether such fair schemes exist for arbitrary non-product distributions on sets of players S, and whether they admit simple algorithmic primitives. Moreover, can we go beyond fair allocation and design such schemes for all possible achievable allocations.

We show that for any arbitrary distribution on sets of players S, and for any achievable allocation, there exist several natural CR schemes that can be succinctly described, are simple to implement and can be efficiently computed to any desired accuracy. We also characterize the space of achievable allocations for any distribution, give algorithms for computing an optimum fair allocation for arbitrary distributions, and describe other natural fair CR schemes for product distributions. These results are based on matrix scaling and various convex programming relaxations.

1 Introduction

Contention resolution (CR) schemes are basic algorithmic primitives that arise naturally, either implicitly or explicitly, in many optimization problems. One of the most basic version of contention resolution was first introduced by Feige and Vondrak [11,12], in the context of allocating items to players, and it deals with the following setting: there is a single item, and there are n players, out of which some subset S might request the item. However, the item can only be allocated to a single player in S and the goal is to decide how to allocate the item in some fair and optimal way.

Part of the work has been done while the author was a postdoctoral fellow at CWI Amsterdam and TU Eindhoven, and was supported by the ERC Consolidator Grant 617951 and the NWO VICI grant 639.023.812.

© Springer Nature Switzerland AG 2021
J. Koenemann and B. Peis (Eds.): WAOA 2021, LNCS 12982, pp. 252–274, 2021.
https://doi.org/10.1007/978-3-030-92702-8_16

More formally, there is an underlying distribution $\mathcal{P}$ over subsets $S \subseteq [n]$ of n players, that specifies the probability p_S that subset S of players request the item. A CR scheme or rule R, specifies for each set S and player $i \in S$, the probability $r_{S,i}$ that player i gets the item. Given such a rule R, player i receives the item with probability $g_i = \sum_{S:i \in S} r_{S,i}\, p_S$. The goal then, is to find a suitable rule R, so that resulting allocation vector $g = (g_1, \ldots, g_n)$ satisfies some desired fairness and optimality properties, depending on the application at hand.

Such questions arise in designing rounding-based algorithms for several problems, like those involving allocation of jobs to machines, or goods to users in combinatorial auctions, labeling problems where a vertex must pick one of several possible labels, assignment problems where a vertex much be matched to one of several neighbors and so on. Here, the solution to the natural linear or convex relaxation gives some distribution $\mathcal{P}$ over possible assignments for each item, out of which exactly one must be chosen. For this reason various CR schemes have received a lot of attention in both offline, online, stochastic and submodular optimization [1, 8, 11, 13, 15, 20].

Here, we focus on the most basic version due to Feige and Vondrak [11] described above.

Feige-Vondrak Scheme for Product Distributions. Feige and Vondrak considered the case where $\mathcal{P}$ is a product distribution corresponding to marginals q_i, i.e., $p_S = \prod_{i \in S} q_i \prod_{i \notin S}(1 - q_i)$. The goal is to allocating the item fairly, i.e., $g_i = \alpha q_i$ (each player receives the item in proportion to its marginal probability of requesting it), with α as large as possible.

Perhaps surprisingly, the solution turns out to be quite non-trivial already for product distributions, and the natural strategies such as allocating players in S uniformly ($r_{S,i} = 1/|S|$), or in proportion to q_i ($r_{S,i} = q_i/(\sum_{i \in S} q_i)$) are sub-optimal [11]. Feige and Vondrak give the following elegant rule (that we call FV scheme): For $S = \{i\}$, $r_{S,i} = 1$, otherwise for all S with $|S| > 1$

$$r_{S,i} = \frac{1}{\sum_{j=1}^n q_j} \left(\sum_{j \in S \setminus \{i\}} \frac{q_j}{|S| - 1} + \sum_{j \notin S} \frac{q_j}{|S|} \right) \qquad \forall i \in S. \qquad (1)$$

They show that this gives a *fair* allocation with $g_i = \alpha q_i$, with the best possible

$$\alpha = (1 - p_\phi)/\left(\sum_i q_i\right) = \left(1 - \prod_{i=1}^n (1 - q_i)\right)/\left(\sum_i q_i\right).$$

Notice that for this α, we have $\sum_i g_i = \alpha(\sum_i q_i) = 1 - p_\phi$, so the rule always allocates an item whenever possible, i.e., $\sum_{i \in S} r_{S,i} = 1$ for all $S \neq \emptyset$.

1.1 Extensions and Motivating Questions

The work of Feige and Vondrak [11] raises several natural questions, which motivate our work.

Question 1. (Arbitrary distributions) Can we go beyond product distributions, and design CR schemes with similar guarantees for *arbitrary* distributions $\mathcal{P}$?

Our starting point for exploring this question is that the need for such CR schemes for general distributions arises while rounding fractional solutions to more complex relaxations such as configuration LPs [3,15,17] or SDPs [2,7], where the underlying variables are correlated, and hence the distribution $\mathcal{P}$ on the resulting sets S is not a product distribution.

Exploring general distributions $\mathcal{P}$ raises several other questions: (i) How is the distribution $\mathcal{P}$ specified? (ii) How does one describe the CR scheme (specifying $r_{S,i}$ explicitly for each S, i could use exponential space)? (iii) Do there exist CR schemes with succinct description, even if $\mathcal{P}$ is completely arbitrary? (iv) Can such CR schemes be found efficiently even if the algorithm only has sampling access to $\mathcal{P}$?

Question 2. (General allocation vectors) What is the space of all possible allocation vectors g achievable by CR schemes? Given arbitrary target allocation g and distribution $\mathcal{P}$, can we efficiently determine whether g is achievable, and find such a scheme for g if it is achievable?

In some applications, one may be interested in allocations g that are not necessarily fair, and it is desirable to design a CR scheme that achieves such a g. It is unclear how to do this even for product distributions $\mathcal{P}$, as the FV-scheme only describes the rule for fair allocations of the form $g = \alpha(q_1, \ldots, q_n)$. The question becomes even more intriguing for arbitrary distributions $\mathcal{P}$.

Questions 1 and 2 above consider general distributions $\mathcal{P}$ and general allocation vectors g. It is also interesting to explore the production distribution $\mathcal{P}$ further.

Question 3. (Other CR schemes for product distributions) Are there other natural CR schemes for product distributions $\mathcal{P}$ that achieve the same allocation as the FV-scheme?

Finally, for general distributions $\mathcal{P}$, the right generalization of a maximally fair allocation is the widely studied notion of max-min fairness, also referred to as lexicographic max-min fairness [4,10,18,19]. We define this formally later below, but intuitively an allocation is max-min fair if no player's share can be increased without decreasing that of another player with *lower* share.

Question 4. Can we efficiently find an optimum max-min fair allocation for an arbitrary distribution $\mathcal{P}$?

In this work, we answer these questions in the affirmative and give several other structural and algorithmic results. These results are described in Sect. 1.2 below. Our results reveal a rich general structure in CR schemes and suggest several new directions for further investigation. Before describing our results we give some notation and basic definitions.

Notation. There are n players denoted by $N = [n]$, and $\mathcal{P}$ specifies an arbitrary distribution on subset of players $S \subseteq [n]$ requesting the single item. We use $\text{supp}(\mathcal{P})$ to denote the support of $\mathcal{P}$, and p_S to denote the probability of set S under $\mathcal{P}$. We use $q_i = \sum_{S \ni i} p_S$ to denote the marginal probabilities of player i requesting the item, and use $\bar{\mathcal{P}} = \langle q_1, \ldots, q_n \rangle$ to denote the product distribution on n players with marginals $q_1, \ldots, q_n$ respectively. A CR scheme R, specifies the probabilities $r_{S,i}$ of giving the item to player $i \in S$ upon seeing the set S. We say that an allocation vector $g = (g_1, \ldots, g_n)$ is *achievable* if there exists some CR scheme R such that for each $i \in [n]$

$$g_i = g_i(R, \mathcal{P}) := \sum_{S \subseteq [n]} r_{S,i} \, p_S.$$

The achievable vectors g form a convex polyhedral set $G = G(\mathcal{P})$ (see Sect. 1.3) and we use $\text{int}(G)$ to denote the interior of G. We call a goal vector *maximal* if an item is assigned whenever possible, i.e., for all $S \neq \emptyset$, $\sum_{i \in S} r_{S,i} = 1$, or equivalently when $\sum_i g_i = 1 - p_\emptyset$. Denote $G^M(\mathcal{P}) = \{g \in G(\mathcal{P}) : \sum_{i \in [n]} g_i = 1 - p_\emptyset\}$ as the set of maximal achievable vectors. While we state our results for maximal achievable vectors $g \in G^M(\mathcal{P})$, they hold for any $g \in G(\mathcal{P})$ by a simple reduction[1].

Max-min Fairness. The FV allocation is both fair and maximal. However, for general distributions $\mathcal{P}$, both these properties need not hold simultaneously[2]. One must allow players to have different $\alpha_i = g_i/q_i$, and the right notion to consider is max-min fairness[3].

For a vector $\alpha = (\alpha_1, \ldots, \alpha_n)$, let $\alpha^\uparrow$ be the vector with entries of α sorted in non-decreasing order. Given vectors $\alpha, \beta \in \mathbb{R}^n$ we say that α is *fairer than* β, denoted as $\alpha \succeq \beta$, if $\alpha^\uparrow$ is lexicographically at least as large $\beta^\uparrow$ (either $\alpha^\uparrow = \beta^\uparrow$, or there is an index $k \in [n]$ such that $\alpha^\uparrow(i) = \beta^\uparrow(i)$ for $i < k$ and $\alpha^\uparrow(k) > \beta^\uparrow(k)$). For an achievable goal vector $g \in G(\mathcal{P})$, let $\alpha(g)$ denote the vector with entries $\alpha_i(g) = g_i/q_i$. Then we say that g is a max-min fair allocation if $\alpha(g) \succeq \alpha(g')$ than any other feasible allocation g'. It is not hard to see that such an allocation is also maximal.

More generally, we will consider max-min fair allocations with respect to any priority values of the players $V = \langle v_1, \ldots, v_n \rangle$, where $v_i \in R^+$ and $\alpha_i^V(g) = g_i/v_i$, and define the fairness relation $(\succeq_V)$ accordingly (the special case above corresponds to $V = q$). That is, an achievable goal vector $g \in G(\mathcal{P})$ is max-min fair with respect to V, if $g \succeq_V g'$ for any achievable vector $g' \in G(\mathcal{P})$.

---

[1] Define a distribution $\tilde{\mathcal{P}}$ on $[n+1]$ players, where for $S \subseteq [n+1]$, we set $\tilde{p}_S = p_{S \setminus \{n+1\}}$ if $n+1 \in S$ and $\tilde{p}_S = 0$ otherwise, and $g_{n+1} = 1 - p_\emptyset - \sum_{i \in [n]} g_i$. Then $\sum_{i \in [n+1]} g_i = 1 - p_\emptyset$ and whenever the item is assigned to player $n+1$ in the CR scheme for $\tilde{\mathcal{P}}$ we do not assign it at all in the scheme for $\mathcal{P}$.

[2] Suppose $n > 3$ and $S = \{1\}$ or $S = \{1, \ldots, n\}$, each with probability $1/2$. Then $q_1 = 1$ and $q_i = 1/2$ for $i \geq 2$. For a *balanced* allocation with $g = \alpha q$, the best g is $(1/(n-1), 1/2(n-1), \ldots, 1/2(n-1))$, but here $\sum_i g_i < 1 - p_\emptyset$.

[3] In the example above, this allows one to achieve the allocation $(1/2, 1/2(n-1), \ldots, 1/2(n-1))$, which is clearly better.

1.2 Our Results

We first show that there exist very natural and succinct to describe CR schemes, even under the most general setting of arbitrary distribution $\mathcal{P}$ and any feasible allocation vector g.

Weight-based Schemes. A CR scheme is *weight-based* if it has the following form: Each player $i \in [n]$ has a (fixed) weight $w_i \geq 0$, and for every non-empty set S and player $i \in S$,

$$r_{S,i} = w_i / \left(\sum_{j \in S} w_j \right).$$

That is, for each set S the rule simply assigns the item to $i \in S$ with probability proportional to w_i. Note that as the weights w_i are independent of S, the CR rule is extremely simple and is succinctly described by only n numbers. Besides simplicity this rule also has several other useful properties, and is widely studied in social discrete choice and econometrics and is referred to as the Multinomial Logit Model [14, 22, 26].

Theorem 1. *For any distribution $\mathcal{P}$ on the players, and any maximal achievable allocation vector in $g \in int(G^M(\mathcal{P}))$, there is a weight-based scheme achieving g.*

We remark that the condition $g \in int(G^M)$ is necessary. E.g., suppose $n = 2$, and $S = \{1\}$ or $\{1, 2\}$ with probability $1/2$ each. Then the allocation $g = (1/2, 1/2)$ is achievable (choose player 1 if $S = \{1\}$ and player 2 otherwise). But in a weight-based scheme as player 2 is only chosen with probability $w_2/(w_1 + w_2)$ when $S = \{1, 2\}$, and as $p_{1,2} = 1/2$, its allocation approaches $1/2$ only as $w_2/w_1 \to \infty$.

We give two proofs of Theorem 1. The first is based on formulating a suitable max-entropy convex program [16, 27] and using convex programming duality [5]. The second proof is based on applying the classic matrix-scaling algorithm [21, 23, 25] to a suitably chosen matrix.

These proofs give algorithms that take $\mathcal{P}$ and g as input and some error parameter $\varepsilon > 0$, and return weights w corresponding to some achievable $\hat{g}$ with $\|g - \hat{g}\| \leq \varepsilon$ (if it exists) in time $\text{poly}(n, \text{supp}(\mathcal{P}), \log(1/\varepsilon))$. However, this is not so useful as the support of $\mathcal{P}$ is exponentially large typically. Using standard sampling based techniques, these can be adapted to run with only access to samples from $\mathcal{P}$. This removes the dependence on $\text{supp}(\mathcal{P})$ at the expense of increasing the dependence on ε from $\log(1/\varepsilon)^{O(1)}$ to $1/\varepsilon^{O(1)}$.

Theorem 2. *Given sampling access to $\mathcal{P}$, there are algorithms based on convex programming and on matrix-scaling, that given a target allocation g and $\varepsilon > 0$, with high probability, find some $\hat{g} \in G(\mathcal{P})$ satisfying $\|g - \hat{g}\|_\infty \leq \varepsilon$, or certify that no such $\hat{g}$ exists. Both algorithms use $O(\varepsilon^{-2} \log n)$ samples and run in time $\text{poly}(n, \varepsilon^{-1})$.*

Permutation Schemes and Characterizing the Set G^M. Next, we consider another natural class of CR schemes based on what we call the *permutation* *scheme.* In a permutation scheme, the players are ordered according to some fixed permutation π, and given a set of players S, the item is assigned to the first player in S according to π.

We show the following general result.

Theorem 3. *For any arbitrary distribution $\mathcal{P}$ and given any feasible goal vector $g \in G^M(\mathcal{P})$, there exists a CR scheme of the following form: A random permutation π is chosen from a (fixed) distribution, depending only on g and $\mathcal{P}$, over at most $n+1$ permutations. Then, given a set S, the item is all allocated according to the permutation scheme given by π.*

Notice that these schemes are succinctly described, requiring only $n+1$ permutations on the n players, and exist in the most general setting possible. The Theorem above follows from the result below which gives an explicit characterization of the set $G^M(\mathcal{P})$ of achievable allocations in the most general setting using permutation schemes.

Theorem 4. *For any arbitrary distribution on players $\mathcal{P}$, the convex set $G^M(\mathcal{P})$ is the convex hull of the goal vectors $g(\pi)$ of permutation schemes.*

This result is based on a non-trivial connection between the allocations achievable by convex combination of permutation schemes and weight-based schemes. Apriori, it is unclear why such schemes exist even for product distributions.

Finally, we give an efficient algorithm to checking whether some given g is achievable, and obtain the corresponding allocation rule.

Theorem 5. *If the distribution $\mathcal{P}$ has the property that the allocation vector $g(\pi)$ can be computed[4] for any permutation π, then for any vector g there is an efficient algorithm to test if $g \in int(G(P))$, and to compute the corresponding CR-scheme.*

The Feige-Vondrak Setting and Sequential Schemes. While the above results hold for arbitrary $\mathcal{P}$ and g, we now revisit the Feige Vondrak setting, where $\mathcal{P}$ is a product distribution and $g = \alpha q$ is the fair allocation. We give another very natural scheme that we call the *Sequential scheme.*

A sequential scheme has a very simple form. We fix an order of players (say $1, \ldots, n$) and compute numbers $\gamma_1, \ldots, \gamma_n$. When S is realized, we go over the players sequentially in the fixed order, and give the item to player i in S with probability γ_i, unless it is the last player in S, in which case it gets the item with probability 1.

As the γ_i are independent of the set S, this gives a succinct scheme with only n parameters. The following result shows that sequential schemes can achieve the same allocation as the FV rule. The proof of this result also gives an explicit expression for the γ_i as a function of $q_1, \ldots, q_n$.

---

[4] It can always be computed efficiently to any desired accuracy $\varepsilon > 0$ using $\text{poly}(n, 1/\varepsilon)$ samples from $\mathcal{P}$.

Theorem 6. *For any product distribution $\mathcal{P} = \langle q, \ldots, q_n \rangle$, and for any ordering of the players, there exists a sequential scheme, that achieves the maximally fair allocation $g_i = \alpha q_i$ for all $i \in [n]$.*

Unfortunately, sequential schemes are not general enough, and we give two examples which show their limitations in a strong way: (i) there exist non-product distributions $\mathcal{P}$, for which the fair allocation g is feasible, but no sequential scheme can achieve it, and (ii) even for $\mathcal{P}$ a product distribution, there exist (non-fair) feasible vectors g, that cannot be achieved by a sequential scheme.

Variance Minimizing Program. Another natural question is whether the FV-scheme in (1), can be viewed more systematically, and obtained as an optimum solution to some natural convex program. We show that this is indeed the case, and it corresponds to a convex program that minimizes a certain weighted squared error objective.

However, this convex program also does not seem too useful for more general settings. In particular, we give examples of (i) general $\mathcal{P}$ for which the fair allocation is feasible, and (ii) where $\mathcal{P}$ is a product distribution for g is not the fair allocation, for which the program does not have a nice structured solution. Finally, we also consider a general class of such convex programs and show their solutions can have a form similar to that of (1).

Efficiently Computing the Max-Min Fair Allocation. For general $\mathcal{P}$, while Theorems 1 and 2 give an efficient test to determine if an allocation vector g is feasible, they do not give an algorithm for computing the max-min fair allocation. Despite extensive work on fairness and in particular on computing max-min fair allocations in both continuous and discrete settings, we are not aware of any work in our setting.

We show the following result for max-min fair allocation with respect to a general *priority* values.

Theorem 7. *For any distribution $\mathcal{P}$ and for any priority vector V, the max-min allocation $g \in G^M(P)$ can be computed exactly in time $poly(n, |supp(\mathcal{P})|)$ using a liner program. If an ϵ additive error is allowed, the time is $poly(n, 1/\varepsilon)$. Finally, an exact computation in $poly(n)$ time is also possible if $g(\pi)$ can be computed efficiently for any permutation π.*

Our algorithm follows a natural approach of finding and removing the most *critical* subset of players and iterating on the residual instance. To do this, we extend an LP-based algorithm of Charikar for computing the densest subgraph in a graph [6], to hypergraphs. For the setting where $g(\pi)$ can be computed efficiently for any permutation, we consider a different LP to compute to the most critical set.

1.3   Preliminaries

We describe some basic properties of the set $G^M(\mathcal{P})$ of maximal achievable goal vectors, that will be useful later. First, we note that $G^M(\mathcal{P})$ is a convex set for an underlying distribution over players $\mathcal{P}$. Let $g, g' \in G^M(\mathcal{P})$ with corresponding CR schemes $R = \{r_{S,i}\}_{S,i}$ and $R' = \{r'_{S,i}\}_{S,i}$. Then for $\alpha \in [0, 1]$, the goal vector $g^{(\alpha)} = \alpha g + (1 - \alpha)g'$ is also valid, as the CR scheme $R^{(\alpha)}$ with $r^{(\alpha)}_{S,i} = \alpha r_{S,i} + (1 - \alpha)r'_{S,i}$ gives that for any $i \in [n]$,

$$g_i^{(\alpha)} = \alpha g_i + (1 - \alpha)g'_i = \alpha \sum_S p_S\, r_{S,i} + (1 - \alpha) \sum_S p_S\, r'_{S,i} = \sum_S p_S\, r^{(\alpha)}_{S,i}.$$

The following gives a test to check whether a given allocation vector g is feasible or not. While the test is not efficient as it checks for exponentially many inequalities, it will be useful for our proofs later. For $S \subseteq [n]$, let $\mathcal{E}(S) = \{T \subseteq [n] : T \cap S \neq \emptyset\}$ be the collection of sets with non-empty intersection with S.

Lemma 1. $g \in G(\mathcal{P})$ iff for all $S \subseteq [n]$, $P(\mathcal{E}(S)) \geq \sum_{i \in S} g_i$, where $P(\mathcal{E}(S)) = \sum_{T \in \mathcal{E}(S)} p_T$.

Proof. For a goal vector g, consider the flow network $F(\mathcal{P}, g) = (V, E)$, with vertices $V = \{s\} \cup \mathrm{supp}(\mathcal{P}) \cup [n] \cup \{t\}$, and edges $E = E_1 \cup E_2 \cup E_3$ with capacity function c, where $E_1 = \{(s, S) : S \in \mathrm{supp}(\mathcal{P})\}$, $c(s, S) = p_S$, $E_2 = \{(i, t) : i \in [n]\}$, $c(i, t) = g_i$ and $E_3 = \{(S, i) : i \in S\}$, $c(S, i) = \infty$.

We claim that g is achievable iff there exists an s-t flow of value $\sum_{i \in N[n]} g_i$. Indeed, given a CR scheme given by $r_{S,i}$ that achieves g, the flow f with $f(s, S) = \sum_{i \in S} r_{S,i} p_S$ for $S \in \mathrm{supp}(\mathcal{P})$, $f(S, i) = r_{S,i} p_S$ for $i \in S$, and $f(i, t) = g_i$ for $i \in [n]$, is feasible and has value $\sum_{i \in N} g_i$. Conversely, given a flow f with value $\sum_i g_i$, setting $r_{S,i} = f(S, i)/p_S$ would achieve g.

Applying max-flow min-cut theorem to this network directly gives the result.

2   Weight-Based Schemes

Recall that in a weight based scheme, $r_{S,i} = w_i/(\sum_{j \in S} w_j)$, for each set S and each $i \in S$. We now prove Theorems 1 and 2. For clarity of exposition, we first focus on showing the existence of weights and assume that $\mathcal{P}$ is given explicitly. The algorithmic version follows from known results for max-entropy convex programs and matrix scaling, and standard ideas based on sampling that are described in Sect. 2.3.

2.1   Convex Programming Based Algorithm

Consider the following convex program with variables $x_{S,i} \geq 0$ for the probability that player $i \in S$ receives the item on outcome S.

$$\max \sum_{S,i} x_{S,i} \ln \frac{e}{x_{S,i}}$$

$$\text{s.t.} \sum_{S:i\in S} x_{S,i} = g_i \qquad \forall i \in [n] \tag{2}$$

$$\sum_{i\in S} x_{S,i} = p_S \qquad \forall S \in \text{supp}(\mathcal{P}) \tag{3}$$

$$x_{S,i} \geq 0 \qquad \forall S \in \text{supp}(\mathcal{P}), i \in S \tag{4}$$

As $g \in \text{int}(G^M)$, $g_i > 0$ for each i, and as the objective function is strictly convex, the optimum solution x^* to the program satisfies $x^*_{S,i} > 0$ for all i, S. As $g \in \text{int}(G^M)$, the program is feasible and also satisfies Slater's condition [5], and by strong duality there is a tight dual solution with the same objective value.

Dual. Let us consider the dual. It has variables $\beta_i \in \mathbb{R}$ for each $i \in [n]$ for the constraints (2), and $\alpha_S \in \mathbb{R}$ for each $S \in \text{supp}(\mathcal{P})$ for the constraints (3). Writing the objective as $\sum_{S,i} x_{S,i}(1 - \ln x_{s,i})$, its partial derivative with respect to $x_{S,i}$ is $(1 - \ln x_{S,i}) - 1 = -\ln x_{S,i}$. As $x^*_{S,i} > 0$ in the primal, by complementary slackness the dual variables for constraints (4) have value 0 and the KKT conditions give that,

$$\beta_i - \alpha_S = \ln x_{S,i} \qquad \forall S, i. \tag{5}$$

Or equivalently, $x_{S,i} = e^{\beta_i}/e^{\alpha_S}$. For a fixed S, summing up (5) over $i \in S$ and using $\sum_{i\in S} x_{i,S} = p_S$ gives $\sum_{i\in S} e^{\beta_i} = e^{\alpha_S} p_S$ and hence

$$x_{S,i} = p_S e^{\beta_i} / \left(\sum_{i\in S} e^{\beta_i} \right).$$

As $r_{S,i} = x_{S,i}/p_S$, this gives that $r_{S,i}$ has the form $w_i/(\sum_{i\in S} w_i)$ for each s, i with $w_i = e^{\beta_i}$.

Running Time and Accuracy. The condition $g \in \text{int}(G^M)$ ensures that g has distance at least ϵ from G^M for some $\epsilon > 0$. The arguments in [24] on the bit-length of solutions for max-entropy convex programs, imply that the bit length of w_i depends as $O(\log 1/\varepsilon)$ on ε and the convex program runs in time polynomial in n, $\text{supp}(\mathcal{P})$ and $\log(1/\varepsilon)$.

2.2 Matrix Scaling Based Algorithm

We now show how to obtain Theorem 1 using matrix scaling. We begin by briefly describing the relevant background on matrix scaling.

Let A be a given $m \times n$ matrix with non-negative entries, and let $r \in (\mathbb{R}^+)^m$ and $c \in (\mathbb{R}^+)^n$ be positive vectors satisfying $\sum_{i\in[m]} r_i = \sum_{j\in[n]} c_j$.

Definition 1. ((r,c)-**scalable**). *A non-negative matrix A is said to be (r,c)-scalable, if there exists non-negative vectors $x \in \mathbb{R}^m$ and $y \in \mathbb{R}^n$ such that $B = \text{Diag}(x) A \text{Diag}(y)$ with entries $b_{ij} = x_i a_{ij} y_j$ has row sums r and column sums c. That is, $\sum_j b_{ij} = r_i$ for $i \in [m]$ and $\sum_i b_{ij} = c_j$ for $j \in [n]$.*

Given an $\epsilon > 0$, we say that A is ε-(r,c) scalable if there is some scaling of A with row sums (exactly) r, and columns sums c' satisfying $\sum_{j=1}^{n}(c'_j - c_j)^2 \leq \varepsilon$. We say that A is almost (r,c)-scalable if it is ϵ-(r,c) scalable for any $\epsilon > 0$.

The following proposition from [23] exactly characterizes approximate scalability for a matrix.

Proposition 1. *A non-negative matrix A is almost (r,c)-scalable iff for every zero minor $Z \times L$ of A (i.e., a submatrix of A with all zero entries), the vectors r and c satisfy $\sum_{i\in[n]\setminus Z} r_i \geq \sum_{j\in L} c_j$.*

Relation to Weight-Based CR Schemes. We now relate matrix scaling to contention resolution.

Given $\mathcal{P}$, let A be a $|\text{supp}(\mathcal{P})| \times n$, incidence matrix for the support of $\mathcal{P}$, defined as $A_{S,i} = 1$ if $i \in S$ and 0 otherwise. The key observation is the following.

Lemma 2. *Let $g \in G^M(\mathcal{P})$, and suppose there exist vectors x and y such that the matrix $\text{diag}(x)A\,\text{diag}(y)$ has row sums p_S for each S and column sums g_j for each j. Then the vector y gives the weights for a weight-based CR scheme that achieves g.*

Proof. By the definition of A, and the properties of the scaling, the sum for row S satisfies

$$p_S = \sum_{j\in[n]} x_S A_{S,j} y_j = \sum_{j\in S} x_S y_j,$$

and hence $x_S = p_S/(\sum_{j\in S} y_j)$. Similarly, the sum for each column j satisfies

$$g_j = \sum_{S} x_S A_{S,j} y_j = \sum_{S:j\in S} x_S y_j = \sum_{S:j\in S} p_S \frac{y_j}{\sum_{j\in S} y_j}.$$

So setting $r_{S,j} = y_j/(\sum_{j\in S} y_j)$ achieves $g_j = \sum_S p_S r_{S,j}$ and $\sum_{j\in S} r_{S,j} = 1$ for all S, giving the claimed weight-based scheme. $\square$

Interestingly, the condition for the existence of the scaling of A in Lemma 2, turns out to be identical to the condition for feasibility of a goal vector $g \in G^M(\mathcal{P})$ as in Lemma 1.

Lemma 3. *For any $\mathcal{P}$, the incidence matrix A is almost (p_S, g)-scalable iff $g \in G^M$.*

Proof. We will show that the condition in Proposition 1 holds for each zero minor $Z \times L$ of A. Fix some $L \subset [n]$. Then it suffices to consider the maximal $Z \times L$ minor of A, which corresponds to $Z = [n]\setminus\mathcal{E}(L)$, the collection all subsets disjoint from L.

For $r = \{p_S\}_S$ and $c = g$, the condition $\sum_{i\in[n]\setminus Z} r_i \geq \sum_{j\in L} c_j$ thus becomes $\sum_{T\in\mathcal{E}(L)} p_T \geq \sum_{i\in L} g_i$, which holds by Lemma 1 for every $L \subseteq [n]$ iff g is achievable. $\square$

It is also known that if A is almost (r,c)-scalable, then an ϵ-(r,c) scaling can be found after $\text{poly}(n, \log(1/\varepsilon))$ iterations [9], giving the claimed running time.

2.3   Handling g at the Boundary and Sampling

Interestingly, the matrix-scaling based algorithm is quite robust and works even if g lies at the boundary of $G^M(\mathcal{P})$. In particular, as the algorithm iteratively computes some row and column scaling x and y, using y as weights at any intermediate step gives some valid and achievable allocation. More precisely, the weights y after $\text{poly}(n, \log 1/\varepsilon)$ iterations will correspond to a goal vector $\hat{g}$ with $\|g - \hat{g}\|_2 \leq \varepsilon$ [9], even for g on the boundary of G^M. The algorithms above assumed access to the entire distribution $\mathcal{P}$. This can be removed using standard arguments that we now sketch. Let q_i be the marginal probability of player i in $\mathcal{P}$. By standard tail bounds, for any $\mathcal{P}$, each q_i can be estimated to within $\pm\varepsilon$ error with high probability, using $O((\log n)/\epsilon^2)$ independent samples from $\mathcal{P}$.

Suppose g is achievable for $\mathcal{P}$ by some CR scheme R given by $r_{S,i}$, i.e., $g_i = \sum_S p_S r_{S,i}$. Take $m = O(\log n/\varepsilon^2)$ samples $\hat{S}$ from $\mathcal{P}$, and consider the (empirical) allocation $\hat{g}_i = \frac{1}{m}\sum_{\hat{S}} r_{\hat{S},i}$ obtained by R on these samples. As $0 \leq r_{S.i} \leq 1$, and as $|q_i - \sum_{\hat{S}} \mathbf{1}(i \in \hat{S})/m| \leq \varepsilon$ for all i, this gives that $\|\hat{g} - g\|_\infty \leq \varepsilon$ with high probability. As $\hat{g}$ is achievable for the sampled distribution $\hat{S}$, running the convex program in proof of Theorem 1 on $\hat{S}$ with goal g (and by using an additional error term ϵ in (2)) will find some $\tilde{g}$, with $\|g - \tilde{g}\|_\infty \leq \|g - \hat{g}\|_\infty + \|\hat{g} - \tilde{g}\| \leq 2\varepsilon$. In particular, this implies the following.

Theorem 8. *There are efficient algorithms, using only sampling access to $\mathcal{P}$, that given g, find an achievable $\hat{g}$ satisfying $\|g - \hat{g}\| \leq \varepsilon$ or certify that no such $\hat{g}$ exists and run in time $\text{poly}(n, 1/\varepsilon)$.*

3   Permutation Schemes and Achievable Goal Vectors

Recall that a given a permutation π of the players, the corresponding permutation scheme allocates the item to $i \in S$ appearing earliest in the ordering π, i.e., for every $S \neq \emptyset$, $r^\pi_{S,i} = 1$ for $i = \arg\min\{\pi(j) : j \in S\}$, and $r_{S,j}(\pi) = 0$ otherwise.

Let Π_n denote the set of all $n!$ permutations on $[n]$. Given a distribution $\mathcal{P}$ on subsets of n players, and a permutation $\pi \in \Pi_n$, let $g(\pi, \mathcal{P})$ denote allocation vector achieved by the permutation scheme given by π.

Theorem 9. *For any $\mathcal{P}$, any maximal achievable allocation $g \in int(G^M(\mathcal{P}))$ is a convex combination of $g(\pi, \mathcal{P})$ for $\pi \in \Pi_n$. Equivalently, the closure of $G^M(\mathcal{P})$ is the convex hull of the vectors $g(\pi, \mathcal{P})$.*

Proof. As all quantities depend on $\mathcal{P}$, we drop it to simplify notation. Fix some $g \in int(G^M)$. It suffices to show that g is some convex combination of $g(\pi)$ for permutations π in Π_n, i.e., $g = \sum_{\pi \in S_n} \alpha(\pi) g(\pi)$ with $\sum_{\pi \in S_n} \alpha(\pi) = 1$ and $\alpha(\pi) \geq 0$ for all π.

As $g \in int(G^M)$, by Theorem 1, g can be achieved by a weight-based scheme given by some weight vector w, i.e., $g_i = \sum_{i \ni S} p_S w_i/(\sum_{j \in S} w_j)$ for all $i \in [n]$.

We define $\alpha(\pi)$ using w as follows. Equip each player i with an independent exponentially distributed random variable X_i with rate $\lambda_i = w_i$ (i.e., with mean

$1/w_i$). For each instantiation of these variables, let π be the ordering of the players in the increasing of their X_i values i.e., $X_{\pi^{-1}(1)} < X_{\pi^{-1}(2)} < \ldots < X_{\pi^{-1}(n)}$ (we ignore ties, as this is a measure zero event). Let α_π denote the probability for the permutations $\pi \in \Pi_n$, and let α denote the resulting distribution on the permutations.

By the memoryless property of exponential random variables, for any set S of players, the probability that player i has minimum X_i value among all the X_j for $j \in S$ is exactly $w_i / \sum_{j \in S} w_j$. In other words,

$$\Pr_{\pi \sim \alpha}[\arg\min\{\pi(j) : j \in S\} = i] = \frac{w_i}{\sum_{j \in S} w_j}.$$

This implies that choosing π according to the distribution α and applying the permutation scheme based on π exactly achieves the goal g.

While the proof may use g exponentially many permutations to write g as a convex combination, as $g \in \mathbb{R}^n$, by Caratheodory's theorem at most $n + 1$ vectors $g(\pi)$ always suffice.

Tightness. We remark that there exist distributions $\mathcal{P}$ for which each of the $n!$ permutations π actually corresponds to a unique extreme point $g(\pi)$ of $G^M(\mathcal{P})$. Consider the product distribution $\mathcal{P}$ with marginals $q_1 = \ldots = q_n = 1/2$. For any permutation π, the player $\pi^{-1}(k)$, which is at position k in the ordering, gets the item with probability exactly 2^{-k}. So $g(\pi) = (2^{-\pi(1)}, \ldots 2^{-\pi(n)})$.

For any π, we claim that $g(\pi)$ cannot be expressed as a convex combination of other $g(\pi')$. Without loss of generality, suppose π is the identity permutation. Then, as $g(1) = 1/2$ and the marginal $q_1 = 1/2$, every permutation π' in the support of the convex combination must have 1 as its first element. Conditioning out the element 1, and applying the argument repeatedly gives that the only permutation in the support of $g(\pi)$ is $(1, 2, \ldots, n)$.

3.1   Alternate Feasibility Test

The characterization above gives another useful test for testing if g is feasible, provided the distribution $\mathcal{P}$ has the property that the goal vector $g(\pi)$ can be efficiently computed for any permutation π. This is true for most natural distributions, or in general $g(\pi)$ can be computed efficiently to desired accuracy by sampling. E.g., for the product distribution $\mathcal{P} = \langle q_1, \ldots, q_n \rangle$, we have the have the explicit expression $g_k(\pi) = q_k \prod_{j : \pi(j) < \pi(k)} (1 - q_j)$.

Theorem 10. *If $\mathcal{P}$ has the property that $g(\pi)$ can be computed for any permutation π, then for any g there is an efficient algorithm to test if $g \in int(G^M(P))$. The algorithm also computes a corresponding CR scheme.*

The proof is based on linear programming. Given a goal vector g, consider the following LP with $n!$ variables x_π for each permutation $\pi \in \Pi_n$.

$$\min \sum_{\pi \in \Pi_n} x_\pi \quad \text{s.t.} \quad \sum_{\pi \in \Pi_n} x_\pi g_i(\pi) \geq g_i, \quad \forall i \in [n], \quad x_\pi \geq 0, \quad \forall \pi \in \Pi_n.$$

Then g is achievable iff the objective $\sum_{\pi \in \Pi_n} x_\pi \leq 1$. As this LP has exponentially many variables, consider the following dual with variables y_i for $i \in [n]$.

$$\max \sum_{i \in [n]} g_i y_i \quad \text{s.t.} \quad \sum_{i \in [n]} y_i g_i(\pi) \leq 1, \quad \forall \pi \in \Pi_n, \quad y_i \geq 0, \quad \forall i \in [n].$$

While the dual has exponentially many constraints, it can be solved efficiently using a *separation oracle*. Recall that for the separation oracle, given a candidate feasible solution y we need to find some permutation π such that $\sum_{i \in [n]} y_i g_i(\pi) > 1$, provided such a permutation exists. The following shows that the such a permutation π is easily obtained by sorting the coordinates of y.

Claim 4. *Given y, let σ be the permutation such that $y_{\sigma(1)} \geq \cdots \geq y_{\sigma(n)}$. Then for any $\mathcal{P}$, the quantity $\sum_i y_i g_i(\pi)$ is maximized for $\pi = \sigma^{-1}$. That is $\sigma^{-1} = \arg\max_\pi (\sum_{i \in [n]} y_i g_i(\pi))$.*

Proof. Let π be some permutation that maximizes $\sum_{i \in [n]} y_i g_i(\pi)$ and $\sum_{i \in [n]} y_i g_i(\pi) > \sum_{i \in [n]} y_i g_i(\sigma^{-1})$. Then $\pi^{-1}(k)$ is the index of player that appears at position k in the ordering π. Suppose for the sake of contradiction that $y_{\pi^{-1}(k)} < y_{\pi^{-1}(k+1)}$. Then, consider the ordering obtained by swapping the players at positions k and $k+1$, i.e. $\pi'(\pi^{-1}(k)) = k+1$ and $\pi'(\pi^{-1}(k+1)) = k$ and $\pi'(i) = \pi(i)$ otherwise. We will show that $(\sum_{i \in [n]} y_i g_i(\pi'))$ is not smaller than $(\sum_{i \in [n]} y_i g_i(\pi))$, giving the desired contradiction. Let

$$\tilde{S} = \{S \subseteq N : \pi^{-1}(k), \pi^{-1}(k+1) \in S, \ k = \min_i \{\pi(i) : i \in S\}\}$$

be the collection of sets containing both $\pi^{-1}(k), \pi^{-1}(k+1)$ and where the player $\pi^{-1}(k)$ is the earliest player in S according to the ordering in π.

The crucial observation is that for any S and $i \in S$, we have $r_{S,i}(\pi) = r_{S,i}(\pi')$, unless $S \in \tilde{S}$ and $i \in \{\pi^{-1}(k), \pi^{-1}(k+1)\}$. Moreover, in this case, $r_{S,\pi^{-1}(k)}(\pi) = 1$, $r_{S,\pi^{-1}(k+1)}(\pi) = 0$, and $r_{S,\pi^{-1}(k)}(\pi') = 0$, $r_{S,\pi^{-1}(k+1)}(\pi') = 1$. This gives that

$$\sum_{i \in [n]} y_i g_i(\pi') - \sum_{i \in [n]} y_i g_i(\pi) = \sum_{S \in \tilde{S}} p_S \cdot (y_{\pi^{-1}(k+1)} - y_{\pi^{-1}(k)})$$

which is non-negative by our assumption.

4   Sequential Schemes for the FV Setting

Recall that a sequential scheme has the following form. There is some fixed order π on the players, and we compute $\gamma_i \in [0,1]$ for each i. When S is realized, we go over the players in the order given by π, and give item to $i \in S$ with probability γ_i, unless i is the last player in S, in which it gets the item with probability 1.

Clearly, this scheme is maximal. We show that it can achieve the same allocation as the FV scheme for any production distribution and for any order on players π.

Theorem 11. *For any product distribution* $\mathcal{P} = \langle q_1, \ldots, q_n \rangle$, *and for any ordering* π *of the players, there exists a sequential scheme, that achieves the maximal fair allocation. Moreover, the* γ_i *are explicitly given as* $\gamma_i = (\alpha_i - R_{i+1})/(1 - R_{i+1})$, *where* $R_k = \prod_{i \geq k}(1 - q_i)$, $\alpha_k = (1 - R_k)/Q_k$ *and* $Q_k = \sum_{i \geq k} q_i$.

Proof. Without loss of generality, we can assume π is the identity permutation, and the players are considered in the order $1, \ldots, n$. Also, given the definition of α_k and R_k, the fairness factor $\alpha = (1 - \prod_{i=1}^n(1 - q_i))/(\sum_{i=1}^n q_i)$ is simply $(1 - R_1)/Q_1 = \alpha_1$. We will show that the desired $g = \alpha q$ is attained by setting $\gamma_k = (\alpha_k - R_{k+1})/(1 - R_{k+1})$ for $k \in [n]$.

We first show that γ_k are well-defined probabilities. Let us recall the Weierstrass' product inequalities.

Fact 5. *Let* $a_1, \ldots, a_n \in [0, 1]$ *and* $S = \sum_i a_i$. *Then* $\prod_i(1 - a_i) \geq 1 - S$ *and* $\prod_i(1 + a_i) \geq 1 + S$.

As $\alpha_k \leq 1$, we clearly have $\gamma_k \leq 1$. To show that $\gamma_k \geq 0$, note that the denominator is non-negative, and the numerator satisfies

$$\alpha_k - R_{k+1} = \frac{1 - R_k}{Q_k} - R_{k+1} = \frac{1}{Q_k}\left(1 - (1 - q_k)R_{k+1} - Q_k R_{k+1}\right)$$

$$= \frac{1}{Q_k}(1 - (1 + Q_{k+1})R_{k+1}) \geq 0,$$

where the second equality uses that $R_k = (1 - q_k)R_{k+1}$ and the third equality uses that $Q_k - q_k = Q_{k+1}$. The inequality follows, as by Fact 5, $1 + Q_{k+1} \leq \prod_{i \geq k+1}(1 + q_i)$, and hence $(1 + Q_{k+1})R_{k+1} \leq \prod_{i \geq k+1}(1 - q_i^2) \leq 1$.

For $i \in [n]$, let $p^{(i)}$ denote the product distribution $\langle q_i, \ldots, q_n \rangle$ on $i, \ldots, n$. That is, for a set $T \subseteq [i, n]$, $p_T^{(i)} = \prod_{j \in T} q_j \prod_{j \geq i, j \notin T}(1 - q_j)$.

Given $\gamma_1, \ldots, \gamma_n$, we define $\tilde{p}_T^{(i)}$ for $i \in [n], T \subseteq [i, n], T \neq \emptyset$, as the probability that each player in T requested the item and that the item is not allocated to any of the players in $[1, i - 1]$ in the sequential scheme.

Claim. For any non-empty subset T of $\{i, \ldots, n\}$, it holds that $\tilde{p}_T^{(i)} = (\alpha_1/\alpha_i)p_T^{(i)}$.

Proof. We first note that $\tilde{p}_T^{(i)} = \prod_{j=1}^{i-1}\left((1 - q_j) + q_j(1 - \gamma_j)\right)p_T^{(i)} = \prod_{j=1}^{i-1}(1 - q_j\gamma_j)p_T^{(i)}$.

This follows as the set T is 'left over' at step i, if and only if, for each player $j \in \{1, \ldots, i-1\}$, either it did not appear in S or it did not pick the item which happens with probability $1 - \gamma_j$ as T is non-empty and hence j was not the last player in S.

So it suffices to show that $1 - q_j\gamma_j = \alpha_j/\alpha_{j+1}$ for each $j \in \{1, \ldots, i-1\}$. To this end, we have

$$1 - q_j\gamma_j = 1 - \frac{q_j(\alpha_j - R_{j+1})}{1 - R_{j+1}} = \frac{1 - R_{j+1} - q_j\alpha_j + q_j R_{j+1}}{1 - R_{j+1}}$$

$$= \frac{1 - R_j - q_j\alpha_j}{1 - R_{j+1}} = \frac{\alpha_j(Q_j - q_j)}{1 - R_{j+1}} = \frac{\alpha_j Q_{j+1}}{1 - R_{j+1}} = \frac{\alpha_j}{\alpha_{j+1}},$$

where the third equality uses that $R_j = (1-q_j)R_{j+1}$, and the fourth (resp. sixth) that $1 - R_j = \alpha_j Q_j$ (resp. $1 - R_{j+1} = \alpha_{j+1}Q_{j+1}$).

We can express probability of player i getting the item as follows.

Claim. The probability the player i gets the item in the sequential scheme is

$$g_i = \tilde{p}_{\{i\}}^{(i)} + \gamma_i \sum_{T \subseteq \{i,\dots,n\}, i \in T, T \neq \{i\}} \tilde{p}_T^{(i)}.$$

Proof. Consider the set of players T left after the first $i - 1$ steps. Then either player i gets the item with probability 1 if $T = \{i\}$, otherwise it gets it with probability γ_i if $i \in T$ and $j \in T$ for some $j > i$.

Noting that $p_{\{i\}}^{(i)} = q_i R_{i+1}$ and that $\sum_{T \subseteq \{i,\dots,n\}, i \in T, T \neq \{i\}} p_T^{(i)} = q_i(1 - R_{i+1})$, by Claim 4 and Claim 4,

$$g_k = \frac{\alpha_1}{\alpha_k} \cdot (q_k R_{k+1} + \gamma_k q_k(1 - R_{k+1})) = \frac{\alpha_1}{\alpha_k} \cdot (q_k R_{k+1} + q_k(\alpha_k - R_{k+1})) = \alpha_1 q_k.$$

Noting that $\alpha_1 = \alpha$ gives the desired result.

Limitations. Unfortunately, sequential schemes are not general enough to work for arbitrary $\mathcal{P}$, or even when $\mathcal{P}$ is a product distribution but g is arbitrary. For $n = 3$, there exists a general distribution $(\Pr[A = \{1\}] = \Pr[A = \{1,2,3\}] = 0.34, \Pr[A = \{2\}] = \Pr[A = \{1,2\}] = 0.16)$ such that a fair allocation that is achievable, but it is not achievable by a sequential scheme (for the order $1, 2, 3$). In addition, there exists a product distribution and achievable vector $(\mathcal{P} = \langle 0.4, 0.1, 0.2 \rangle, g = (0.356, 0.077, 0.135))$ which cannot be achieved by a sequential scheme (for the order $1, 2, 3$). Understanding the class of allocations g, and the class of distributions $\mathcal{P}$ for which such schemes work might be an interesting question for further investigation.

5   Convex Program for the FV Scheme

Recall that the FV scheme considers the product distribution $\mathcal{P} = \langle q_1, \dots, q_n \rangle$ with $g = \alpha q$ where $\alpha = (1 - \prod_i (1 - q_i))/(\sum_i q_i)$. Consider the following program with variables $r_{S,i}$, that tries to finds a rule $r_{S,i}$ to minimize the (weighted) quadratic variation about $1/|S|$, while satisfying the global constraints.

$$\min \sum_{i \in S} (|S| - 1) p_S \left(r_{S,i} - \frac{1}{|S|} \right)^2$$

$$\text{s.t.} \quad \sum_{S \ni i} p_S r_{S,i} = \alpha q_i \qquad\qquad \forall i \in [n] \qquad (6)$$

$$\sum_{i \in S} r_{S,i} = 1 \qquad\qquad \forall S \qquad (7)$$

$$r_{S,i} \geq 0 \qquad\qquad \forall S, \forall i \in S \qquad (8)$$

Theorem 12. *For product distributions, the optimum solution to the program above is the FV scheme* (1).

Proof. We first note that after rearranging the terms, the FV scheme can be written as

$$r_{S,i} = \frac{1}{|S|} + \frac{(\overline{q}_S - q_i)}{(|S| - 1)(\sum_{i=1}^{n} q_i)} \qquad \text{for} |S| > 1, i \in S \qquad (9)$$

where $\overline{q}_S = \sum_{i \in S} q_i / |S|$. Moreover, $r_{S,i} = 1$ for $|S| = 1$.

We wish to show that this is the optimum solution to the convex program. To do this, we construct a feasible dual solution and show that the primal-dual pair satisfies the KKT conditions, and use strong duality for convex programs (and that Slater's condition is satisfied).

Let us define the dual variables $\beta_i \in \mathbb{R}$ for $i \in [n]$ for constraints (6), $\gamma_S \in \mathbb{R}$ for all S in (7), and $\delta_{S,i} \geq 0$ for (8). The complementary slackness conditions for (8) give $r_{s,i}\delta_{i,S} = 0$. Taking the Lagrangian and the partial derivatives with respect to $r_{S,i}$ gives the condition

$$2(|S| - 1) \, p_S \cdot (r_{S,i} - 1/|S|) - \gamma_S - \beta_i \, p_S - \delta_{i,S} = 0 \qquad \forall S, \forall i \in S. \qquad (10)$$

Consider the following dual solution:

$$\beta_i = -2q_i/(\sum_{i=1}^{n} q_i), \quad \gamma_S = (\sum_{i \in S} \beta_i) \, p_S/|S| = -2\overline{q} \, p_S/(\sum_{i=1}^{n} q_i) \quad \text{and} \quad \delta_{S,i} = 0.$$

We show that this primal-dual pair satisfies the KKT conditions. First, $r_{s,i}\delta_{i,S} = 0$ holds trivially as $\delta_{S,i} = 0$. So (10) becomes

$$2(|S| - 1) \, p_S \cdot (r_{S,i} - 1/|S|) - \gamma_S - \beta_i \, p_S = 0. \qquad (11)$$

For S with $|S| = 1$, this holds easily as the first term above becomes 0, and our choice satisfies $\gamma_S = -\beta_i p_S$ for $S = \{i\}$.

For S with $|S| > 1$, plugging the values of γ_S and β_i, cancelling p_S and re-arranging, (11) simplifies to

$$r_{S,i} = \frac{1}{S} + \frac{1}{2(|S| - 1)}(\beta_i - \gamma_S) = \frac{(\overline{q}_S - q_i)}{(|S| - 1)(\sum_{i=1}^{n} q_i)},$$

which exactly corresponds to the FV scheme. The primal feasibility of $r_{S,i}$ follows from the (somewhat tedious) calculations in [11], which show that $r_{S,i}$ is a valid and maximally fair scheme.

6   Extending the Variance Minimization Program

More generally, for any distribution $\mathcal{P}$ and any goal vector g, one can consider a more general family of convex programs with arbitrary weights w_S for set S in the objective (the program for the FV scheme has $w_S = |S| - 1$).

$$\min \sum_{i \in S} w_S \, p_S \, (r_{S,i} - 1/|S|)^2 \quad \text{s.t.} \quad \sum_{S \ni i} p_S \cdot r_{S,i} = g_i, \forall i; \quad \sum_{i \in S} r_{S,i} = 1, \forall S.$$

Suppose there is some natural choice of weights w_S (possibly depending of $\mathcal{P}, g$), so that the constraints $r_{S,i} \geq 0$ were automatically satisfied by the optimum solution to this convex program. Then, we claim that the resulting CR rule given by $r_{S,i}$ has a very succinct representation, similar to FV scheme.

Dual. As before, consider the dual of this program with dual variables $\beta_i, \gamma_S \in \mathbb{R}$. Then taking the Lagrangian and partial derivatives, gives the KKT conditions:

$$2w_S \, p_S \, (r_{S,i} - \frac{1}{|S|}) - \gamma_S - \beta_i \, p_S = 0, \forall S, i.$$

For a fixed S, summing this up over all $i \in S$ and using that $\sum_{i \in S} r_{S,i} = 1$, the first term becomes 0 and we get $\gamma_S = -\sum_{i \in S} \beta_i p_S / |S|$. Let us denote $\overline{\beta}_S = \sum_{i \in S} \beta_i / |S|$. Then, this gives that

$$r_{S,i} = \frac{1}{|S|} + \frac{\beta_i - \overline{\beta}_S}{2w_S}. \tag{12}$$

Let us call a CR scheme a β-scheme, if there is some natural choice of weights w_S, such that the optimum solution to the convex program has the form above. By our discussions above, the FV scheme is a β-scheme for $w_S = |S| - 1$ with $\beta_i = -2q_i/(\sum_{i=1}^n q_i)$. It is an interesting question to explore if β-schemes exists for more general distributions or even for more general allocation vectors and product distributions. The following two examples show that this is not true for the choice $w_S = |S| - 1$, that worked for the FV scheme.

Computing β efficiently. We remark that if a β-scheme exists, for some choice of w_S, then given any goal vector g, the corresponding $\beta_i(g)$ can be computed in time polynomial in n, assuming suitable access to $\mathcal{P}$. In particular, as $g_i = \sum_{S \ni i} r_{S,i} p_S$, and $r_{S,i}$ is given by (12), g is linear in β and hence $g = J\beta + v_0$ where v_0 is some affine shift and J is the $n \times n$ Jacobian matrix with entries $J_{ij} = \partial g_i / \partial \beta_j$. Suppose J (the entries of which only depend on $\mathcal{P}$) and w_S can be computed explicitly, and that the goal vector $g(0)$ for $\beta = \mathbf{0}$ (that has entries $g_i(0) = \sum_{S \ni i} p_S / |S|$) can be computed. Then $\beta(g) = J^+(g - g(0))$, where J^+ is the pseudoinverse of J.

Limitations. For $n = 3$, there exists a general distribution ($\Pr[\{1\}] = 0.55$, $\Pr[\{3\}] = 0.2$, $\Pr[\{1,3\}] = 0.05$, $\Pr[\{1,2,3\}] = 0.25$) such that a a fair allocation is achievable, but is not achievable using a β-scheme with the choice of weights $w_S = |S| - 1$. And there exists a product distributions and an achievable goal vector g ($\mathcal{P} = \langle 0.28, 0.59, 0.52 \rangle$, and $g = (0.25437, 0.29448, 0.309446)$) which cannot be achieved using a β-scheme with the choice of weights $w_S = |S| - 1$.

7 Computing the Max-Min Allocation

Recall that given $\mathcal{P}, V$ our goal is to compute a feasible $g \in G(\mathcal{P})$, such that for any $g' \in G(\mathcal{P})$ we have $g \succeq_V g'$. In other words, the vector $\alpha^V(g)$ with entries

g_i/v_i satisfies that $\alpha^V(g)^\uparrow$ is lexicographically the largest among all other feasible allocations g'.

For any set S, the maximum probability that the item is allocated to players in S is $\sum_{T:T\cap S\neq\emptyset} p_T$. So by averaging, under any allocaton g, some player $i \in S$ always has $\alpha_i^V(g) \leq (\sum_{T:T\cap S\neq\emptyset} p_T)/(\sum_{i\in S} v_i)$. The algorithm proceeds by finding the set S will the smallest such value, called the critical value, and ensures that g_i/v_i equals this value for players in S. It then iterates on the residual instance. Below, we show that the critical probabilities can only increase, and that the resulting allocation in the fairest with respect to V. Later we show how to compute the critical set S.

We omit the superscript V in α^V for convenience, whenever it is clear from the context. For $S \subseteq [n]$, let $V(S) = \sum_{i\in S} v_i$ and let $E^A(S) = \{T \subseteq A, T \cap S \neq \emptyset\}$ be the collection of subsets of A intersecting S. For a collection of sets $\mathcal{S}$, let $P(\mathcal{S}) = \sum_{S\in\mathcal{S}} p_S$.

Consider the following algorithm.

Algorithm CompuleFair$(\mathcal{P}, V)$:

1. Init: $A_1 \leftarrow [n]$, $k \leftarrow 1$
2. while $A_k \neq \emptyset$
   (a) $S_k \leftarrow \arg\min_{S\subseteq A_k} \left\{ \frac{P(E^{A_k}(S))}{V(S)} \right\}$, $\alpha_k \leftarrow \frac{P(E^{A_k}(S_k))}{V(S_k)}$
   (b) $g_i \leftarrow v_i \cdot \alpha_k$, for $i \in S_k$, $A_{k+1} \leftarrow A_k \setminus S_k$, $k \leftarrow k+1$

Theorem 13. *For any distribution $\mathcal{P}$ and for any priority vector V, the fairest allocation $g \in G(P)$ can be computed exactly in time $\text{poly}(n, |supp(\mathcal{P})|)$ using a liner program. If an ϵ additive error is allowed, the running time is $\text{poly}(n, 1/\epsilon)$. An exact computation in $\text{poly}(n)$ time is also possible if $g(\pi)$ can be computed efficiently for any permutation π.*

First, we prove the correctness of the algorithm, let K be the number of iterations. The following simple claim shows that α_i can only increase over the iterations.

Claim 6. *For all $i \in [1, K-1]$, $\alpha_i \leq \alpha_{i+1}$.*

Proof. For the sake of contradiction, suppose that $\alpha_i > \alpha_{i+1}$ for some $i \in [K-1]$. Then for $S' = S_i \cup S_{i+1}$,

$$\frac{P(E^{A_i}(S'))}{V(S')} = \frac{P(E^{A_i}(S_i)) + P(E^{A_{i+1}}(S_{i+1}))}{V(S_i) + V(S_{i+1})} < \frac{P(E^{A_i}(S_i))}{V(S_i)} = \alpha_i,$$

where the first equality follows from $E^{A_i}(S_i) \cap E^{A_{i+1}}(S_{i+1}) = \emptyset$, and the inequality follows from $\frac{a+x}{b+y} < \frac{a}{b}$ if $\frac{x}{y} < \frac{a}{b}$, for $a, b, x, y > 0$. This contradicts that S_i is the critical subset at step i.

Next, we prove that the output of **ComputeFair**$(\mathcal{P}, V)$ is achievable.

Claim 7. *For any distribution* $\mathcal{P}$ *and priority* V, *the output* g *of* ***ComputeFair***$(\mathcal{P}, V)$ *lies in* $G(\mathcal{P})$.

Proof. (Proof of Claim 7). By Lemma 1, we need to show that for all S we have $P(\mathcal{E}(S)) \geq \sum_{i \in S} g_i$, where $\mathcal{E}(S) = \{T \subseteq [n] : T \cap S \neq \emptyset\}$. This follows as

$$\sum_{i \in S} g_i = \sum_{j \in [K]} \sum_{i \in S \cap S_j} g_i = \sum_{j \in [K]} \sum_{i \in S \cap S_j} \alpha_j v_i$$

$$\leq \sum_{j \in [K]} \sum_{i \in S \cap S_j} v_i \cdot \frac{P(E^{A_j}(S \cap S_j))}{V(S \cap S_j)} = \sum_{j \in [K]} P(E^{A_j}(S \cap S_j)) \leq P(\mathcal{E}(S)),$$

where the first inequality follows from the minimality of S_j, and the second inequality follows from $E^{A_j}(S \cap S_j) \cap E^{A_{j'}}(S \cap S_{j'}) = \emptyset$, for $j \neq j'$.

Claim 8. *For any* $\mathcal{P}$ *and* V, *let* g *the output of* ***ComputeFair***$(\mathcal{P}, V)$, *then* $g \succeq_V g'$ *for all* $g' \in G(\mathcal{P})$.

Proof. For the sake of contradiction, consider some $\mathcal{P}, V$ and $g' \in G(\mathcal{P})$ such that $g' \succ_V g$, with the fewest number of players n. Let r' be the rule corresponding to g', i.e., $g'_i = \sum_{S \subseteq [n]} r'_{S,i} p_S$. Consider the following two cases depending on the first critical set $\bar{S}_1$.

Case 1. $g'_i = g_i$ for all $i \in S_1$. In this case, we claim that for all $S \in E(S_1)$, we have $\sum_{i \in S_1} r'_{S,i} = 1$ as

$$\sum_{S \in E(S_1)} p_S \cdot \sum_{i \in S_1} r'_{S,i} = \sum_{i \in S_1} g'_i = \alpha_1 \cdot \sum_{i \in S_1} v_i = \sum_{S \in E(S_1)} p_S.$$

So, for all $S \in E(S_1)$, we have $\sum_{i \in S \setminus S_1} r'_{S,i} = 0$.

Consider the distribution $\hat{\mathcal{P}}, \hat{V}$ on players $A_2 = [n] \setminus S_1$, defined as $\hat{p}_S = p_S$ for $S \subseteq A_2$ and $\hat{v}_i = v_i$ for $i \in A_2$. Let $\hat{g}$ be the output of **ComputeFair**$(\hat{\mathcal{P}}, \hat{V})$, and note that by the design of the algorithm $\hat{g}_i = g_i$ for all $i \in A_2$. Moreover, let $\tilde{g}$ the vector which corresponds to the assignment rule $\tilde{r}$, where $\tilde{r}_{S,i} = r'_{S,i}$, for $S \subseteq A_2$. Since for all $S \in E(S_1)$ we have $\sum_{i \in S \setminus S_1} r'_{S,i} = 0$, therefore $\tilde{g}_i = g'_i$ for all $i \in A_2$. Hence, $\tilde{g} \succ_{\hat{V}} \hat{g}$ as well, with a smaller number of players, which contradicts the minimality of $\mathcal{P}$.

Case 2. $g'_j \neq g_j$ for some $j \in S_1$. Suppose $g'_i < g_i$ for some $i \in S_1$, then as $g'_i/v_i < g_i/v_i = \alpha_1$, and as α_k are non-decreasing by Claim 6, it follows that $g'_i/v_i < g_r/v_r$ for all $r \in [n]$, contradicting that $g' \succeq_V g$. Thus, we have $g'_j > g_j = \alpha_1 \cdot v_j$, and $g'_i \geq g_i = \alpha_1 \cdot v_i$ for all $i \in S_1$, but this is impossible as

$$\alpha_1 \cdot V(S_1) = \alpha_1 \cdot v_j + \sum_{i \in S_1 \setminus \{j\}} \alpha_1 \cdot v_i \leq \alpha_1 \cdot v_j + \sum_{i \in S_1 \setminus \{j\}} g'_i < \sum_{i \in S_1} g'_i \leq P(E(S_1)) = \alpha_1 \cdot V(S_1).$$

7.1 Computing the Critical Subset

We now give an algorithm that given $\mathcal{P}$, computes the critical subset of players. For clearer exposition, we assume that $\mathcal{P}$ is given explicitly and show that the algorithm runs in $\text{poly}(n, |\text{supp}(\mathcal{P})|)$. The sampling argument in Sect. 2 directly gives the ε-approximate version, which runs in time $\text{poly}(n, 1/\varepsilon)$.

The algorithm is based on a result of Charikar [6] for computing dense subgraphs in graphs, but for our setting we need to extend it to hypergraphs and to handle the weights v_i.

Consider the following LP with variables x_S for each set S in the support of $\mathcal{P}$ and y_i for $i \in [n]$.

$$\min \sum_S p_S x_S \quad \text{s.t.} \quad x_S \geq y_i, \forall i \in S; \quad \sum_{i \in [n]} v_i y_i \geq 1; \quad x_S, y_i \geq 0, \forall i, S. \quad \text{(LP}\alpha\text{)}$$

Given a solution, let us define $T(r) = \{i \in [n] : y_i \geq r\}$. The algorithm solves LPα, and outputs $T(r^*)$ where $r^* = \arg\min_r \frac{P(E(T(r)))}{V(T(r))}$. Note that it suffices to only check for $r \in \{y_1, \ldots, y_n\}$.

Lemma 9. *The optimal value for LPα is $\min_T \frac{P(E(T))}{V(T)}$. The algorithm also outputs the critical set T.*

Proof. First, note that for any set T, there exists a feasible solution with value of $P(E(T))/V(T)$, by setting $y_i = 1/V(T)$ for all $i \in T$, and $x_S = 1/V(T)$ for all $S \in E(T)$, and all the other variables to 0. This satisfies all the constraints and has the claimed objective value.

Given a solution to LPα with value u, we show how to find an integral solution T with value u. First, we can assume that for all sets S we have $x_S = \max\{y_i : i \in S\}$. Consider a set T and a collection of sets $\mathcal{S}$ parameterized by $r \geq 0$ as follows: $T(r) = \{i : y_i \geq r\}$ and $\mathcal{S}(r) = \{S : x_S \geq r\}$.

We claim that $E(T(r)) = \mathcal{S}(r)$ for all $r \geq 0$. Fix some r. As $x_S \geq y_i$ for all $i \in S$, if $i \in T(r)$, then $S \in \mathcal{S}(r)$ for all S containing i. This implies that $\mathcal{S}(r)$ contains $E(T(r))$. Conversely, as $x_S = \max\{y_i : i \in S\}$, if $S \in \mathcal{S}(r)$ then there exists some $i \in S$ such that $y_i \geq r$ and hence $i \in T(r)$.

We claim that there exists some r such that $\frac{P(E(T(r)))}{V(T(r))} \leq u$. Indeed if not, then using $\int_0^\infty 1_{\{x \geq r\}} dr = x$, we have the following contradiction

$$u = \sum_S p_S x_S \int_0^\infty \Big(\sum_{S \in \mathcal{S}(r)} p_S \Big) dr > u \cdot \int_0^\infty \Big(\sum_{i \in T(r)} v_i \Big) dr = u \Big(\sum_i v_i y_i \Big) \geq u.$$

7.2 Computing the Critical Subset Using an Oracle

One can also compute the critical subset using oracle access to the allocation vectors $g(\pi)$ for permutations, using the linear program in Theorem 9, and thus avoid knowing $\mathcal{P}$ exactly. Consider the following LP.

$$\min \sum_{\pi \in \Pi} x_\pi \quad \text{s.t.} \quad \sum_{\pi \in \Pi} x_\pi g_i(\pi), \forall i \in [n]; \qquad x_\pi \geq 0, \forall \pi \in \Pi.$$

$$\text{(LP-Perm-}\alpha)$$

The dual of this LP is the following.

$$\max \sum_{i \in [n]} v_i \cdot y_i \quad \text{s.t.} \quad \sum_{i \in [n]} y_i g_i(\pi) \leq 1, \forall \pi \in \Pi; \qquad y_i \geq 0, \forall i \in [n].$$

$$\text{(LP-}\alpha\text{-Perm-Dual)}$$

As before, let $T(r) = \{i \in [n] : y_i \geq r\}$. The algorithm solves LP-α-Perm-Dual by using a separation oracle as defined in Claim 4, and outputs $T(r^*)$ where $r^* = \arg\max_r \frac{V(T(r))}{P(E(T(r)))}$. As previously, it suffices to consider $r \in \{y_1, \ldots, y_n\}$.

Lemma 10. *The optimal value of the LP is* $\max_T \frac{V(T)}{P(E(T))}$. *The algorithm also outputs the critical set* T.

Proof. For any $S \subseteq [n]$ and $\pi \in \Pi$, note that $\sum_{i \in S} g_i(\pi) \leq P(E(S))$, as $P(E(S))$ is the total probability mass of all sets that contain some element of S. Moreover, for S is a prefix of π, $\sum_{i \in S} g_i(\pi) = P(E(S))$.

Let u the optimal value of LP-α-Perm-Dual, and let $T = \arg\max_S(V(S)/P(E(S)))$. First we show that $v \geq V(T)/P(E(T))$. Setting $y_i = 1/P(E(T))$, for any $\pi \in \Pi$, we have $\sum_{i \in [n]} y_i g_i(\pi) = (\sum_{i \in T} g_i(\pi))/P(E(T)) \leq 1$, and the objective is $(\sum_{i \in T} v_i)/P(E(T)) = V(T)/P(E(T))$. On the other hand, given a solution y, let π_y be order according to non-decreasing values of y (i.e., $y_{\pi_y(i)} \geq y_{\pi_y(i+1)}$ for all $i \in [n-1]$). Let $S(r) = \{i \in N : y_i \geq r\}$, note that $S(r)$ is a prefix of π_y for any r. By the constraint for π_y in the dual we have: $1 \geq \sum_{i \in [n]} y_i g_i(\pi_y) = \int_0^\infty \left(\sum_{i \in S(r)} g_i(\pi_y)\right) dr = \int_0^\infty P(E(S(r))) dr$ Now, an r such that $V(S(r))/P(E(S(r))) \geq u$ exists, otherwise we get the contradiction that

$$u = \sum_{i \in [n]} v_i \cdot y_i = \int_0^\infty V(S(r)) dr < u \int_0^\infty P(E(S(r))) dr \leq u.$$

References

1. Adamczyk, M., Włodarczyk, M.: Random order contention resolution schemes. In: Foundations of Computer Science (FOCS), pp. 790–801 (2018)
2. Bansal, N., Srinivasan, A., Svensson, O.: Lift-and-round to improve weighted completion time on unrelated machines. In: Symposium on Theory of Computing, STOC, pp. 156–167 (2016)
3. Bansal, N., Sviridenko, M.: The santa claus problem. In: Symposium on Theory of Computing, STOC, pp. 31–40. ACM (2006)
4. Bertsekas, D., Gallagher, R.: Data Networks. Prentice Hall, Hoboken (1992)

5. Boyd, S., Vandenberghe, L.: Convex Optimization. Cambridge University Press, New York (2004)
6. Charikar, M.: Greedy approximation algorithms for finding dense components in a graph. In: APPROX, pp. 84–95 (2000)
7. Charikar, M., Makarychev, K., Makarychev, Y.: Near-optimal algorithms for unique games. In: Symposium on Theory of Computing, STOC, pp. 205–214 (2006)
8. Chekuri, C., Vondrák, J., Zenklusen, R.: Submodular function maximization via the multilinear relaxation and contention resolution schemes. SIAM J. Comput. **43**(6), 1831–1879 (2014)
9. Cohen, M.B., Madry, A., Tsipras, D., Vladu, A.: Matrix scaling and balancing via box constrained newton's method and interior point methods. In: Symposium on Foundations of Computer Science, FOCS, pp. 902–913 (2017)
10. Dresher, M.: Theory and Applications of Games of Strategy. RAND Corporation, Santa Monica (1951)
11. Feige, U., Vondrák, J.: Approximation algorithms for allocation problems: improving the factor of 1–1/e. In: Symposium on Foundations of Computer Science, FOCS, pp. 667–676 (2006)
12. Feige, U., Vondrák, J.: The submodular welfare problem with demand queries. Theor. Comput. **6**(1), 247–290 (2010)
13. Feldman, M., Svensson, O., Zenklusen, R.: Online contention resolution schemes. In: Symposium on Discrete Algorithms, SODA, pp. 1014–1033 (2016)
14. Hensher, D.A., Rose, J.M., Greene, W.H.: Applied Choice Analysis. Cambridge University Press, Cambridge (2015)
15. Im, S., Shadloo, M.: Weighted completion time minimization for unrelated machines via iterative fair contention resolution. In: Symposium on Discrete Algorithms, SODA, pp. 2790–2809 (2020)
16. Jaynes, E.T.: Information theory and statistical mechanics. Phys. Rev. **106**, 620–630 (1957)
17. Karmarkar, N., Karp, R.M.: An efficient approximation scheme for the one-dimensional bin-packing problem. In: Foundations of Computer Science, FOCS, pp. 312–320 (1982)
18. Kelly, F.P., Maulloo, A.K., Tan, D.K.: Rate control for communication networks: shadow prices, proportional fairness and stability. J. Oper. Res. Soc. **49**(3), 237–252 (1998)
19. Kohlberg, E.: On the nucleolus of a characteristic function game. SIAM J. Appl. Math. **20**(1), 62–66 (1971)
20. Lee, E., Singla, S.: Optimal online contention resolution schemes via ex-ante prophet inequalities. In: European Symposium on Algorithms, ESA, volume 112 of LIPIcs, pp. 57:1–57:14 (2018)
21. Linial, N., Samorodnitsky, A., Wigderson, A.: A deterministic strongly polynomial algorithm for matrix scaling and approximate permanents. Combinatorica **20**(4), 545–568 (2000)
22. Luce, R.D.: Individual Choice Behavior: A Theoretical Analysis. Wiley, New York (1959)
23. Rothblum, U.G., Schneider, H.: Scalings of matrices which have prespecified row sums and column sums via optimization. Linear Algebra Appl. **114**, 737–764 (1989)
24. Singh, M., Vishnoi, N.K.: Entropy, optimization and counting. In: Symposium on Theory of Computing, STOC, pp. 50–59 (2014)

25. Sinkhorn, R.: A relationship between arbitrary positive matrices and doubly stochastic matrices. Ann. Math. Stat. **35**(2), 876–879 (1964)
26. Train, K.E.: Discrete Choice Methods with Simulation. Cambridge University Press, Cambridge (2009)
27. Wainwright, M.J., Jordan, M.I.: Graphical models, exponential families, and variational Inference. Found. Trends® Mach. Learn. **1**(1–2), 1–305 (2008)

Author Index

Printed in the United States
by Baker & Taylor Publisher Services